THE

Bee-Keeper's Guide;

—OR—

MANUAL OF THE APIARY,

—BY—

A, J, COOK,

Late Professor of Entomology in the Michigan State Agricultural College,
Professor of Zoology Pomona College, Claremont, California,

AUTHOR OF

"Injurious Insects of Michigan," "Maple Sugar and the
Sugar Bush," and "Silo and Silage."

EIGHTEENTH EDITION.

Revised, Enlarged, Re=Written and Beautifully Illustrated.

TWENTIETH THOUSAND.

CHICAGO, ILL.
GEORGE W. YORK & COMPANY,
PUBLISHERS.
1904

The Beekeeper's Guide
or Manual of the Apiary

© A. J. Cook

ISBN 978-1-912271-01-6

Published by Northern Bee Books, 2017
Scout Bottom Farm
Mytholmroyd
Hebden Bridge HX7 5JS (UK)

Printed by Lightning Source, UK

TO THE

REVEREND L. L. LANGSTROTH,

THE

INVENTOR OF THE MOVABLE-FRAME HIVE,

THE HUBER OF AMERICA,

AND ONE OF THE GREATEST MASTERS OF PURE AND APPLIED

SCIENCE, AS RELATING TO APICULTURE,

IN THE WORLD,

THIS MANUAL IS GRATEFULLY DEDICATED

BY

THE AUTHOR.

PREFACE.

In 1876, in response to a desire frequently expressed by my apiarian friends, principally my students, I published an edition of 3000 copies of the little, unpretending "Manual of the Apiary." This was little more than the course of lectures which I gave annually at the Michigan Agricultural College. In less than two years this was exhausted, and the second edition, enlarged, revised, and much more fully illustrated, was issued. So great was the sale that in less than a year this was followed by the third and fourth editions, and, in less than two years, the fifth edition (seventh thousand) was issued.

In each of the two following years, another edition was demanded. In each of these editions the book has been enlarged, changes made, and illustrations added, that the book might keep pace with our rapidly advancing art.

So great has been the demand for this work, not only at home and in Europe, but even in more distant lands, and so great has been the progress of apiculture—so changed the views and methods of our best bee-keepers—that the author feels warranted in thoroughly revising and entirely recasting this eighth edition (tenth thousand). Not only is the work re-written, but much new matter, and many new and costly illustrations, are added.

The above I quote directly from the preface of the eighth edition, published in 1883. Since then four editions have appeared, each revised as the progress of the art required.

In electrotyping the eighth edition, through an accident very poor work was done, so that the impressions of the last three editions have been far from satisfactory. This has led me wholly to revise the present, or thirteenth edition. In doing this I have thought it wise to add largely, especially to the scientific portion, as the intelligence of our bee-keepers demands the fullest information. I have thus added one hundred and fifty pages and more than thirty illustrations. All this has involved so

PREFACE.

much expense that I am forced, though very reluctantly, to increase the price of the work.

As our bee-keepers know, I have permitted wide use of the illustrations prepared expressly for this work, believing heartily in the motto, "greatest good to the greatest number;" so I have drawn widely from others. I am greatly indebted to all these, and have given credit with the illustration.

Since the above was penned three editions have appeared, the last, sixteenth, in 1899. Each has been revised. Both the science and practice have so advanced that I now recast entirely this, the seventeenth edition.

I wish again to express my thanks and gratitude to our wide-awake American apiarists, without whose aid it would have been impossible to have written this work. I am under special obligation to Messrs. Cowan, York and Root, and to my students who have aided me, both in the apiary and laboratory.

As I stated in the preface to the eighth edition, it is my desire and determination that this work shall continue to be the exponent of the most improved apiculture; and no pains will be spared, that each succeeding edition may embody the latest improvements and discoveries wrought out by the practical man and the scientist, as gleaned from the excellent home and foreign apiarian and scientific periodicals.

The above was prefaced to the Eighteenth one thousand published in 1900. This Nineteenth one thousand has been wholly revised, about 80 pages and 75 engravings added. We believe it is now at the front in bee-keeping science and practice.   A. J. COOK.

Pomona College, Claremont, California, 1902.

CONTENTS.

INTRODUCTION.

WHO MAY KEEP BEES.

SPECIALISTS.

Any person who is cautious, observing and prompt, will succeed in bee-keeping. He must expect to work with full energy through the busy season, and persist though discouragement and misfortune both confront him. I need not mention capital or location, for men of true metal—men whose energy of body and mind bespeak success in advance—will solve these questions long before their experience and knowledge warrant their assuming the charge of large apiaries.

AMATEURS.

Bee-keeping is specially to be recommended as an avocation. Bees are of great value in fertilizing fruits, grains and vegetables; they also save millions of pounds of most wholesome food which would otherwise go to waste; and experience amply proves that they may be kept in city, village and country at a good profit, and so any person, possessed of the proper ability, tact and energy, may adopt bee-keeping, and thus do good, gain pleasure, and often receive profit, as experience has shown, more than is derived from the regular occupation. The late Mr. C. F. Muth, of Cincinnati, long kept bees very profitably on his store, in the very heart of the city. Hundreds of our most successful bee-keepers live in small towns and villages, and add bee-culture to their work in shop, office, or study, and receive health, pleasure, and money as a reward. Ladies all over our country are finding in this pursuit pleasure, and opportunity to exercise in the pure air, which means health, and money. Farmers are adding bee-keeping to their farms, to find not infrequently that the bees are their most profitable property. Orchardists, especially, need and must have bees to pollinate the fruit-blossoms, and insure a crop. The time required will of course depend upon the number of colonies kept; but with wise management, this time may be given at any time of the day or week, and thus not interfere with the regular business. Thus residents of country, village or city, male or female, who enjoy the society and study of natural objects, and wish to add to their income and pleasure, will find here an ever waiting opportunity.

WHO SHOULD NOT KEEP BEES.

There are occasionally persons to whom the venom of the bee is a serious poison. If such persons are stung anywhere their eyes swell so they can not see, the skin blotches, and serious irritation is felt over the entire body. Such persons are often overcome with fever for several days, and, though very rarely, the sting sometimes proves fatal. It goes without saying that such persons should not keep bees.

It is a well known fact that the sting of the honey-bee becomes less and less poisonous the more one is stung. The system becomes inoculated against the poison. My own experience proves this most conclusively. Every bee-keeper will receive occasional stings, but these become more and more rare, and soon occasion neither fear nor anxiety.

INDUCEMENTS TO BEE-KEEPING.

RECREATION.

I name this first, as it was the pleasure in store that led me to the art of keeping bees, though I was terribly afraid of bees at the beginning. There is a rare fascination in the study of nature. Insect life is ever presenting the most pleasurable surprises to those who study it. Bees, from their wonderful instincts, curious structure and habits, and the interesting relations which they sustain to vegetable life, are most fascinating objects of study. The observant and appreciative bee-keeper is ever the witness of exhibitions that incite wonder and admiration. This is why bee-keepers are always enthusiasts. I know of no class of laborers who dwell more fondly on their work and business than do bee-keepers. A thorough study of the marvelous economy of the honey-bee must, from its very nature, bring delight and admiration. A farmer once said to me, "Were it not for the generous profits of the business, I would still keep bees for the real pleasure I receive in the business." I once asked a hard worked teacher why he kept bees. I felt like saying amen to his answer: "For the restful pleasure which the work gives." I have often gone to the bees tired and nervous, and after an hour's labor, felt refreshed, as by sound sleep. I have been deeply gratified many times by the letters thanking me for having turned the writers' attention towards bee-keeping. I often think that if a person does embark in bee-keeping, commencing in a small way—*and no person should begin in any other way*—the knowledge gained and consequent pleasure received will prove ample remuneration, even should no practical results follow. The man is broadened by the study, and better fitted to enjoy life.

Some years since my old friend and college classmate, O. Clute, visited me. Of course, I must show him the bees. He was delighted, took this "Manual" home with him, purchased some bees at once, and became enraptured with the work, and the result of all this was another first-class bee-keeper and that most fascinating work of fiction, "Blessed Bees."

PROFITS.

The profits in bee-keeping offer strong inducements towards its adoption as a pursuit. I believe few manual-labor occupations offer so large returns, if we consider the capital invested. True, bee-keeping requires hard work, but this is only for a portion of the year, and in winter there is almost no work, especially if the bee-keeper buys all his hives, sections, etc., which is usually wiser than to make them. The cautious, prompt and skillful bee-keeper will often be able to secure an annual average of seventy-five pounds per colony, besides doubling the number of his colonies. This will give $10.00 per colony at least, which is almost as much as the colony, with required apparatus, is worth. Of course, poor years will confront the bee-keeper. Winter losses will be experienced by the beginner. Some will fail entirely. The fickle, careless, indolent man will as surely fail in bee-keeping as in any other calling. Yet if one studies the science and art, and commences bee-keeping in a small way, as all should, he will be no great loser, even if he find that he is not suited to the business. He knows more and is a broader man for this study and experience. My brother, who is a good farmer, with a fertile and well-stocked farm, commenced bee-keeping more to interest his boys than aught else. He has met very little loss in wintering—for years together none at all. For three successive years his sixty colonies of bees gave him more profits than all the balance of his farm. As he said at one of the Michigan State Conventions : "I find my bees the pleasantest and most profitable part of my farm." He added the surprising remark, "Nothing on my farm bears neglect better than my bees." I might add that neglect is rarely seen on his farm.

Adam Grimm, James Heddon, G. M. Doolittle, E. J. Oatman, and many others, have made much money in this pursuit. Mr. Hetherington keeps thousands of colonies of bees, and has received over $10,000 cash receipts in a single year. Mr. Clute, an able clergyman, has often received more money from his bees than from his salary as a preacher. All over our country men are gaining a livelihood in this industry, and often earning as much more in other pursuits. The opportunity to make money, even with hardships and privations, is attractive and seldom disregarded. What shall we say then of this opportunity, if the labor which it involves, brings in itself healthful recreation and constant delight ? Dr. C. C. Miller gave up a $2500 salary to engage in bee-keeping. Though a specialist, and though his profits some years, owing to the drouth, are nothing, yet he is contented with the business, and has no idea of changing for any other.

EXCELLENCE AS AN AMATEUR PURSUIT.

After twenty years of experience, I am persuaded that no business offers more as an avocation. Indeed, I think bee-keeping may ofttimes best serve as a second business. We have already seen that bees are a blessing, and I would have every person, whatever his leading business, keep a few colonies of bees, unless by taste, nature or temperament, he be unfitted for the work. Bee-keeping offers additional funds to the poorly paid ; outdoor air to clerk and office-hand ; healthful exercise to the person of sedentary habits, opportunity for the poor to reap what would otherwise go to waste, and superior recreation to the student, teacher and professional man, especially to him whose life-work is of that dull, hum-drum, routine order that seems to rob life of all zest.

The labor required in bee-keeping, especially if but few colonies are kept, can, with thought and management, be so arranged as not to infringe upon the time demanded by the regular occupation. Even the farmer, by wise foresight, can arrange so that his bees will not interfere greatly with his regular farm work. I have never received more hearty thanks than from persons whom I had influenced to add the care of bees to their other duties.

ADAPTATION TO WOMEN.

Apiculture may also bring succor to those whom society has not been over-ready to favor—our women. Widowed mothers, dependent girls, the weak and the feeble, all may find a blessing in the easy, pleasant and profitable labors of the apiary. Of course, women who lack vigor and health can care for but very few colonies, and must have sufficient strength to bend over and lift the small-sized frames of comb when loaded with honey, and to carry empty hives. With the proper thought and management, full colonies need never be lifted, nor work done in the hot sunshine. Yet, right here let me add, and emphasize the truth, *that only those who will let energetic thought and skillful plan, and above all promptitude and persistence, make up for physical weakness, should enlist as apiarists.* Usually a stronger body and improved health, the result of pure air, sunshine and exercise, will make each successive day's labor more easy, and will permit a corresponding growth in the size of the apiary for each successive season. One of the most noted apiarists, not only in America, but in the world, sought in bee-keeping her health, and found not only health, but reputation and influence. Some of the most successful apiarists in our country are women. Of these, many were led to adopt the pursuit because of waning health, grasping at this as the last and successful weapon with which to vanquish the grim monster.

That able apiarist, and terse writer on apiculture, Mrs. L. Harrison, states that the physicians told her that she could not live; but apiculture did for her what the physicians could not do—restored her to health, and gave her such vigor that she has been able to work a large apiary for years.

Said "Cyula Linswik"—whose excellent and beautifully

written articles have so often charmed the readers of the bee-journals, and who has had many years of successful experience as an apiarist—in a paper read before the Michigan convention in March, 1887 : "I would gladly purchase exemption from indoor work, on washing-day, by two days' labor among the bees, and I find two hours' labor at the ironing-table more fatiguing than two hours of the severest toil the apiary can exact." I repeat, that apiculture offers to many women not only pleasure but profit

Mrs. L. B. Baker, of Lansing, Mich., who had kept bees very successfully for four years, read an admirable paper before the same convention, in which she said : " But I can say, having tried both (keeping boarding-house and apiculture), I give bee-keeping the preference, as more profitable, healthful, independent and enjoyable. * * * I find the labors of the apiary more endurable than working over a cook-stove indoors, and more pleasant and conducive to health. * * * I believe that many of our delicate and invalid ladies would find renewed vigor of body and mind in the labors and recreations of the apiary. * * * By beginning in the early spring, when the weather was cool and the work light, I became gradually accustomed to outdoor labor, and by midsummer found myself as well able to endure the heat of the sun as my husband, who has been accustomed to it all his life. Previously, to attend an open-air picnic was to return with a headache. * * * My own experience in the apiary has been a source of interest and enjoyment far exceeding my anticipations." Although Mrs. Baker commenced with but two colonies of bees, her net profits the first season were over $100 ; the second year but a few cents less than $300 ; and the third year about $250. " The proof of the pudding is in the eating ;" and such words as those above show that apiculture offers special inducements to our sisters to become either amateur or *professional* apiarists. At the present time almost every State has women bee-keepers, whose success has won attention. True it is, that in neatness and delicacy of manipulation, the women far surpass the men. The nicest honey produced in Michigan, year after year, comes from the apiary of two ladies who I believe are peers of any bee-keepers in our country.

IMPROVES THE MIND, THE OBSERVATION, AND THE HEART.

Successful apiculture demands close and accurate obser-vation, and hard, continuous thought and study, and this, too, in the wondrous realm of nature. In all this, the apiarist re-ceives manifold and substantial advantages. In the cultiva-tion of the habit of observation a person becomes constantly more able, useful and susceptible to pleasure—results which also follow as surely on the habit of thought and study. It is hardly conceivable that the wide-awake apiarist who is so frequently busy with his wonder-working comrades of the hive, can ever be lonely, or feel time hanging heavily on his hands. The mind is occupied, and there is no chance for *ennui.* The whole tendency of such thought and study, where nature is the subject, is to refine the taste, elevate the desires, and ennoble manhood. Once get our youth, with their sus-ceptible natures, engaged in such wholesome study, and we shall have less reason to fear the vicious tendencies of the street, or the luring vices and damning influences of the saloon. Thus apiculture spreads an intellectual feast that even the old philosophers would have coveted; furnishes the rarest food for the observing faculties, and, best of all, by keeping its votaries face to face with the matchless creations of the *All Father,* must draw them toward Him "who went about doing good," and "in whom there was no guile."

YIELDS DELICIOUS FOOD.

A last inducement of apiculture, certainly not unworthy of mention, is the offering it brings to our tables. Health, yea our very lives, demands that we eat sweets. It is a truth that our sugars, and especially our commercial syrups, are so adul-terated as to be often poisonous. The apiary in lieu of these, gives us one of the most delicious and wholesome of sweets, which has received merited praise, as food fit for the gods, from the most ancient time to the present day. Ever to have within reach the beautiful, immaculate comb, or the equally grateful nectar, right from the extractor, is certainly a bless-ing of no mean order. We may thus supply our families and friends with a food element, with no cloud of fear from vile, poisonous adulterations. We now know that if we eat cane, sugar—the common sugar of our tables—it is converted by the digestive fluids into a glucose-like sugar, which is probably nearly or quite identical with honey-sugar. The bees do the same with the nectar, which is dilute cane-sugar, of flowers. Thus we may reason that honey is our most wholesome sugar, for here the bees have in part digested our food for us.

BRINGS THE SECOND BLADE OF GRASS.

We now know that bees do most valuable work in pollinating the fruit-blossoms. No orchard will give full fruitage without the visits of nectar-loving insects. Of these valued friends, no other is at all comparable to the honey-bee, in the value of its service. I know of California orchards whose productiveness has been immensely increased by the introduction of an apiary. Every orchard should have an apiary in its near vicinity.

ADDS TO THE NATION'S WEALTH.

An excellent authority placed the number of colonies of bees in the United States, in 1881, at 3,000,000, and the honey-production for that year at more than 20,000,000 pounds. The production for that year was not up to the average, and yet the cash value of the year's honey crop exceeded $30,000,000. We may safely add as much more as the value of the increase of colonies, and we have a grand total of $60,000,000—nearly enough to pay the interest on the national debt, were the bonds all refunded. Mr. Root, in his excellent "A B C of Bee-Culture," estimates, from sections sold, that 125 million pounds of honey are produced annually and sold for $10,000,000. And yet all this is but gathered nectar, which would go to waste were it not for the apiarist and his bees. We thus save to the country that which would otherwise be a total loss. Apiculture, then, in adding so immensely to the productive capital of the country, is worthy, as an art, to receive the encouragement and fostering care of the State. And the thought that he is performing substantial service to the State, may well add to the pleasure of the apiarist, as he performs his daily round of labor. When we add to this the vastly greater indirect benefit which comes through the agency of bees in fertilizing flowers —a benefit which can hardly be computed—we then understand the immense value which comes from bees. Truly, the bee-keeper may feel proud of the grand part which his bees perform in the economy of that part of nature which most concerns man and most generously ministers to man's wants.

WHAT SUCCESSFUL BEE-KEEPING REQUIRES.

MENTAL EFFORT.

No one should commence this business who is not willing to read, think, and study. To be sure, the ignorant and unthinking may stumble on success for a time, but sooner or later failure will set her seal upon their efforts. Those of our apiarists who have studied the hardest, observed the closest, and thought the deepest, have even passed the late terrible winter with but slight loss. Those who fail, often fail because of just this lack of mental preparation.

Of course the novice will ask, "How and what shall I study?"

EXPERIENCE NECESSARY.

Nothing will take the place of real experience. Commence with a few colonies, even one or two is best, and make the bees your companions at every possible opportunity. Note every change, whether of the bees, their development, or work, and then by earnest thought strive to divine the cause.

LEARN FROM OTHERS.

Great good will also come from visiting and even working for a time with other bee-keepers. Note their methods, hives, sections, etc. Strive by conversation to gain new and valuable ideas, and gratefully adopt whatever is found, by comparison, to be an improvement upon your own past system and practice.

AID FROM CONVENTIONS.

Attend conventions whenever distance and means render this possible. Here you will not only be made better by social intercourse with those whose occupation and study make them sympathetic and congenial, but you will find a real conservatory of scientific truths, valuable hints, and improved instruments and methods. And the apt attention—rendered possible by your own experience—which you will give to essays, discussions, and private conversations, will so enrich your mind that you will return to your home encouraged and able to do better work, and to achieve higher success. I have attended nearly all the meetings of the Michigan Bee-Keepers' Association, many of those of California, and several of the meetings of the National Bee-Keepers' Association, and never yet when I was not well paid for all trouble and expense by the many, often very valuable, suggestions which I received.

AID FROM BEE-PERIODICALS.

Every apiarist should take and read at least one of the many excellent bee-periodicals that are issued in our country. It has been suggested that Francis Huber's blindness was an advantage to him, as he thus had the assistance of two pairs of eyes, his wife's and servant's, instead of one. So, too, of the apiarist who reads the bee-publications. He has the aid of the eyes, and the brains, of hundreds of intelligent and observing bee-keepers. Who is it that squanders his money on worse than useless patents and fixtures? He who "*can not afford*" to take a bee-paper.

It would be invidious and uncalled for to recommend any one of these valuable papers to the exclusion of the others. Each has its peculiar excellencies, and all who can may well call to his aid two or more of them.

AMERICAN BEE JOURNAL.—This is the oldest American bee-paper, and the only weekly journal devoted exclusively to bee-keeping in the United States. It was founded in 1861, by the late Samuel Wagner, whose breadth of culture, strength of judgment, and practical and historical knowledge of bee-keeping, were remarkable. Even to-day those early volumes of this paper are very valuable parts of any bee-keeper's library. Under the able management of Mr. Thomas G. Newman, the late editor, the paper made great and continuous advancement. The contributors to the "American Bee Journal" are the successful bee-keepers of America, and so it has a wide influence. It is now edited by George W. York, whose skill, enterprise, and ability, are no whit behind those who founded and raised this journal to its present proud place. The publishers are George W. York & Co., 334 Dearborn St., Chicago, Ill. Subscription price, $1.00 a year.

GLEANINGS IN BEE-CULTURE.—This semi-monthly journal, which has just completed its 28th volume, has shown great vigor and energy from its very birth. Its editor is an active apiarist, who is constantly experimenting; a terse, able writer, and brimful of good-nature and enthusiasm. I am free to say that in practical apiculture I am more indebted to Mr. A. I. Root than to any other one person, except Rev. L. L. Langstroth. I also think that, with few exceptions, he has done more for the recent advancement of practical apiculture than any other person in our country or the world. This sprightly and beautifully illustrated journal is edited by E. R. Root, Medina, Ohio. Price, $1.00 a year.

CANADIAN BEE JOURNAL.—This excellent periodical, though published across the line, is worthy of high praise and patronage. Mr. D. A. Jones was its founder, and his ability, enterprise, and long and successful experience gave this paper great prestige. Perhaps no bee-keeper in the world has sacrificed more in the way of time and money, and received less for it, than has Mr. Jones, This is a monthly journal, and is published by the Goold, Shapley & Muir Co., Ltd., Brantford, Ont., at $1.00 a year. W. J. Craig is its editor.

BEE-KEEPERS' REVIEW.—Although the Bee-Keepers' Review has less of years, it is already away up to the front, and an indispensable adjunct to every live apiarist. Its success has been quite phenomenal. The ability, energy, and successful experience of the editor, both as a writer and as a bee-keeper, fit him most admirably for his work. Not only has he won success in all departments of bee-keeping, but he has long been esteemed as one of the most able of our American apicultural writers. Published by W. Z. Hutchinson, Flint, Mich., at $1.00 a year.

AMERICAN BEE-KEEPER.—The ability, enterprise and long and successful experience of Harry E. Hill, editor of this paper, are all well-known. It is a 20-page monthly magazine, neatly edited and well illustrated. It is published by W. T. Falconer Mfg. Co., Jamestown, N. Y., at 50 cents a year.

PROGRESSIVE BEE-KEEPER.—This is one of the later bee-papers, but it shows wonderful progress and great promise of usefulness. Its present editor, R. B. Leahy, is noted for his ability, enterprise, and pushing business ways. It is published monthly by Leahy Mfg. Co., Higginsville, Mo. Price, 50 cents a year.

ROCKY MOUNTAIN BEE JOURNAL.—This latest journal is edited by H. C. Morehouse, and is published monthly by him at Boulder, Colo. It shows vigor, and gives promise of long life and great usefulness. Its locality is very fortunate. Price, $1.00 a year.

BOOKS FOR THE APIARIST.

Having read many of the books treating of apiculture, American and foreign, I can freely recommend such a course to others. Each book has peculiar excellencies, and may be read with interest and profit.

LANGSTROTH ON THE HONEY-BEE.—This treatise will ever remain a classic in bee-literature. I can not over-estimate the benefits which I have received from a study of its pages. The style of this work is so admirable, the subject-matter so replete with interest, and the entire book so entertaining, that it is a desirable addition to any library, and no thoughtful, studious apiarist can well be without it. It is especially happy in detailing the work of experimentation, and in showing with what caution the true scientist establishes principles or deduces conclusions. The work is wonderfully free from errors, and had the science and practice of apiculture remained stationary, there would have been little need of another work. We are happy to state, however, that this work is now revised by no less able authorities than Chas. Dadant & Son, which places it high among our bee-books of to-day. Price, $1.20.

A B C OF BEE-CULTURE.—This work is by the editors of "Gleanings in Bee-Culture." It is arranged in the convenient form of our cyclopedias, is printed in fine style, on beautiful paper, and is very fully illustrated. I need hardly say that the style is pleasing and vigorous. The subject matter is fresh, and embodies the most recent discoveries and inventions pertaining to bee-keeping. Price, $1.20.

FORTY YEARS AMONG THE BEES.—This book is written by Dr. C. C. Miller, of Marengo, Ill., who is an authority on practical bee-keeping. It contains 328 pages, with 112 beautiful, original illustrations, taken by the author himself. It shows in minutest detail just how Dr. Miller does things with bees and makes a great success with them. Price, $1 00.

SCIENTIFIC QUEEN-REARING.—This work is by that well-known and thoroughly practical bee-keeper, G. M. Doolittle. It is invaluable, treating, as it does, of a method by which the very best queen-bees are reared in accord with nature's way. Price, $1.00.

ADVANCED BEE-CULTURE.—This is a full and plain explanation of the successful methods practiced by the author, W. Z. Hutchinson. Price, 50 cents.

FOREIGN PUBLICATIONS.

The BRITISH BEE JOURNAL, as the exponent of British methods and practices, is interesting and valuable to American bee-keepers. It shows that in many things, as in the method of organizing and conducting conventions, so as to make them highly conducive to apicultural progress, we have much to learn from our brothers in Britain. The editor is one of the best informed bee-keepers of the world. The best way for Americans so secure this journal is through the editors of our American bee-papers.

FOREIGN BOOKS.

The best of these, indeed one of the best ever published, is THE HONEY-BEE, by Thomas W. Cowan, of London, England. It is the recognized authority in Europe, as it may well be. It is not only beautiful, but full, accurate, and scientific. As a history of scientific discovery in relation to bees, it is of special interest. It deserves a place in every bee-keeper's library. Price, $1.00.

A more pretentious book is BEES AND BEE-KEEPING, by Frank Cheshire. In workmanship and illustration it is most admirable. It is a compilation from Schiemenz, Girard, Wolff, and others. Many of the pages and many of the finest illustrations are taken bodily, and, we are pained to say, with no credit. As we should expect, the work is not as reliable as the smaller work of Mr. Cowan. Price, $5.50.

As practical guides, I do not think the foreign works superior to our own. Indeed, I think the beginner would profit most by studying our American books. The advanced bee-keeper will gain much in discipline and knowledge by a careful reading of the foreign works on bee-keeping. Foreign scientists, especially the Germans, are at the head, but no nation is quicker to discern the practical bearing and utilize the facts and discoveries in science than are Americans. The Germans had hardly shown how centrifugal force could be used to separate honey from the comb before the Americans had given us our beautiful extractors. The same is true of comb-foundation machines. The Germans pointed out the true nature of "foul brood," and discovered the germicides for its cure, yet I believe ten times as many Americans as foreigners profit by this knowledge.

PROMPTITUDE.

Another absolute requirement of successful bee-keeping is prompt attention to all its varied duties. Neglect is the rock on which many bee-keepers, especially farmers, find too often they have wrecked their success. I have no doubt that more colonies die from starvation than from all the bee-maladies known to the bee-keeper. And why is this? Neglect is the apicide. I feel sure that the loss each season by absconding colonies is almost incalculable, and what must we blame? Neglect. The loss every summer by enforced idleness of queen and workers, just because room is denied them, is very great. Who is the guilty party? Plainly, Neglect. If we would be successful, Promptitude must be our motto. Each colony of bees requires but very little care and attention. Our every interest requires that this be not denied, nor even granted grudgingly. The very fact that this attention is slight, renders it more liable to be neglected; but this neglect always involves loss—often disaster. True, with thought and management the time for this care can be arranged at pleasure and the amount greatly lessened, but the care must *never be neglected*.

ENTHUSIASM.

Enthusiasm, or an ardent love of its duties, is a very desirable, if not an absolute, requisite to successful apiculture. To be sure, this is a quality whose growth, with only slight opportunity, is almost sure. It only demands perseverance. The beginner, without either experience or knowledge, may meet with discouragements—unquestionably will. Swarms will be lost, colonies will fail to winter, and the young apiarist will become nervous, which fact will be noted by the bees with great disfavor, and, if opportunity permits, will meet reproof more sharp than pleasant. Yet, with PERSISTENCE, all these difficulties quickly vanish. Every contingency will be foreseen and provided against, and the myriad of little workers will become as manageable and may be fondled as safely as a pet dog or cat, and the apiarist will minister to their needs with the same fearlessness and self-possession that he does to his gentlest cow or favorite horse. *Persistence, in the face of all these discouragements which are so sure to confront inexperience, will surely triumph.* In sooth, he who appreciates the beautiful and marvelous, will soon grow to love his companions of the hive, and the labor attendant upon their care and management. Nor will this love abate until it has been kindled into enthusiasm.

True, there may be successful apiarists who are impelled by no warmth of feeling, whose superior intelligence, system and promptitude, stand in lieu of, and make amends for, absence of enthusiasm. Yet I believe such are rare, and certainly they work at great disadvantage.

PART FIRST.

NATURAL HISTORY

OF THE

HONEY-BEE.

Natural History of the Honey-Bee.

CHAPTER L

THE BEE'S PLACE IN THE ANIMAL KINGDOM.

It is estimated by eminent naturalists that there are more than 1,000,000 species of living animals. It will be both interesting and profitable to look in upon this vast host, that we may know the position and relationship of the bee to all this mighty concourse of life.

BRANCH OF THE HONEY-BEE.

The great French naturalist, Cuvier, a cotemporary of Napoleon I, grouped all animals which exhibit a ring structure into one branch, appropriately named Articulates, as this term indicates the jointed or articulated structure which so obviously characterizes most of the members of this group.

The terms "joint" and "articulation," as used here, have a technical meaning. They refer not to the hinge or place of union of two parts, but to the parts themselves. Thus, the parts of an insect's legs are styled "joints" or "articulations." All apiarists who have examined carefully the structure of a bee, will at once pronounce it an Articulate. Not only is its body, even from head to sting, composed of joints, but by close inspection we find the legs, the antennæ, and even the mouth-parts, likewise jointed.

The worms, too, are Articulates, though in some of these, as the leech, the joints are very obscure. The bee, then, which gives us food, is distantly related to the dreaded tape-worm, with its hundreds of joints, which, mayhaps, robs us of the same food after we have eaten it; and to the terrible pork-worm, or trichina, which may consume the very muscles we have developed in caring for our pets of the apiary.

In classifying animals, the zoologist has regard not only to the morphology—the gross anatomy—but also to the embryology, or style of development before birth or hatching. On both embryological and morphological grounds, Huxley and other recent authors are more than warranted in separating the Vermes, or worms, from the Articulates of Cuvier as a separate phylum. The remaining classes are now included in the branch Arthropoda. This term, which means jointed feet, is most appropriate, as all of the insects and their allies have jointed feet, while the worms are without such members.

The body-rings of these animals form a skeleton, firm, as in the bee and lobster, or more or less soft, as in most larvæ. The hardness of the crust is due to the deposit within it of a hard substance called chitine, and the firmness of the insect's body varies simply with the amount of this chitine. This skeleton, unlike that of Vertebrates, or back-bone animals, to which man belongs, is outside, and thus serves to protect the inner, softer parts, as well as to give them attachment, and to give strength and solidity to the animal.

This ring structure, so beautifully marked in our golden-banded Italians, usually makes it easy to separate, at sight, animals of this branch from the Vertebrates, with their usually bony skeleton ; from the less active Molluscan branch, with their soft, sack-like bodies, familiar to us in the snail, the clam, the oyster, and the wonderful cuttle-fish—the devil-fish of Victor Hugo—with its long, clammy arms, strange ink-bag, and often prodigious size ; from the branch Echinodermata, with its graceful star-fish and sea-stars, and elegant sea-lilies ; from the Cœlenterata with its delicate but gaudy jelly-fish, and coral animals, the tiny architects of islands and even continents ; from the lowly Porifera or sponges which seem so little like an animal ; and from the lowest, simplest, Protozoan branch, which includes animals often so minute that we often owe our very knowledge of them to the microscope, and so simple that they have been regarded as the bond which unites plants with animals.

CLASS OF THE HONEY-BEE.

The honey-bee belongs to the class Hexapoda, or true insects. The first term is appropriate, as all have in the imago, or last stage, six legs. Nor is the second term less applicable, as the word "insect" comes from the Latin, and means to cut in, and in no other Arthropod does the ring-structure appear so marked upon merely a superficial examination.   More than this, the true insects when fully developed have, unlike all other Arthropods, three well-marked divisions of the body,

FIG. 1.

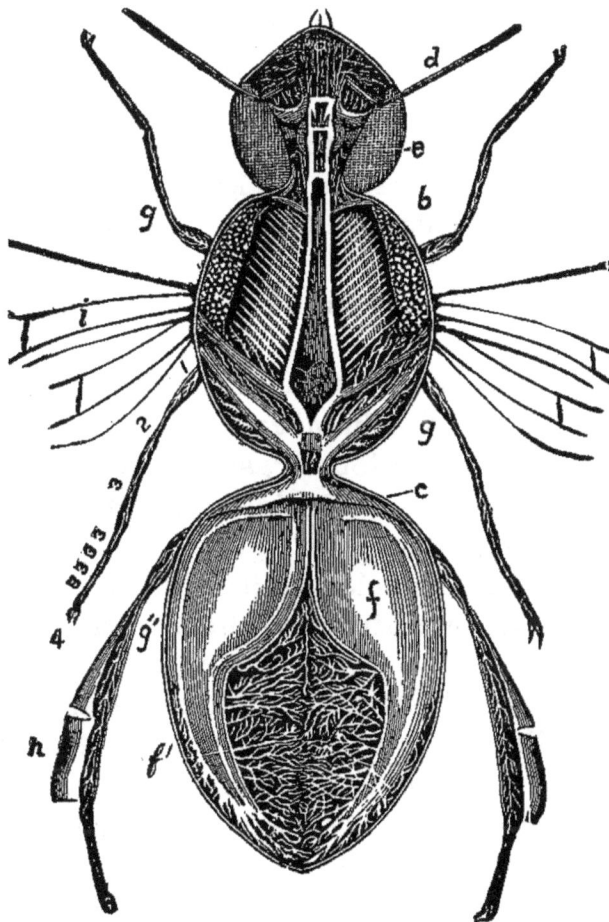

Respiratory Apparatus of Bee, magnified—After Duncan.

a Head, *b* Thorax, *c* Abdomen, *d* Antennæ, *e* Compound Eyes, *f* Air-sacs, *g g' g"* Legs, *f'* Tracheæ.

(Fig. 1), namely: the head (Fig. 1, *a*), which contains the antennæ (Fig. 1, *d*), the horn-like appendages common to all insects; eyes (Fig. 1, *e*), and mouth organs; the thorax (Fig. 1, *b*), which bears the legs (Fig. 1, *g*), and wings, when they are present; and lastly, the abdomen (Fig. 1, *c*), which, though usually without appendages, contains the ovipositor, and, when present, the sting. Insects undergo a more striking metamorphosis than do most other animals. When first hatched they are worm-like, and called "larvæ" (Fig. 39, *f*), which means masked; afterward they are frequently quiescent, and would hardly be supposed to be animals at all. They are then known as pupæ (Fig. 39, *g*). At last there comes forth the mature insect or imago (Fig. 1), with compound eyes, antennæ and wings. In some insects the transformations are said to be incomplete, that is, the larva, pupa, and imago differ little except in size, and that the latter possesses wings. The larvæ and pupæ of such insects are known as nymphs. We see in our bugs, lice, locusts and grasshoppers, illustrations of insects with incomplete transformations. In such cases there is a marked resemblance from the newly-hatched larva to the adult.

The other classes of the phylum Arthropoda, are the Crustacea, Myriapoda, and Arachnoidea. The Crustaceans include the jolly cray-fish and the lobster, so indifferent as to whether they move forward, backward or sidewise; the shorter crab, the sow-bug, lively and plump, even in its dark, damp home under old boards; and the barnacles, which fasten to the bottom of ships, so that vessels are often freighted with life, without, as well as within.

The myropods are the so-called "Thousand-Legged Worms." These are wormlike in form. The body is hardly differentiated at all. The name comes from the numerous legs, which though never a thousand may reach one-fourth that number. Myriapods have only simple eyes, and all have antennæ. Of the Myriapoda the Millipeds have numerous segments, often as many as sixty, have four legs to each joint, are cylindrical, and are often pests in the garden, as they are vegetable eaters. The Centipeds have fewer joints, may be no more than thirty, only one pair of legs to each segment, and feed on insects, etc. Their bite is venomous, and the bite of the larger ones may prove harmful even to man himself.

The Class Arachnida includes the spider group. These animals all have, when mature, eight legs. They never have but two parts to the body, the head-thorax and abdomen.

Their eyes are simple, and they are without antennæ. The wee mites belong here. They have hardly any divisions to the body. The mouth-parts form a mere portico. When first hatched they have only six legs. The so-called red spider (red mite), so destructive in the orchard, belongs here, as do also the sugar, cheese, flour, and chicken mites. The ticks are but colossal mites. Of these, the Texas Cattle Tick (Boophilus bovis) causes the Texas fever in cattle. The cause of the fever is a protozoan animal, Pyrosoma bigeminum. This is in the blood of Texas cattle, but is harmless. Carried by the tick to other cattle, it brings disease and death. The scorpions are also Arachnids. One of these stings as does the bee, and the sting is often quite venomous. The whip scorpion of Florida is named from its caudal appendages. It is entirely harmless. The Datames, which I call the "California bee-killer" (Fig. 292), and which is described among the bee-enemies, belongs here. Grandfather Graybeard also belongs in the scorpion order. It is only useful in pointing the way to lost cows, etc. Its legs point every way. The spiders are the highest Arachnids. They differ from mites in possessing two well-marked divisions of the body, and in always having eight legs, and from the scorpions in never having the abdomen jointed. The spiders have a poisonous bite, but rarely inflict injury to man. Their silk and spinning habits are exceedingly interesting. Spiders are almost as marvelous in their life history as are the bees. Like the Datames, to be spoken of as a bee-enemy, spiders often kill our pets of the hive.

ORDER OF THE HONEY-BEE.

The honey-bee belongs to the order Hymenoptera (from two Greek words meaning membrane and wing), which also includes the wasps, ants, ichneumon-flies, gall-flies, and saw-flies. This group contains insects which possess a tongue by which they may suck (Figs. 16 and 54), and strong jaws (Fig. 65) for biting. Thus, the bees can sip the honeyed sweets of flowers, and also gnaw away mutilated comb. They have, besides, four wings, and undergo complete transformations.

There are among insects strange resemblances. Insects of one order will show a marked likeness to those of another.

This is known as mimicry, and sometimes is wonderfully striking between very distant groups. Darwin and Wallace have shown this to be a developed peculiarity, not always possessed by the ancestors of the animal, and that it comes through the laws of variation and natural selection to serve the purpose of protection. Right here we have a fine illustration of this mimicry. Just the other day I received, through Mr. A. I. Root, an insect which he and the person sending it to him supposed to be a bee, and he desired to know whether it was a malformed honey-bee, or some other species. This insect, though looking in a general way much like a bee, had only two wings, had no jaws, and its antennæ were close together in front, and mere stubs. In fact, it was no bee at all, but belonged to the order Diptera, or two-winged flies. I have received several similar insects, with like inquiries. Among Diptera there are several families, as the Œstridæ, or bot-flies, some of the Asilidæ, or robber-flies (Fig. 268), which are often fierce enemies of our bees, the Syrphidæ—a very useful family, as the larvæ or maggots often live on plant-lice—whose members are often seen sipping sweets from flowers, or trying to rob the honey from other bees—the one referred to above belonged to this family—and the Bombyliidæ, which in color, form, and hairy covering, are strikingly like wild and domesticated bees. The maggots of some of these feed on the larvæ of various of our wild bees, and of course the mother fly must steal into the nests of the latter to lay her eggs. So in these cases there is seeming evidence that the mimicry may serve to protect these fly-tramps as they steal in to pilfer the coveted sweets, or lay the fatal eggs. Possibly, too, they may have a protective scent, as they have been seen to enter a hive in safety, though a bumble-bee essaying to do the same found the way barricaded with myriad simitars, each with a poisoned tip.

Some authors have placed Coleoptera, or beetles, as the highest of insects, others claim for Lepidoptera, or butter-flies and moths, a first place, while others, and with the best of reasons, claim for Hymenoptera the highest position. The larger brain, wondrous habits, and marvelous differentiation of mouth-organs, legs, etc., more than warrant placing them at the head. The moth is admired for the glory of its coloring and elegance of its form, and the beetle for the luster and brilliancy of its elytra, or wing-covers; but these insects only revel in Nature's wealth, and live and die without labor or purpose. Hymenoptera, usually less gaudy, often quite plain and unattractive in color, are yet the most highly endowed among insects. They live with a purpose in view, and are the best models of industry to be found among animals. Our bees practice a division of labor; the ants are still better political economists, as they have a specially endowed class in the community which are the soldiers, and thus are the defenders of each ant-kingdom. Ants also conquer other communities, take their inhabitants captive, and reduce them to abject slavery— requiring them to perform a large portion, and sometimes the whole, of the labor of the community. Ants tunnel under streams, and in the tropics some leaf-eating species have been observed to show no mean order of intelligence, as some ascend trees to cut off the leafy twigs, while others remain below and carry these branches through their tunnels to their underground homes. Indeed, the Agricultural ant, of Texas, actually clears land and grows a special kind of plant on which it feeds. (See McCook's Ants.)

The parasitic Hymenoptera are so-called because they lay their eggs in other insects, that their offspring may have fresh meat not only at birth, but so long as they need food, as the insect fed upon generally lives till the young parasite, which is working to disembowel it, is full-grown; thus this steak is ever fresh as life itself. These parasitic insects show wondrous intelligence, or sense-development, in discovering their prey. I have caught ichneumon-flies—a family of these parasites—boring through the bark and a thin layer of solid beech or maple wood, and upon examination I found the prospective victim further on in direct line with the insect auger, which was to intrude the fatal egg. I have also watched ichneumon-flies depositing eggs in leaf-rolling caterpillars, so surrounded with tough hickory leaves that the fly had to pierce several thicknesses to place the egg in its snugly-ensconced victim. Upon putting these leaf-rolling caterpillars in a box, I reared, of course, the ichneumon-fly and not the moth. Is it instinct or reason that enables these flies to gauge the number of their eggs to the size of the larva which is to receive them, so that there may be no danger of famine and starvation? For true it is that while small caterpillars will receive but few eggs, large ones may receive several. Even the honey-bee some-times falls victim to such parasites, as I shall show in speak-ing of enemies of bees. How strange the habits of the saw-fly, with its wondrous instruments, more perfect than any saws of human workmanship, and the gall-flies, whose poison-ous stings, as they fasten their eggs to the oak, rose, or other leaves, cause the abnormal growth of food for the still un-hatched young. In the South it is reported that bees often obtain no small amount of nectar from species of oak-galls.

The providing and caring for their young, which are at first helpless, is peculiar among insects, with slight exception, to the Hymenoptera, and among all animals is considered a mark of high rank. Such marvels of instinct, if we may not call it intelligence, such acumen of sense perception, such wonderful habits, all these, no less than the compact structure, small size and specialized organs of nicest finish, more than warrant that grand trio of American naturalists—Agassiz, Dana, and Packard—in placing Hymenoptera first in rank among insects. As we shall detail the structure and habits of the *highest* of the high—the bees—in the following pages, I am sure no one will think to degrade the rank of these wonders of the animal kingdom.

FAMILY OF THE HONEY-BEE.

The honey-bee belongs to the family Apidæ, of Leach, which includes not only the hive-bee, but all insects which feed their helpless larvæ on pollen, pollen or honey, or food digested by the adult bees.

Many authors separate the lower bees, principally because of their shorter tongues, from the others, under the family name, Andrenidæ. In this case all the bees are grouped as Mellifera or Anthophila Latr. I shall group all bees in the one family Apidæ, and regard the Andrenæ and their near relatives as a sub-family. The insects of this family all have branched or plumose hairs on some portion of the body, broad heads, elbowed antennæ (Fig. 1, *d*), which are thirteen jointed in the males, and only twelve jointed in the females. The jaws or mandibles (Fig. 65) are strong and usually toothed. The tongue or ligula is very long and slim in the higher genera, but short and flattened in the lower ones. The second jaws or maxillæ (Fig. 54, *m x*) are long and prominent, and ensheath the tongue, with which they are folded back when not in use, once or more under the head. All the insects of this family have, on the four anterior legs, a stiff spine on the end of the tibia (Fig. 69), the fourth joint of the leg from the body—called the tibial spur, and all except the genus Apis, which includes the honey-bee, in which the posterior legs are without tibial spurs, have two tibial spurs on the posterior legs. Nearly all bees (the parasitic genera are exceptions) have the first joint of the tarsus of the posterior legs much broadened (Fig. 71), and this, together with the broad tibia, is hollowed out (Fig. 70), forming quite a basin or basket—the corbicula—on the outer side, in the species of Apis and Bombus, which basket is deepened by long, stiff hairs. These receptacles, or pollen-baskets, are found only in such bees as gather much pollen. A few of the Apidæ—thieves by nature—cuckoo-like, steal unbidden into the nests of others, and here lay their eggs. As their young are fed and fostered by another, such bees gather no pollen, and so, like drone-bees, need no organs for collecting it. These parasites illustrate mimicry, already described, as they look so like the foster-mothers of their own young that unscientific eyes would often fail to distinguish them. Probably the bees thus imposed upon are no sharper, or they would refuse ingress to these merciless vagrants.

The larvæ (Fig. 39, *f*) of all insects of this family are maggot-like, wrinkled, footless, tapering at both ends, and, as already stated, have their food prepared for them. They are helpless, and thus all during their babyhood (the larva state)—the time when all insects are most ravenous, and the only time when many insects take food; the time when all growth in size, except such enlargement as is required by egg-development, occurs—these infant bees have to be fed by their mothers or elder sisters. They have a mouth with soft lips, and weak jaws, yet it is doubtful if all or much of their food is taken in at this opening. There is much reason to believe that the honey-bees especially, like many maggots—such as the Hessian-fly larvæ—absorb much of their food through the body walls. From the mouth leads the alimentary canal, which has no anal opening. So there are no excreta other than gas and vapor, except the small amount which remains in the stomach and intestine, which are shed with the skin at the time of the last molt. What commendation for their food, nearly all capable of nourishment, and thus assimilated !

To this family belongs the genus of stingless bees, Melipona, of Mexico and South America, which store honey not only in the hexagonal brood-cells, but in great wax-reservoirs. They, like the unkept hive-bee, build in hollow logs. They are exceedingly numerous in each colony, and it has thus been thought that there was more than one queen. They are also very prodigal of wax, and thus may possess a prospective commercial importance in these days of comb foundation. In this genus the basal joint of the tarsus is triangular, and there are two submarginal cells, not three, to the front wings. They are also smaller than our common bees, and have wings that do not reach the tip of their abdomens. Mr. T. F. Bingham, inventor of the bee-smoker, bought a colony of the stingless bees from Mexico to Michigan. The climate seemed unfavorable to them, as soon the bees all died. I now have some of the bees, and their great black honey and pollen cells in our museum. The corbiculæ, or pollen-baskets, are specially well marked, and the posterior tibial spur is wanting in these small bees.

Another genus of stingless bees, the genus Trigona, have the wings longer than the abdomens, and their jaws toothed. These, unlike the Melipona, are not confined to the New World, but are met with in Africa, India, and Australasia. These build their combs in tall trees, fastening them to the branches much as does the Apis dorsata, soon to be mentioned.

Of course insects of the genus Bombus—our common bumble-bees—belong to this family. Here the tongue is very long, the bee large, and the sting curved, with the barbs very short and few. Only the queen survives the winter. In spring she forms her nest under some sod or board, often in a deserted mouse-nest, hollowing out a basin in the earth, and after storing a mass of bee-bread she deposits several eggs in the mass. The larvæ are soon hatched out and develop in large, coarse cells, not unlike the queen-cells of our hive-bees. When the bees issue from these cells the latter are strengthened with wax. Later in the season, these coarse wax-cells, which contain much pollen, become very numerous, serving both for brood and honey. At first, in spring, the queen has all to do, hence the magnificent bumble-bees, the queens, seen about the lilacs in early spring. Soon the smaller workers become abundant, and relieve the queen, which then seldom leaves the nest. Later, the drones and the smaller, because yet unimpregnated and non-laying, queens appear. Thus, the bees correspond with those of the hive. The young queens mate in late summer, and are probably the only ones that survive the winter. Mating is performed on the wing. I once saw a queen Bombus fall to the earth, dragging a male from which she would have torn loose had I not captured both. The bumble-bee drones are often seen collected about shady places at the mating season in August.

Bees of the genus Xylocopa much resemble bumble-bees, though they are usually black, less hairy, and are our largest bees. They have not the corbiculæ. These are among our finest examples of boring insects. With their strong bidentate jaws they cut long tunnels, often two or more feet long, in sound wood. These burrows are partitioned by chips into cells, and in each cell is left an egg and bee-bread for the larva, soon to hatch. These bees do no slight damage by boring into cornices, window-casings, etc., of houses and out-buildings. At my suggestion, many people thus annoyed have plugged these tunnels with a mixture of lard and kerosene, and have speedily driven the offending bees away. These are the bees which I have discovered piercing the base of long tubular flowers, like the wild bergamot. I have seen honey-bees visiting these slitted flowers, the nectar of which was thus made accessible to them. I have never seen honey-bees biting flowers. I think they never do it. Xylocopa Californica is very common here at Claremont. The females are

black, and the males light yellow. My students told me the
females would not sting. I said that was strange, and picked
one up. I threw it down very quickly, and have not repeated
the experiment.

The mason-bees—well named—construct cells of earth,
which, by aid of their spittle, they cement so that these cells
are very hard. There are several genera of these bees, the
elegant Osmia, the brilliant Augochlora, the more sober but
very numerous Andrena—the little black bees that often steal
into the hives for honey, etc. Some burrow in sand, some
build in hollowed-out weeds, some build mud cells in crevices,
even small key-holes not being exempt, as I have too good
reason to know. The Yale locks in our museum have thus
suffered. Here the lard and kerosene mixture again comes in
play.

The tailor, or leaf-cutting, bees, of the genus Megachile,
make wonderful cells from variously shaped pieces of leaves.
These are always mathematical in form, usually circular and
oblong, are cut—the insect making scissors of its jaws—from
various leaves, the rose being a favorite. I have found these
cells made almost wholly of the petals or flower-leaves of the
rose. The cells are made by gluing these leaf-sections in con-
centric layers, letting them overlap. The oblong sections form
the walls of the cylinder, while the circular pieces are crowded
into the tubes as we press circular wads into our shot-guns,
and are used at the ends, or for partitions where several cells
are placed together. When complete, the single cells are in
form and size much like a revolver cartridge. When several
are placed together, which is usually the case, they are
arranged end to end, and in size and form are quite like a small
stick of candy, though not more than one-third as long.
These cells I have found in the grass, partially buried in the
earth, in crevices, and in one case knew of their being built in
the folds of a partially-knit sock, which a good house-wife had
chanced to leave stationary for some days. These leaf-cutters
often have yellow hairs underneath their bodies, which aid
them in carrying pollen. I have noticed them each summer,
for some years, swarming on the Virginia creeper, often called
woodbine, while in blossom, in quest of pollen, though I have

rarely seen the hive-bee on these vines. The tailor-bees often cut the foliage of the same vines quite badly. The males of these bees have curiously modified, and broadly fringed anterior legs. I have found these tailor-bees as common in California as in Michigan.

I have often reared beautiful bees of the genera Osmia and Augochlora, which, as already stated, are also called mason-bees. Their glistening colors of blue and green possess a luster and reflection unsurpassed even by the metals themselves. These rear their young in cells of mud, in mud-cells lining hollow weeds and shrubs, and in burrows which they dig in the hard earth. In early summer, during warm days, these glistening gems of life are frequently seen in walks and drives intent on gathering earth for mortar, or digging holes, and will hardly escape identification by the observing apiarist, as their form is so much like that of our honey-bees. They are smaller, yet their broad head, prominent eyes, and general form, are very like those of the equally quick and active, yet more soberly attired, workers of the apiary. The beautiful— often beautifully striped—species of Ceratina look much like those of Osmia, but they nest in hollows in stems of various plants, which, in some cases, they themselves form. In southwestern Michigan they do no little harm by boring the blackberry canes. They have simple hind legs.

Other bees—the numerous species of the genus Nomada, and of Apathus—are the black sheep in the family Apidæ. These tramps, already referred to, like the English cuckoo and our American cow-blackbird, steal in upon the unwary, and, though all unbidden, lay their eggs ; in this way appropriating food and lodgings for their own yet unborn. Thus these insect vagabonds impose upon the unsuspecting foster-mothers in their violated homes, and these same foster-mothers show by their tender care of these merciless intruders, that they are miserably fooled, for they carefully guard and feed infant bees, which, with age, will in turn practice this same nefarious trickery. The Apathus species are parasites on the Bombus ; the Nomada species, which are small bees, often beautifully ringed, on the small black Andrenæ.

The species of Andrena, Halictus, the red Sphecodes, and

others of the Andrenidæ of some authors, have short, flat tongues, with equi-jointed labial palpi. These bees have been little studied, and there are very numerous undescribed species.

I reluctantly withhold further particulars of this wonderful bee-family. When first I visited Messrs. Townley and Davis, of Michigan, I was struck with the fine collection of wild-bees which each had made. Yet, unknowingly, they had incorporated many that were not bees. Of course, many apiarists will wish to make such collections, and also to study our wild bees. I hope the above will prove both a stimulus and aid. I hope, too, that it will stimulate others, especially youth, to the valuable and intensely interesting study of these wonders of nature. I am glad to open to the reader a page from the book of nature so replete with attractions as is the above. Nor do I think I have taken too much space in revealing the strange and marvelous instincts, and wonderfully varied habits, of this brightest of insect families, at the head of which stand our own fellow-laborers and companions of the apiary.

I shall be very glad to receive specimens of wild bees from every State in our country. To send bees or other insects, kill with gasoline or chloroform, wrap with cotton or tissue paper, so as to prevent injury, and mail in a strong box.

THE GENUS OF THE HONEY-BEE.

The genus Apis includes all bees that have no tibial spurs on the posterior legs, and at the same time have three cubital or sub-costal cells (6, 7, 8, Fig. 2)—the second row from the costal or anterior edge—on the front or primary wings. The marginal cell (Fig. 2, 5) is very long. On the inner side of the posterior basal tarsus, opposite the pollen-baskets, in the neuters or workers, are rows of hairs (Fig. 71), which are used in collecting pollen. In the males, which do no work except to fertilize the queens, the large compound eyes meet above, crowding the simple eyes below (Fig. 3), while in the workers (Fig. 4) and queens these simple eyes (called ocelli) are above, and the compound eyes wide apart. The compound eyes are in all cases hairy (Figs. 3, 4). The drones and queens have weak jaws, with a rudimentary tooth (Fig. 65, *a b*), short tongues, and no pollen-baskets, though they have the broad tibia and wide basal tarsus (Fig. 48, *t*, *s*).

There is some doubt as to the number of species of this

FIG. 2.

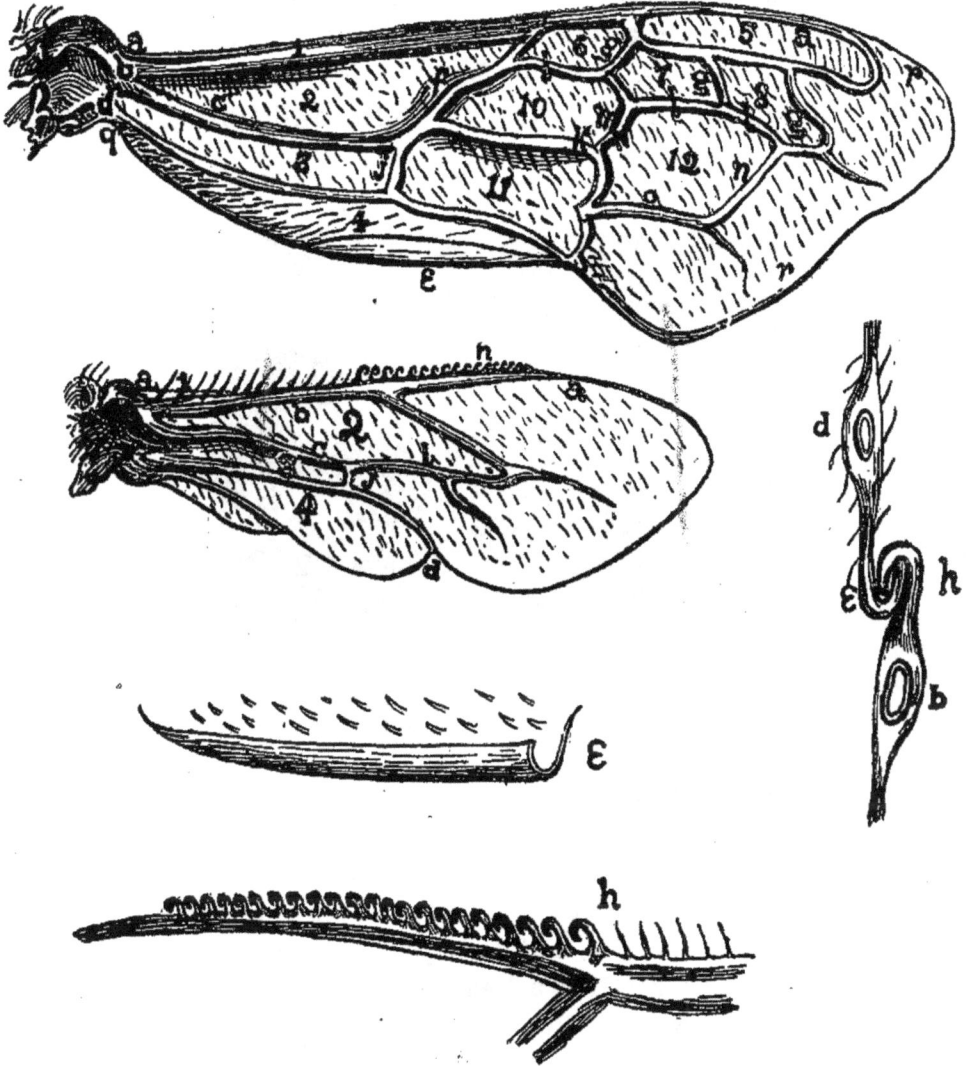

Anterior and Posterior Wings of Bee.—Original.

h Hooklets much magnified.
a Costal vein.
b Sub-costal vein.
c Median vein.
d Anal vein.
q Posterior margin.
e Fold where hooklets catch.
h Hooklets.

1 Costal cell.
2 Median cell.
3 Sub-median cell.
4 Anal cell.
5 Marginal cell.
6, 7, 8, Sub-marginal cells.
10, 11 and 12, Discoidal cells.

genus. It is certain that the Italian bee, the Egyptian bee, the Cyprian bee, and the bees of Syria, of which Mr. Benton states that there are at least two distinct races, are only races of the Apis mellifera, which also includes the Tunisian or Punic bees, the Carniolan, and the German or black bee.

Mr. F. Smith, an able entomologist of England, considers Apis dorsata of India and the East Indies, Apis zonata of the Philippine Islands, Apis indica of India and China, and Apis

FIG. 3.

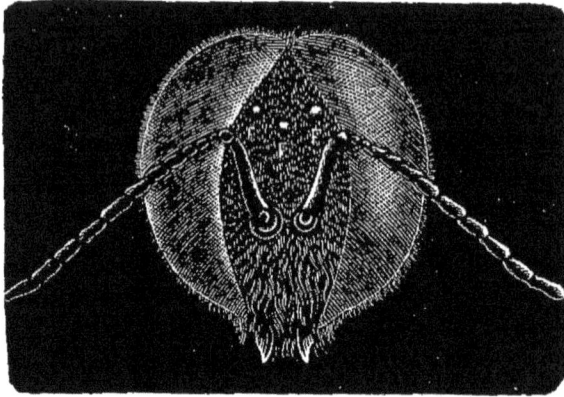

Head of Drone, magnified.—Original.

Antennæ, Compound Eyes, Simple Eyes.

florea of India, Ceylon, China, and Borneo, as distinct species. He thinks, also, that Apis adonsoni and Apis nigrocincta are distinct, but states that they may be varieties of Apis indica. Others think them races of dorsata. Some regard Apis unicolor as a distinct species, but it is probably a variety of Apis mellifera. As Apis mellifera has not been found in India, and is a native of Europe, Western Asia, and Africa, it seems quite possible, though not probable, that several of the above may turn out to be only varieties of Apis mellifera. If there are only color and size to distinguish them, and, indeed, one may add habits, then we may suspect, with good reason, the validity of the above arrangement. If there be structural difference, as Mr. Wallace says there is, in the male dorsata, then we may call them different species. The Italian *certainly* has a longer tongue than the German, yet that is not sufficient to separate them as species. Apis zonata of the East Indies, and Apis unicolor, are said to be very black.

I append the following chart, which I think represents pretty accurately the species, races and varieties of the genus Apis. (See page 48.)

Where a race is followed by an interrogation point, there is a question if it should not be considered a variety of the last preceding race not thus marked. Some of the races, like the Italian, Cyprian, Greek, etc., Vogel considers had their origin in a cross between the yellow and black races. Vogel's conclusion was reached from a long series of experiments, crossing Italian and German bees, and then breeding from such crosses. It seems likely that through the law of variation

FIG. 4.

Head of Worker, magnified.—Original.

Antennæ, Compound Eyes, Simple Eyes.

each race might have originated independently, or possibly all, as varieties of the Egyptian bee.

In the autumn of 1879, Mr. D. A. Jones, of Beeton, Ontario, Canada, inaugurated the grandest enterprise ever undertaken in the interest of apiculture. This was nothing less than to visit Cyprus, Syria, and the more distant India and the East Indies, for the purpose of securing and introducing into America such species and races of bees as gave promise of superior excellence. Mr. Jones procured the services of Mr. Frank

SPECIES.	RACES.	VARIETIES.
Apis Indica, Fab.	A. dorsata nigripennis, Latr.	
Apis florea, Fab.	A. dorsata bicolor, Klug.	
Apis dorsata, Fab.	A. dorsata zonata, Smith.	
	A. mellifera nigra. German Bee.	Carniolan or Krainer. Heath. Hungarian. Dalmatian. Herzegovinian. Smyrnian. Tunisian. Common black.
	A. mellifera fasciata, Egyptian Bee.	
	Syrian (?)	
	South Palestine (?)	
Apis mellifera.	Cyprian (?)	
	Italian (?)	
	Greek (?)	
	Bonnat (?)	
	Caucasian (?)	
	A. mellifera unicolor Latr. Madagascar.	
	A. mellifera adonsoni, African Bee.	

Benton, a graduate of the Michigan Agricultural College, a
fine linguist and skilled apiarist, to aid in his undertaking.
After visiting the principal apiaries of Europe, these gentle-
men located at Larnica, in the island of Cyprus, where they
established a large apiary composed of Cyprian and Syrian
bees. The Cyprian bees were purchased on the island, while
the Syrians were procured personally by Mr. Jones in Syria.
The following June Mr. Jones returned to America with several hundred queens of these two races. Mr. Benton remained
at Larnica to rear and ship more queens to Europe and America. The following winter Mr. Benton visited Ceylon, Farther
India, and Java, as Mr. Jones was determined to ascertain if
there were better bees than those we already had, and if so to
secure them. Apis dorsata (Figs. 5, 6) was the special object of
the quest, and as this bee was known as the "great bee of
Java," Mr. Benton visited that island, in hopes to procure
these bees. But to the sore disappointment not only of those
who had the enterprise in charge, but of all progressive apiarists, the bees in question were not to be found on that island.
Mr. Benton learned at a great cost that this bee is rare in

Java, but common in the jungles of Ceylon, Hindoostan, Farther India, Sumatra, Borneo, and Timor. In Ceylon, Mr.

FIG. 5.

A. dorsata Worker, X2.
(From Department of Agriculture.)

Benton saw many colonies, most of which were in inaccessible places, though he secured, after great labor and hardship, four colonies.

FIG. 6.

A. dorsata Drone, X2.
(From Department of Agriculture.)

These bees usually suspend their great combs, which are often six feet long and four feet wide, to overhanging rocks, or to horizontal branches of trees. In one case, Mr. Benton found them in the crevice of a rock, nearly surrounded by the same. This indicates that they may be kept in hives. The combs hang side by side, so do those of our common bees, but are one-half inch apart. Mr. Benton found the tops of the combs, which contain the honey, from three to six inches thick, while those where brood is reared, are one and one-half inches thick. Drones and workers are all reared in the same cells, which are about the size of the drone brood-cells of our honey-comb, The worker-bees, some specimens of which I have received from

FIG. 7.

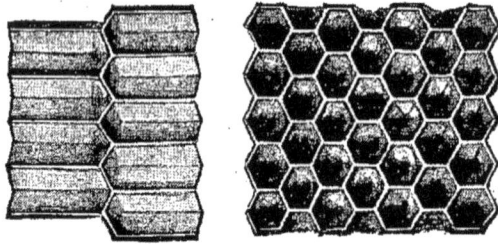

Worker-Cells.—A. indica.
(From Department of Agriculture.)

Mr. Jones, in size and general appearance much resemble our Italian queens. They have blue-black wings, black bodies, which are ringed very much as are our Italians, only the yellow largely predominates. Mr. Benton writes me that in form and style of flight they much resemble wasps. They are the same size as the drones, varying from three-fourths to seven-eighths of an inch in length. They are easily handled by aid of smoke, and are very clumsy in their attempts to sting. Their sting is no larger than that of our common bees, while the pain from their sting, Mr. Benton says, is not so great. The drones are dark brown, marked with yellow. Strangely enough, they only fly, unless disturbed, after sundown. This is unfortunate, as with the same habits we might hope to mate them with our common bees, and thus procure a valuable cross. This may be a developed peculiarity, to protect them from birds, and so might very likely disappear with domestication. The queens are leather-colored, and smaller, as compared with the workers, than are our common queens. The queens are more restless than are the workers while being

handled. While procuring these bees, Mr. Benton was pros-
trated with a fever, and so the bees, during their long voyage
to Syria, were neglected. Strange to say, one colony survived
the long confinement, but perished soon after reaching Syria.
We can not call this journey a failure, as we now have the
information that will render a second attempt surely successful. What has been learned will make the enterprising bee-
keeper more desirous than ever to secure these bees. Their
large size, and immense capabilities in the way of wax-secretion, as well as honey-storing, give us reason to hope for substantial benefits from their importation.

Mr. Benton also found A. indica and A. florea on the
Island of Ceylon. I have received some of the bees and comb
of the former species. The comb is very delicate, the cells
(Fig. 7) being only one-sixth of an inch in diameter. The
workers are less than one-half of an inch long, brown in color,
and their entire abdomens are beautifully ringed with brown
and yellow. The drones are black, and very small. The one
I have measures an eighth of an inch less in length than does
the worker. The queens are leather-colored, and very large as
compared with the workers. They are as large as are our

FIG. 8.

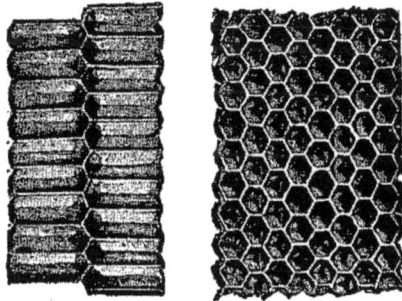

Worker-Cells.—A. Florea.
(From Department of Agriculture.)

common queens. These bees are very quick, and are domesticated on the Island of Ceylon. The workers of A. florea are
also banded, and are more beautiful even than those of A.
indica. They are very small. The combs are not larger than
one's hand, and so diminutive are the cells (Fig. 8) that 100

bees are produced to the square inch. The color is blue-black, with the basal third of the abdomen orange.

The sting of these two species is very small. From the small amount of stores which they gather, the tendency which they have to swarm out, and their inability to stand the cold, these two species promise little of value except from a scientific point of view. One colony of A. florea was brought by Mr. Benton to Cyprus, but it swarmed out and was lost.

It seems strange that the genus Apis should not have been native to the American continent. The "large brown bee" which some of our bee-keepers think native to America, is undoubtedly but a variety of the common black, or German, bee. Without doubt there were no bees of this genus here till introduced by the Caucasian race. It seems more strange, as we find that all the continents and islands of the Eastern Hemisphere abound with representatives. It is one more illustration of the strange, inextricable puzzles connected with the geographical distribution of animals.

SPECIES OF OUR HONEY-BEES.

The bees at present domesticated are all of one species—A pis mellifera. The character of this species will appear in the next chapter, as we proceed with their anatomy and physiology. As before stated, this species is native exclusively to the Eastern Hemisphere, though it has been introduced wherever civilized man has taken up his abode.

RACES OF THE HONEY-BEE.

German or Black Bees.

The German or black bee is the race best known, as through all the ages it has been most widely distributed. The name "German" refers to locality, while the name "black" is a misnomer, as the bee is a gray-black. The queen, and, in a less degree, the drones, are darker, while the legs and under_ surface of the former are brown, or copper-colored, and of the latter light gray. The tongue of the black worker I have found, by repeated dissections and comparisons made both by myself and by my pupils, is shorter than that of the Italian worker, and generally less hairy. The bees are more irritable, and so more likely to sting than the Italians. They are also wont to keep flying before one's face in threatening mien for hours, until killed. The wise apiarist will dispatch such quarrelsome workers at once. The black bees have been known no longer than the Italians, as we find the latter were known both to Aristotle, the fourth century B. C., and to Virgil, the great Roman poet, who sang of the variegated golden bee, the first century B. C.; and we can only account for the wider distribution of the German bee by considering the more vigorous, pushing habits of the Germanic races, who not only over-ran and infused life into Southern Europe, but have vitalized all christendom.

Ligurian or Italian Bee.

The Italian bee is characterized as a race, not only by difference of color, habits, and activity, but also by possessing a little longer tongue. These bees were first described as distinct from the German race by Spinola, in 1805, who gave the name " Ligurian " bee, which name prevails in Europe. The name comes from a province of Northern Italy, north of the Ligurian Gulf, or Gulf of Genoa. This region is shut off from Northern Europe by the Alps, and thus these bees were kept apart from the German bees, and in warmer, more genial Italy, was developed a distinct race—our beautiful Italians. It seems to me quite reasonable to suppose from the appearance of the bees, and also from the migrations of the human race, that the Italian bee is an off-shoot from the Cyprian, and quite likely both of these of the Syrian race.

In 1843, Von Baldenstein procured a colony of these bees, which he had previously observed as peculiar, while stationed as a military captain in Italy. He published his experience in 1848, which was read by Dzierzon, who became interested, and through him the Italian became generally introduced into Germany. In 1859—six years after Dzierzon's first importation—the Italian bee was introduced into England by Neighbour. The same year, Messrs. Wagner and Colvin imported the Italians from Dzierzon's apiary into America ; and in 1860,

Mr. S. B. Parsons brought the first colonies that were imported direct from Italy.

The Italian worker-bee is quickly distinguished by the bright yellow rings at the base of the abdomen. Perhaps golden would be a better term, as these bands are often bright orange. If the colony be pure, every bee will show three of these golden girdles (Fig. 9, A, B. C). The first two segments

FIG. 9.

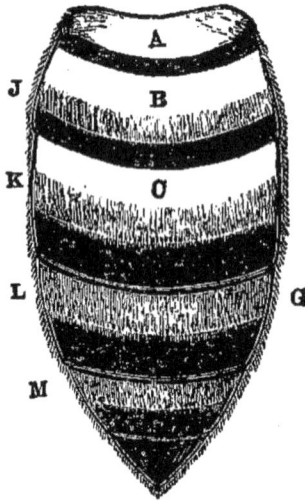

Abdomen of Italian Worker.
(From A. I. Root Co.)

or rings of the abdomen, except at their posterior border, and also the base or anterior border of the third, will be of this orange-yellow hue. The rest of the back or dorsal surface will be much as in the German race. Underneath the abdomen, except for a greater or less distance at the tip, will also be yellow, while the same color appears more or less strongly marked on the legs. The workers have longer ligulæ or tongues (Fig. 54) than the German race, permitting them to gather nectar from long flower-tubes, which is inaccessible to our common bees, and their tongues are also a little more hairy than are those of the black bees. They are also more active, and less inclined to sting. The queen has the entire base of her abdomen, and sometimes nearly the whole of it, orange-yellow. The variation as to amount of color is quite striking. Sometimes very dark queens are imported right from the Ligurian hills, yet all the workers will wear the badge of purity—the three golden bands.

The drones are quite variable. Sometimes the rings and patches of yellow will be very prominent, then, again, quite indistinct. But the under side of the body is always, so far as I have observed, mainly yellow.

A variety of our Italian bees, which is very beautiful and gentle, has the rows of white hairs (Fig. 9, J, K, L, M) unusually distinct, and is being sold in the United States under the name of Albinos. They are not a distinct race. In fact, I have often noticed among Italians the so-called Albinos several times, and have not found them superior, or even equal, I think, to the average Italian.

THE SYRIAN AND CYPRIAN RACES.

Through the enterprise of Messrs. D. A. Jones and Frank Benton, we now have these races in our country, and have proved the truth of the assertion of noted European apiarists, that the Cyprian is a distinct race of bees.

Mr. Benton, than whom no one is better fitted to express a correct opinion, thinks that the Cyprian bees are the offspring of the Syrian. This opinion is strengthened by the close resemblance of the two races, and by the fact that migrations of all kinds have gone westward. A similar argument would make it presumable that the Cyprians gave rise to the Italians.

The Cyprian bees resemble the Italians very closely. They may be distinguished by the bright leather-colored lunulc which tips their thorax posteriorly, and by the fact that the under side of their bodies is yellow to the tip. They are more active than are the Italians, and the queens are more prolific.

The good qualities of the Italians seem all to be exaggerated in the Cyprians, except the trait of amiability. The Cyprian bees are second only to the Egyptian in irritability. That they will become less cross with handling is to be expected.

The Syrian bees are from Asiatic Turkey, north of Mount Carmel, and are a very well marked race. The Syrian queens are remarkably uniform. Their abdomens above are, like the little A. indica, beautifully banded with yellow and black. They are very quick and remarkably prolific. They do not cease laying even when the honey-flow ceases. They are often

FIG. 10.

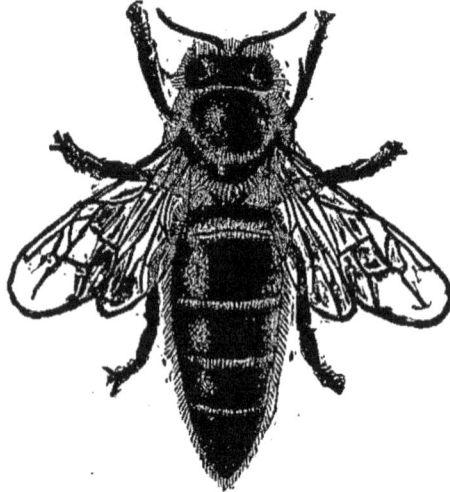

Carniolan Queen.—X2.
(From Department of Agriculture.)

kept prisoners in the cells longer than are queens of other races, and so may fly out at once upon emerging. They

FIG. 11.

Carniolan Worker.—X2.
(From Department of Agriculture.)

emerge from the cells at about the same time, so that often all the queens may emerge from the cells within a few hours, or even one hour. The workers closely resemble those of the

Italian race, only that they were more yellow beneath, and when first from the cells, or newly hatched, they are very dark, owing to the fact that the body-rings seemed pushed together. They are admirable in the way they defend their hives against robbers, the ease with which they are shaken from the combs, their great activity, their great tendency to remain in the hive on very windy days, the wonderful fecundity of the queen, her persistence in laying during a dearth of nectar-secretion, and their great euperiority for queen-rearing.

FIG. 12.

Carniolan Drone.— X2.

(From Department of Agriculture.)

often starting fifty or more good queen-cells. Neither the Cyprian nor Syrian has found favor in America, and have largely been given up.

OTHER RACES.

The Egyptian bees are very yellow, intensely cross, and frequently have laying workers. These are probably the bees which are famous in history, as having been moved up and down the Nile, in rude boats or rafts, as the varying periods of nectar-secreting bloom seemed to demand.

The heath bees of Northern Germany are much like the common German bees, of which they are a strain, except that they are far more inclined to swarm.

The Carniolan bees (Figs. 10, 11, 12) of Southwestern

Austria, also called Krainer bees, from the mountainous region of Krain, Austria, are praised as a very hardy variety. They are black with white rings—a sort of albino—German bee. They are like the heath variety, but are specially noted for their very gentle dispositions. Some European bee-keepers claim that this strain or variety is much superior to the common German bees. Mr. Benton, I think, holds strongly to this opinion. After a brief trial I am pleased with these bees.

The Hungarian bees are longer than the typical German race, and are covered with gray hairs. During the poor season of 1875 in Europe, these bees, like the Carniolans, were found superior even to the Italians.

The beautiful Dalmatian bees are slim, wasp-like, and very black. The rings of their abdomens are banded with lightish yellow. Their honey is even more white and beautiful than that of the German race. Some of the best European bee-keepers claim that they are superior to the Italian bees.

Akin to the Dalmatian bees are the Herzegovinian variety, which comes from the mountainous region of Bosnia, bordering on the Adriatic Sea. A better marked variety—the Smyrnian bees—from Western Asia, and also much praised by some of the noted Austrian bee-keepers, as are also the Caucasian, from the Caucasus Mountains, which are said to be very active and amiable.

The Tunisian bees, from Tunis in the north of Africa, are said to be even darker than the black or German bee. They are described as quite irritable. These were the " Punic bees " sold in the United States some years since. They did not keep in favor. It is stated that there is a race of bees which are domesticated in the south of Africa. From the descriptions I should think them quite like our Albinos in appearance. They are said to be excellent honey-producers, and to work even by moonlight. It is quite likely that some of these varieties might be found to endure our severe winters better than the pure German type, or the Italians. Now that we are to have an experimental station in each State, we may expect that all these races will be imported, that we may prove them and know which is the best.

BIBLIOGRAPHY.

It would be a pleasing duty, and not an unprofitable one, to give in this connection a complete history of entomology so far as it relates to Apis mellifera. But this would take much space, and as there is quite a full history in books, that I shall recommend to those who are eager to know more of this interesting department of natural history, I will not go into details.

Aristotle wrote of bees more than three hundred years B. C. About three hundred years later, Virgil, in his fourth Georgic, gave to the world the views then extant on this subject, gathered largely from the writings of Aristotle. The poetry will ever be remarkable for its beauty and elegance— would that as much might be said for the subject matter, which, though full of interest, is full of errors. A little later, Columella, though usually careful and accurate in his observations, still gave voice to the prevailing errors, though much that he wrote was valuable, and more was curious. As Mr. Langstroth once said to me, Columella wrote as one who had handled the things of which he wrote; and not like Virgil, as one who was dealing with second-hand wares. Pliny, the elder, who wrote in the second century, A. D., helped to continue the erroneous opinions which previous authors had given, and not content with this, he added opinions of his own, which were not only without foundation, but were often the perfection of absurdity.

After this, nearly two thousand years passed with no progress in natural history; even for two centuries after the revival of learning, we find nothing of note. Swammerdam, a Dutch entomologist, in the middle of the seventeenth century, wrote a general history of insects; also, "The Natural History of Bees." He and his English contemporary, Ray, showed their ability as naturalists by founding their systems on insect transformations. They also revived the study and practice of anatomy, which had slept since its first introduction by Aristotle, as the great stepping-stone in zoological progress. I never open the grand work of Swammerdam, with its admirable illustrations, without feelings of the most profound respect and admiration. Though a very pioneer in anatomy, and one of the founders of Natural Science, and possessed of lenses of very inferior quality, yet he wrote with an accuracy, and illustrated even minute tissues with a correctness and elegance that might well put to the blush many a modern writer. His description of the bee's tongue is more accurate than that even of the last edition of the Encyclopedia Britanica.

Ray also gave special attention to Hymenoptera, and was much aided by Willoughby and Lister. At this time Harvey, so justly noted for his discovery of the circulation of the blood, announced his celebrated dictum, "Omnia ex ovo"—all life from eggs—which was completely established by the noted Italians, Redi and Malpighi. Toward the middle of the eighteenth century, the great Linnæus—"the brilliant Star of the North"—published his "Systema Naturæ," and threw a flood of light on the whole subject of natural history. His division of insects was founded upon presence, or absence, and characteristics, of wings. This, like Swammerdam's basis, was too narrow, yet his conclusions were remarkably correct. Linnæus is noted for his accurate descriptions, and especially for his gift of the binomial method of naming plants and animals, giving in the name the genus and species, as Apis mellifera, which he was first to describe. He was also the first to introduce classes and orders, as we now understand them. When we consider the amount and character of the work of the great Swede, we can but place him among the first, if not as the first, of naturalists. Contemporary with Linnæus (also written Linne) was Geoffroy, who did valuable work in defining new genera. In the last half of the century appeared the great work of a master in entomology, DeGeer, who based his arrangement of insects on the character of wings and jaws, and thus discovered another of Nature's keys to aid him in unlocking her mysteries. Kirby well says, "He *united* in himself the highest merit of an anatomist, a physiologist and as the observant historian of the habits and economy of insects, he is above all praise. What a spring of self-improvement, enjoyment and public usefulness, is such an ability to observe as was possessed by the great DeGeer.

Contemporary with Linnæus and DeGeer was Reaumur, of

France, whose experiments and researches are of special interest to the apiarists. Perhaps no entomologist has done more to reveal the natural history of bees. Especially to be commended are his method of experimenting, his patience in investigation, the elegance and felicity of his word-pictures, and, above all, *his devotion to truth*. We shall have occasion to speak of this conscientious and indefatigable worker in the great field of insect life frequently in the following pages. Bonnet, of Geneva, the able correspondent of Reaumur, also did valuable work, in which the lover of bees has a special interest. Bonnet is specially noted for his discovery and elucidation of parthenogenesis—that anomalous mode of reproduction—as it occurs among the Aphides or plant-lice, though he did not discover that our bees, in the production of drones, illustrate the same doctrine. Though the author of no system, he gave much aid to Reaumur in his systematic labor.

At this same period systematic entomology received great aid from Lyonnet's valuable work. This author dissected and explained the development of a caterpillar. His descriptions and illustrations are wonderful, and will proclaim his ability as long as entomology is studied.

We have next to speak of the great Dane, Fabricius—a student of Linnæus—who published his works from 1775 to 1798, and thus was revolutionizing systematic entomology at the same time that we of America were revolutionizing government. He made the mouth organs the basis of his classification, and thus followed in the path which DeGeer had marked out; though it was scarcely beaten by the latter, while Fabricius left it wide and deep. His classes and orders are no improvement on—in fact, are not nearly as correct as—his old master's. In his description of genera—where he pretended to follow nature—he has rendered valuable service. In leading scientists to study parts, before little regarded, and thus to better establish affinities, he did a most valuable work. His work is a standard, and should be thoroughly studied by all entomologists.

Just at the close of the last century appeared the "greatest Roman of them all," the great Latreille, of France, whose name we have so frequently used in the classification of the

honey-bee. His is called the Elective System, as he used
wings, mouth-parts, transformations, in fact, all the organs,
the entire structure. He gave us our Family Apidæ, our Genus
Apis, and, as will be remembered, he described several of the
species of this genus. In our study of this great man's work,
we constantly marvel at his extensive researches and remarkable talents. Lamarck, of this time, did very admirable work.
So, too, did Cuvier, of Napoleon's time, and the learned Dr.
Leach, of England. Since then we have had hosts of workers
in this field, and many worthy of not only mention but praise;
yet the work has been to rub up and garnish rather than to
create. Of late, Mr. F. T. Cresson, of Philadelphia, has given
a synopsis of the Hymenoptera of North America, together
with a list of the described species. This is one of the many
valuable publications of the American Entomological Society.

I will close this brief history with a notice of authors who
are very serviceable to such as may desire to glean farther of
the pleasures of systematic entomology; only remarking that
at the end of the next chapter I shall refer to those who have
been particularly serviceable in developing the anatomy and
physiology of insects, especially of bees.

VALUABLE BOOKS FOR THE STUDENT OF ENTOMOLOGY.

For mere classification, no work is equal to Westwood on
Insects—two volumes. In this the descriptions and illustrations are very full and perfect, making it easy to study the
families, and even genera, of all the orders. This work and
the following are out of print, but can be got with little trouble
at second-hand book-stores. Kirby and Spence—" Introduction to Entomology "—is a very complete work. It treats of
the classification, structure, habits, general economy of insects,
and gives a history of the subject. It is an invaluable work,
and a great acquisition to any library. Dr. Packard's " Guide
to the Study of Insects " is a valuable work, and being American is specially to be recommended. His later " Text-Book of
Entomology " is invaluable to the student. "Injurious Insects " is the title of two valuable books, one by Dr. T. Harris,
and the other by Mary Treat. The Reports of Dr. T. Harris,

Dr. A. Fitch, and Dr. C. V. Riley, the Illinois Entomological Reports, and the Entomological Reports of the Departments of the Interior, and of Agriculture, will also be found of great value and interest. Cresson's Synopsis, already referred to, will be indispensable to every student of bees or other Hymenopterous insects. Smith's Entomology and Comstock's Entomology are indispensable to every person at all interested in Entomology. The Reports of the several Experiment Stations, especially New Jersey and Cornell, are of great value. (See "Bibliography" at the close of the first part of this volume.)

CHAPTER II.

ANATOMY AND PHYSIOLOGY.

In this chapter I shall give first the general anatomy of insects; then the anatomy, and still more wonderful physiology, of the honey-bee.

ANATOMY OF INSECTS.

In all insects the body is divided into three well-marked portions (Fig. 1); the head (Figs. 3 and 4), which is strengthened by cross-pieces or braces (Fig. 13, 14), containing the mouth

FIG. 13.

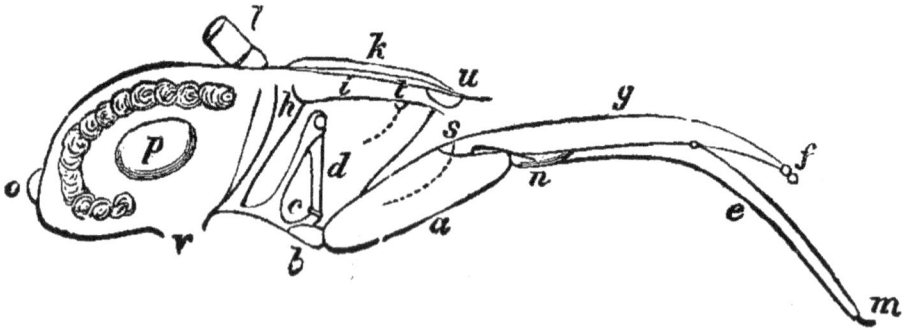

Longitudinal Section Bees' Head (from Cowan.)

a Mentum.	e Ligula.	k Clypeus.	p Brain.
b Sub-mentum	f, g Labialpalpus.	m Funnel.	r Occiput.
c Rods.	h Head-brace.	n Paraglossa.	s Duct from glands.
d Lora.	i Pharynx.	o Ocellus.	

t Duct from lower head-glands.   t, t Labrum.

organs, the eyes, both the compound and, when present, the simple, and the antennæ; the thorax, which is composed of three rings, and gives support to the one or two pairs of wings, when present, and to the three pairs of legs; and the abdomen, which is composed of a variable number of rings, and gives support to the external sex-organs, and, when present, to the sting. Within the thorax (Fig. 25) there are little more than muscles, as the concentrated strength of insects, which enables

FIG. 15.

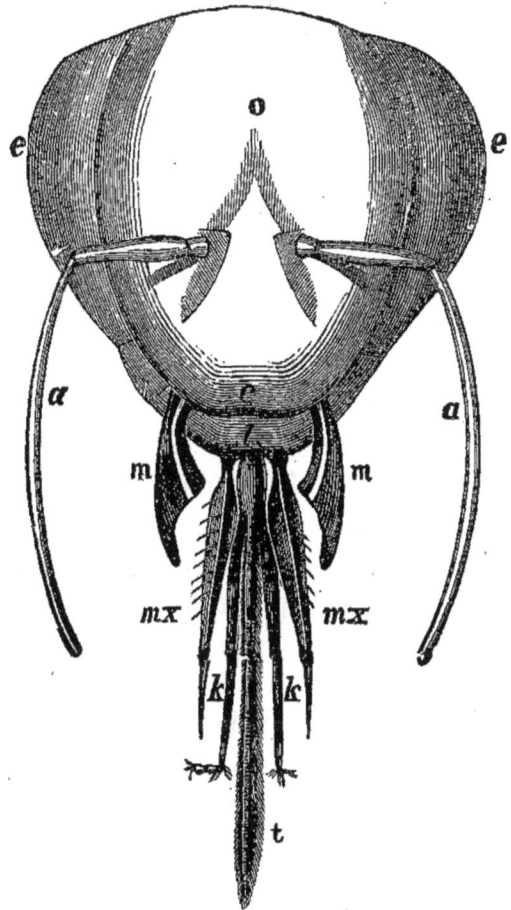

FIG. 14.

Cross Section of Head Showing Braces (After Macloskie.)

c, c Chitinous rods, which support the cardines.
h, h Strengthening rods.

Head of Bee much magnified.—Original.

o Epicranium,   c Clypeus.
e,e Compound eyes.   l Labrum.
a.a Antennæ.   m Jaws or mandibles.
mx 2d Jaws or maxillæ. t Ligula.
k,k Labial palpi.

them to fly with such rapidity, dwells in this confined space. Within the abdomen, on the other hand, are the sex-organs, by far the greater and more important portions of the alimentary canal, and other important organs.

ORGANS OF THE HEAD.

Of these the mouth organs (Fig. 15) are most prominent. These consist of an upper lip—labrum, and under lip—labium, and two pairs of jaws which move sidewise; the stronger, horny jaws, called mandibles, and the more membranous, but usually longer, named maxillæ. The labrum (Fig. 15, *l*) is well described in the name upper lip. It is attached, usually, by a movable joint to a similarly shaped piece above it, called the clypeus (Fig. 15, *c*), and this latter to the broad epicranium (Fig. 15, *o*), which carries the antennæ, the compound, and, when present, the simple eyes (Fig. 3).

The labium (Fig. 16) is not described by the name under lip, as its base forms the floor of the mouth, and its tip the tongue. The base is usually broad, and is called the mentum, and from this extends the ligula (Fig. 15, *t*), which in bees is a sucking organ or tongue.

On either side, near the junction of the ligula and mentum, arises a jointed organ, rarely absent, called the labial palpus (Fig. 15, *k*, *k*), or, together, the labial palpi. Just within the angle formed by these latter and the ligula arise the para-glossæ (Fig. 16, *n*, *n*,) one on either side. These are often wanting, though never in bees.

The jaws or mandibles (Fig. 15, *m*, *m*) arise one on either side just below and at the side of the labrum, or upper lip. These work sidewise instead of up and down, as in the higher animals, are frequently very hard and sharp, and sometimes armed with one or more teeth. A rudimentary tooth (Fig. 65, *a*, *b*) is visible on the jaws of drone and queen bees.

Beneath the jaws or mandibles, and inserted a little farther back, are the second jaws, or maxillæ (Fig. 15, *mx*), less dense and firm than the mandibles, but far more complex. Each maxilla arises by a small joint (Fig. 16, *c*), the cardo; next this is a larger joint (Fig. 16, *k*), the stipes; from this extends on the inside the broad lacinia (Fig. 16, *h*, *h*,) or blade, usually fringed with hairs on its inner edge, towards the mouth; while on the outside of the stripes is inserted the—from one to several jointed—maxillary palpus. In the honey-bee the maxillary palpi (Fig. 16, *i i*) are very small, and consist of two joints, and in some insects are wholly wanting. Sometimes, as in some of the beetles, there is a third piece running from the stipes between the palpus and lacinia, called the galea. The

FIG. 16.

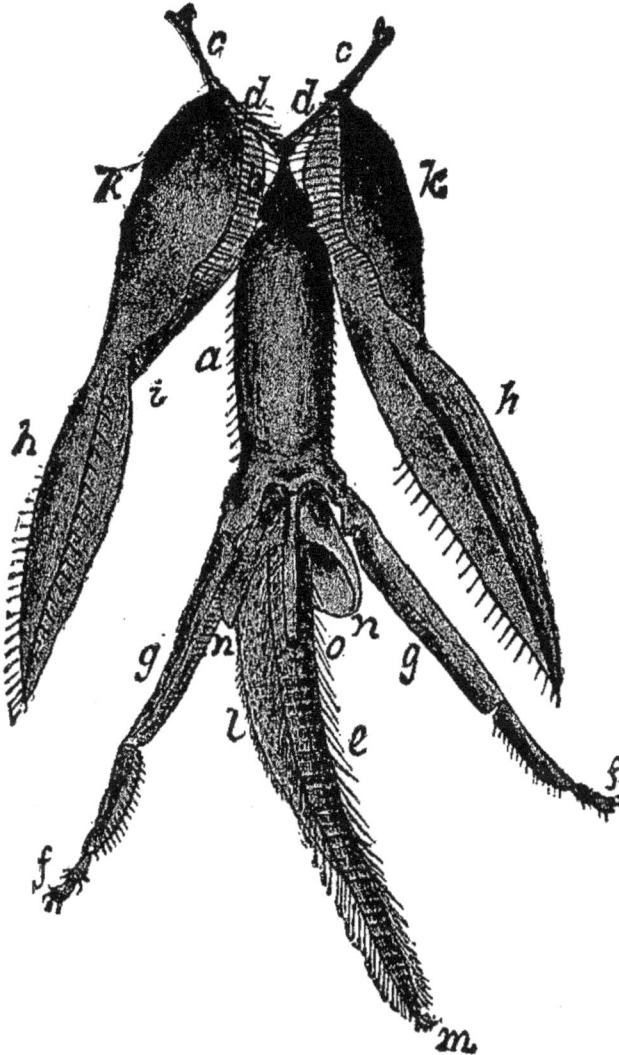

Tongue of Bee.—From Cowan.

a Mentum.	*f, g* Labial palpi.	*m* Funnel of tongue.
b Sub-mentum.	*h, h* L-Lacinia.	*n, n,* Paraglossa.
c, c Cardines.	*i, i,* Maxillary palpi.	*o* Opening of tongue.
d, d Lora.	*k, k* Stipes.	
e Ligula hairs.	*l* Ligula.	

maxillæ also move sidewise, and probably aid in holding and turning the food while it is crushed by the harder jaws, though in some cases they, too, aid in triturating the food.

These mouth-parts are very variable in form in different insects. In butterflies and moths, two-winged flies and bugs, they are transformed into a tube, which in the last two groups forms a hard, strong beak or piercer, well exemplified in the mosquito and bedbug. In all the other insects we find them much as in the bees, with the separate parts varying greatly in form, to agree with the habits and character of their possessors. No wonder DeGeer and Fabricius detected these varying forms as strongly indicative of the nature of the insect, and no wonder that by their use they were so successful in forming a natural classification.

If, as seems certain, the " Doctrine of Natural Selection " is well founded, then a change in habit is the precursor of a

FIG. 17.

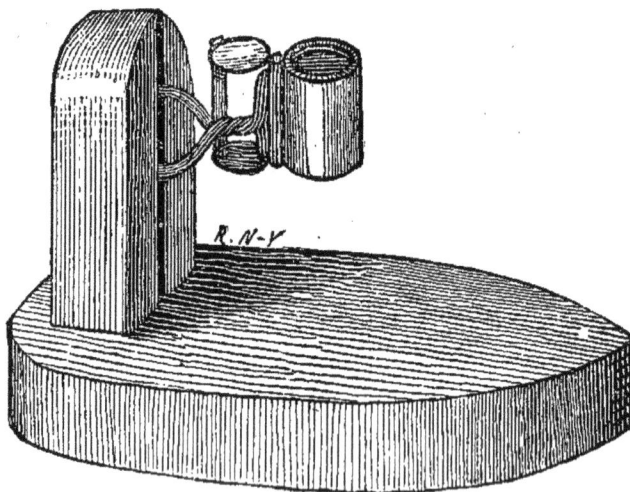

Microscope Mounted for Dissecting.—Original.

change in structure. But what organs are so intimately related to the habits of animals, as the mouth and other organs that have to do with food-taking and food-getting ?

Every bee-keeper will receive great benefit by dissecting these parts and studying their form and relations for himself. By getting his children interested in the same, he will have conferred upon them one of the rarest of blessings.

To dissect these parts, first remove the head and carefully pin it to a cork, passing the pin through, well back between the eyes. Now separate the parts by two needle-points, made by inserting a needle for half its length into a wooden pen-holder, leaving the point projecting for three-fourths of an inch. With one of these in each hand commence operations. The head may be either side up. Much may be learned in dissecting large insects, like our largest locusts, even with no

FIG. 18.

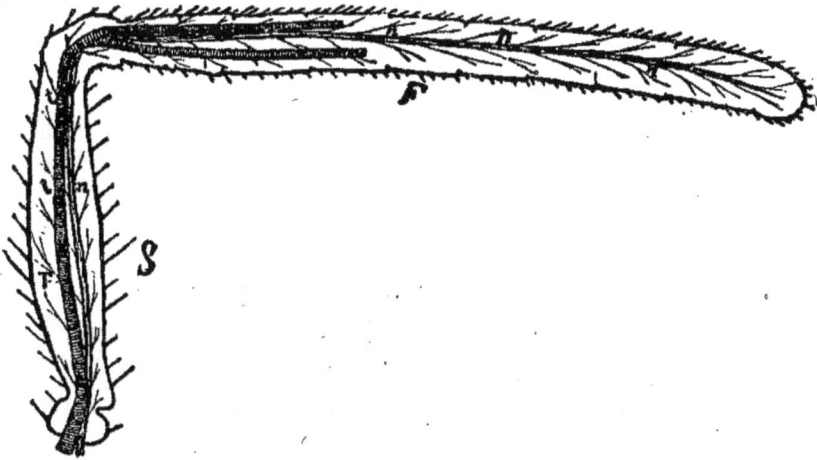

Antenna of Bee much magnified.—Original.

s Scape.   t Tracheæ.
f Flagellum.   n Nerves.

glass; but in all cases, and especially in small insects, a good lens will be of great value. The best lens now in the market is the Coddington lens, mounted in German silver. These are imported from England. They can be procured of any optician, and cost only $1.50. These lenses can be mounted in a convenient stand (Fig. 17), which may be made in twenty minutes. I think one of these more valuable than a large compound microscope, which costs many times as much. Were I obliged to part with either, the latter would go.

I require my students to do a great deal of dissecting, which they enjoy very much, and find very valuable. I would much rather that my boy would become interested in such study than to have him possessor of infinite gold rings, or even a huge gold watch with a tremendous charm. Let such pleasing recreation gain the attention of our boys, and they will ever contribute to our delight, and not sadden us with anxiety and fear.

The antennæ (Fig. 15, *a, a*) are the horn-like jointed organs situated between, or below and in front of, the large compound eyes of all insects. They are sometimes short, as in the house-fly, and sometimes very long, as in crickets and green grasshoppers. They may be straight, curved, or elbowed. In form they are very varied, as thread-like, tapering, toothed, knobbed, fringed, feathered, etc. The antennæ of many Hymenopterous insects are elbowed (Fig. 18). The long first joint in this case is the scape, the remaining joints (Fig. 18, *F*) the flagellum. A large nerve (Fig. 18, *n*) and a

FIG. 19.

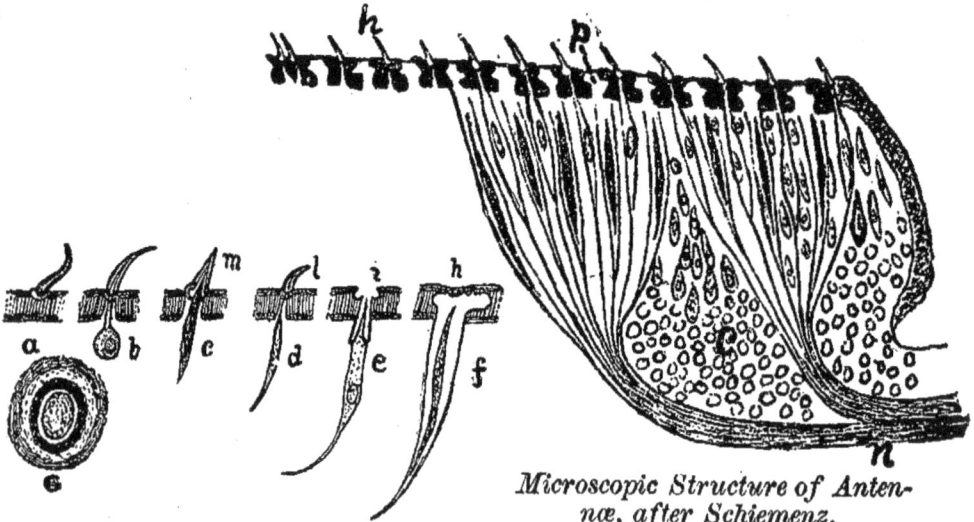

Microscopic Structure of Antennæ, after Schiemenz.

Antennal Hairs.—Original.

n Nerves.	*h* Tooth hairs.	*b* Hairs of scape.
c Cells.	*e, p* Pits or pori.	*b, c* Hairs of scape and flagellum.

trachea (Fig. 18, *t*) enter the antenna. The function of the antennæ is now pretty well, if not wholly, understood. That they often serve as most delicate touch-organs no observing apiarist can doubt. Tactile nerve-ending hairs are often found in great numbers. With the higher insects, like most Hymenopterons, this tactile sense of the antennæ is doubtless very important.

It is now fully demonstrated that the sense of smell is located in the antennæ. Sulzer, in the eighteenth century, suggested that an unknown sense might exist in the antennæ. Reaumur, Lyonet, Bonnet, etc., thought this might be the sense of smell. Dumeril, Lehrmann, who said that a nerve vessel and muscle entered the antennæ, and Cuvier, etc., thought the sense of smell was located in the spiracles or breathing-mouths. Huber thought the organ of smell was located in the mouth. Latreille and Newport, of the last century, believed the antennæ contained the organs of hearing. Strauss-Durckheim located them in the spiracles, while Wolff wrote a beautiful monograph to prove that the sense of smell was situated in the hypo-pharynx beneath the labrum. Erichson, in 1848, discovered pits in the antennæ—pori—covered with a membrane (Fig. 19, *p*), which he thought organs of smell. The next year Burmeister found hairs in these pits in beetles, which varied according as the beetle ate plant-food or carrion.

Leydig, in 1855, showed that Erichson was correct, that there were pits also on the antennæ and pegs (Fig. 19, *p*), or tooth-like hairs, perforated at the end—olfactory teeth. It remained for Hauser (1880) to complete the demonstration. He experimented with insects by the use of carbolic acid, turpentine, etc. He found that this greatly disturbed the insects when their antennæ were intact, and that even after he had withdrawn the offensive substance the insect would continue to rub its antennæ as if to remove the disturbing odor—a sort of holding its nose. He then cut off the antennæ to find that the insect was now insensible to the irritant. He next put food before the insects, which was quickly found and appropriated; but after the antennæ were cut off the food was found with difficulty, if at all. Experiment showed that in mating the same was true. Insects often find their mates when to us it would seem impossible. Thus, I have known hundreds of male moths to enter a room by a small opening in a window, attracted by a female within the room. I have also known them to swarm outside a closed window lured by a female within. Male insects have even been known to reach their mates by entering a room through a stovepipe. Yet Hauser found that this ability was gone with the loss of the antennæ. Kraepelin and others have since proved the correctness of

Hauser's conclusions. So that we now know that the antennæ, in most insects at least, contain the organs of smell. Histologically this apparatus is found to consist of nerves (Fig. 18, *n*) which run from the brain to the antennæ, and at the outer, sensitive end, contain a cell (Fig. 19) with one or more nuclei. These nerves may end in perforated, tooth-like hairs on the antennæ (Fig. 19, *h, b, c, d*) in pegs which have no chitinous sheath, which push out from the bottom of pits—pori—which exist often in great numbers in the antennæ (Fig. 19, *i, e, f.*) While Erichson first discovered the pits (Fig. 19, *p, l*) in the antennæ, Burmeister discovered the sensitive, nerve-ending hairs (Fig. 19, *a, l, m, d, h*) at their bottom, and Leydig the perforated pegs, or tooth-like hairs. We may state, then, that the antennal organ of smell consists of a free or sunken hair-like body which opens by a pore or canal to a many nucleated ganglionic mass. We thus understand how the bee finds the nectar, the fly the meat, and the drone and other male insects their mates. Similar structures in and about the mouth are proved by Kraepelin and Lubbock to be organs of taste. Mr. Cheshire speaks of small pits in the antennæ which he regards as organs of hearing. He gives, however, no proof of this, and the pits that he describes are not at all ear-like in their structure. Dr. Packard says that there is no proof that any insects except crickets and locusts have real organs of hearing. He here refers to the ear-like organs situated on the sides of the body of these insects. Similar organs on the legs of the katydid are also probably auditory. Dr. C. S. Minot, in reviewing Graber's work, says that it has not been demonstrated that even these tympanal organs are auditory, and adds that all attempts to demonstrate the existence of an auditory organ in insects has failed. There is little doubt but that this is a correct statement. That insects are conscious of vibrations which with us cause sound, I think no observing person can doubt. It is proved by the love-note of the katydid, the cicada and the cricket. Every apiarist has noticed the effect of various sounds made by the bees upon their comrades of the hive; and how contagious is the sharp note of anger, the low hum of fear, and the pleasant tone of a new swarm as it commences to enter its new home. Now, whether insects take note of these vibrations, as we recognize pitch, or whether they just distinguish the tremor, I think no one knows. There is some reason to believe that their delicate touch-organs may enable them to discriminate between vibrations, even more accurately than can we by the use of our ears. A slight jar will quickly awaken a colony of hybrids, while a loud noise will pass unnoticed. If insects can appreciate with great delicacy the different vibratory conditions of the air by an excessive development of the sense of touch, then undoubtedly the antennæ may be great aids. Dr. Clemens thought that insects could

FIG. 20.

FIG. 21.

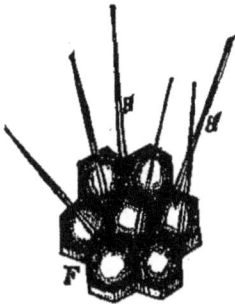

Facets of Compound Eyes,

after Dujardin.

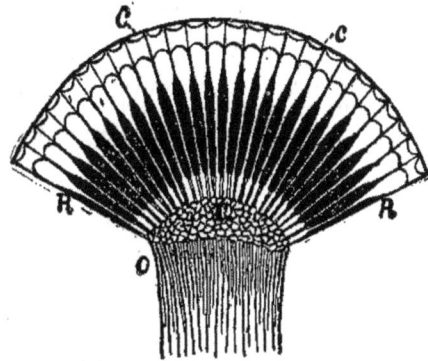

Section of Compound Eye, after Gagenbower.

| F Facets. | c Cornea. | C Cells. |
| H Hairs. | R Rods. | O Nerve. |

only detect atmospheric vibrations. So, too, thought Linnæus and Bonnet. Mayer has proved that the hairs on the antennæ of mosquitoes vibrate to different sounds. From our present knowledge, this view seems the most reasonable one, for nothing answering in the least to ears, structurally, has yet been discovered.

The eyes are of two kinds, the compound, which are always present in mature insects, and the ocelli or simple eyes, which may or may not be present. When present there are usually three of these ocelli (Fig. 3), which, if joined by lines, will describe a triangle, in the vertices of whose angles are the ocelli. Rarely there are but two ocelli, and very rarely but one.

The simple eyes (Fig. 3, *fff*) are circular, and possess a cornea, lens or cone, and retina, which receives the nerve of sight.

From the experiments of Reaumur and Swammerdam, which consisted in covering the eyes with varnish, they con-

FIG. 22.

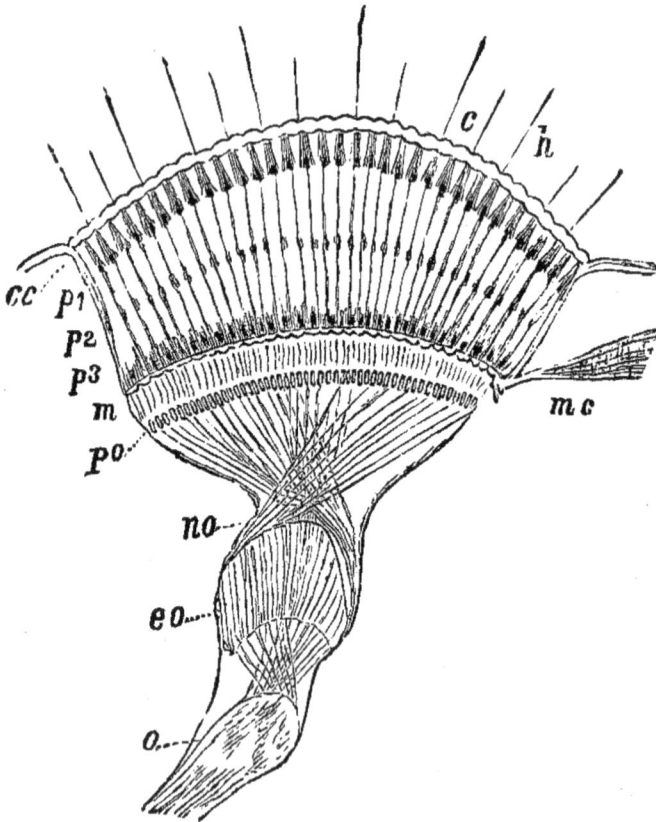

Longitudinal Section Eye.—From Cowan.

c Facet.	*cc* Lenses.	*m* Basilar membrane.
h Hair.	$p^1p^2p^3$ Rods.	*o* Optic nerve.

cluded that vision with these simple eyes is very indistinct, though by them the insect can distinguish light. Some have thought that these simple eyes were for vision at slight distances. Lubbock, Forel, and others, are doubtless correct in the view that the ocelli are for near vision, and for use in dark places. Larvæ, like spiders and most myriapods, have only simple eyes.

The compound eyes (Fig. 3) are simply a cluster of simple eyes, so crowded that they are hexagonal (Fig. 20). The cornea or facet (Fig. 20) is transparent, modified, chitinous skin. Just

FIG. 23.

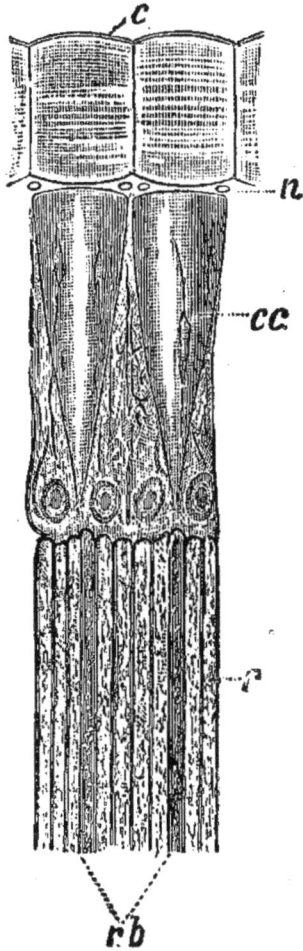

Longitudinal Section of part
of Eye, after Cowan.

c Facet.   cc Cones.
n Nuclei.   r Retinulæ.

FIG. 24.

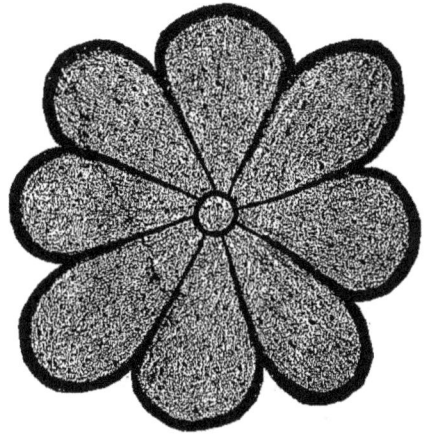

Retinulæ of Eye, after Dujardin.

Rods much magnified,
after Dujardin.

within each facet is the crystalline lens (Fig. 22, cc) or crystalline cone back of which extend the rods (Fig. 21, R, Fig. 23, cc) which consist of chitinous threads. Each rod is surrounded by rounded columns, eight in bees (Fig. 24)—retinulæ—which are enclosed by pigment membranes. This serves in the black lining of our own eyes and of optical instruments, to limit or absorb the rays of light. At the base of the rods is spread the nervous termination of the great optic nerves (Fig. 21), which extend from very near the brain, and which, before reaching the eye, passes through the three ganglionic enlargements (Fig. 22). Unlike the same in vertebrate eyes, the rods point forward.

It is thought that the optic nerve is very short, and that the retina of other higher animals is represented by the three enlargements (Fig. 22), which, as in higher animals, are fibrous cellular and ganglionic, and by the central rods of the reti-nulæ. The sensitive portion is doubtless the end of these rods. Insects, like bees, have a well-developed crystalline cone (Fig. 23), and such eyes are called eucone ; others have this less de-veloped, and their eyes are called pseudocone.

The old theory of Leeuwenhoek, Gottsche, and Platean, that each of the parts of a compound eye, each ommatidium, forms a distinct image, and these together make a compound whole, as, do our two eyes, the images overlapping, is now abandoned for the mosaic theory of Muller. Lubbock argues strongly for this view, and nearly all now accept it as true. Each of the ommatidia give a direct, not reverse, image, as do the ocelli, and each an image of only a point. Thus, the image is a true mosaic, as Muller called it. The crystalline cone covered with black pigment permits only a point to be imag-ined, and so each of the separate eyes or ommatidia images a separate point of the object seen, and all the entire object. Lubbock argues that the compound eyes do not determine form, but only motion, and that is what would be useful to protect the insect. Delicate tracheæ pass into the eyes between the rods.

The color of eyes varies very much, owing to the pigment. In some of the bees, wasps and Diptera, or two-winged flies, the coloration is exceedingly beautiful. Girschner thinks that insects with highly colored eyes do not see as well as others. Often the irridescence or play of colors, as the angle of vision changes, is wonderfully rich.

The form, size, and position of eyes vary much, as seen by noticing the eyes (Fig. 3, 4) of drones and workers. Sometimes, as in bees (Fig. 3, 4), the eyes are hairy, the hairs aris-ing from between the facets. These hairs are protective, and very likely tactile. Usually the eyes are naked. The number of simple eyes which form the compound eye is often pro-digious. There may be 25,000 in a single compound eye. There are 4,000 or 5,000 in the worker-bee.

The compound eyes are motionless, but from their size and sub-spherical shape they give quite a range of vision. It is not likely that they are capable of adjustment to accord with different distances, and it has been supposed, from the direct, darting flight of bees to their hives, and the awkward work they make in finding a hive when moved only a short distance, that their eyes are best suited to long vision.

Sir John Lubbock has proved, by some interesting experi-ments with strips of colored paper, that bees can distinguish colors. Honey was placed on a blue strip, beside several others of various colors. In the absence of the bees he changed the position of this strip, and upon their return the bees went to the blue strip rather than to the old position. Our practical apiarists have long been aware of this fact, and have con-formed their practice to this knowledge, in giving a variety of colors to their hives. Apiarists have frequently noted that bees have a rare faculty of marking positions, but for slight distances their sense of color will correct mistakes which would occur if position alone were their guide. Platean argues that insects are little guided by color, as they find flowers with no color, or the color obscured. This does not prove that color is not an aid, but that another sense—evidently of smell— supplements the sense of sight.

Lubbock's experiments prove that ants and wasps also distinguish colors. This is doubtless true of all insects that love sweets and are attracted by flowers. I have noticed a curious blunder made by bees in case of two houses which are just alike, but five rods apart. Honey placed on one porch is scarce found by bees before the corresponding porch of the other house will be swarming with bees also, though no honey is near it. The bees are simply fooled. This experiment has been tried several times, so there can be no mistake. It shows that sight, not mere position, nor yet odor, is guide, even at long distances. This disproves the general view that insects can see but at very short range.

Within the head is the large brain (Fig. 27, b), which will be described as we come to speak of the nervous system. There are also chitinous bars (Fig. 14) and braces within, which serve greatly to strengthen this portion of the insect.

APPENDAGES OF THE THORAX.

The organs of flight are the most noticeable appendages of the thorax. The wings are usually four, though the Diptera have but two, and some insects—as the worker-ants—have none. The front or primary wings (Fig. 2) are usually larger than the secondary or hind wings, and thus the mesothoracic

FIG. 25.

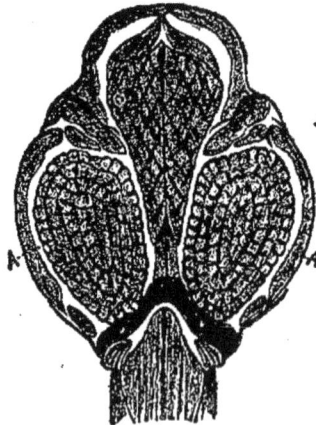

Muscles of Thorax, after Wolff.

L Muscles to raise front wing.   D Muscles to lower front wing.
A Muscles of hind wing.

or middle ring of the thorax, to which they are attached, is usually larger than the metathorax or third ring. The wings consist of a broad frame-work of veins (Fig. 2), covered by a thin, tough membrane. The main ribs or veins are variable in number, while towards the extremity of the wing are more or less cross-veins, dividing this portion of the wings into more or less cells. In the higher groups these cells are few, and quite important in classifying. Especially useful in grouping bees into their families and genera are the cells in the second row, from the front or costal edge of the primary wings, called the sub-costal cells. Thus, in the genus Apis there are three such cells (Fig. 2—6, 7, 8), while in the Melipona there are only two. The ribs or veins consist of a tube within a tube, the inner one forming an air-tube, the outer one carrying blood. On the costal edge of the secondary wings we often find hooks (Fig. 2, *h*) to attach them to the front wings.

The wings are moved by powerful muscles, compactly located in the thorax (Fig. 25), the strength of which is very great. The rapidity of the vibrations of the wings when flight is rapid, is almost beyond computation. Marey found by his

FIG. 26.

Hairs of Bees.—Original.

ingenious and graphic method that they number in the bee 190 in a second. This may be far from the maximum. Think of a tiny fly out-stripping the fleetest horse in the chase, and then marvel at this wondrous mechanism.

The legs (Fig. 1, *g, g, g*) are six in number in all mature insects, two on the lower side of each ring of the thorax. These are long or short, weak or strong, according to the habit of the insect. Each leg consists of the following joints or parts: The coxa (Fig. 67, *c*)*, which moves like a ball-and-socket joint in the close-fitting coxal cavities of the body-rings. Next to this follow in order the broad trochanter (Fig. 67, *T*,) which is double in several families of Hymenoptera like the very valuable ichneumon and chalcid flies, the large, broad femur (Fig. 66, *F*), the long, slim tibia (Fig. 67, *T*), frequently bearing strong spines at or near its end, called tibial spurs, and followed by the from one to five jointed tarsi (Fig. 67, 1, 2, 3, 4, 5). All these parts move freely upon each other, and will vary in form to agree with their use. At the end of the last tarsal joint are two hooked claws (Fig. 68), between which are the pulvilli, which are not air-pumps as usually described, but rather glands, which secrete a sticky substance which enables insects to stick to a smooth wall, even though it be above them. The legs, and in fact the whole crust, are more or less dense and hard, owing to the deposit within the structure of chitine.

The hairs of insects (Fig. 26) are very various in form, development and function. Some are short, others long ; some simple, others beautifully feathered ; some are tactile, like those of the eyes of the bees, some are protective and for warmth, and some are used as brushes, combs, and for collecting, transferring and carrying pollen.

INTERNAL ANATOMY OF INSECTS.

The muscles of insects are usually whitish. Sometimes I have noticed quite a pinkish hue about the muscles of the thorax. They vary in form and position to accord with their use. The mechanism of contraction is the same as in higher animals. The ultimate fibers of the voluntary muscles, when highly magnified, show the striæ or cross-lines, the same as do the voluntary muscles of vertebrates, and are very beautiful as microscopic objects. The fibers of each separate muscle are not bound together by a membrane, as in higher animals. In insects the muscles are widely distributed, though, as we should expect, they are concentrated in the thorax and head. In insects of swiftest flight, like the bee, the thorax (Fig. 25) is almost entirely composed of muscles ; the œsophagus, which carries the food to the stomach, being very small. At the base of the jaws (Fig. 65) the muscles are large and firm. The number of muscles is astounding. Lyonet counted over 3,000 in a single caterpillar, nearly eight times as many as are found in the human body. The strength, too, of insects is prodigious. There must be quality in muscles, for muscles as large as those of the elephant, and as strong as those of the flea, would hardly need the fulcrum which the old philosopher demanded in order to move the world. Fleas have been made to draw miniature cannon, chains and wagons many hundred times heavier than themselves.

The nerves of insects are in no wise peculiar, so far as known, except in position. Each nerve consists of a bundle of fibers, some of which are sensitive, and some motor. As in

FIG. 27.

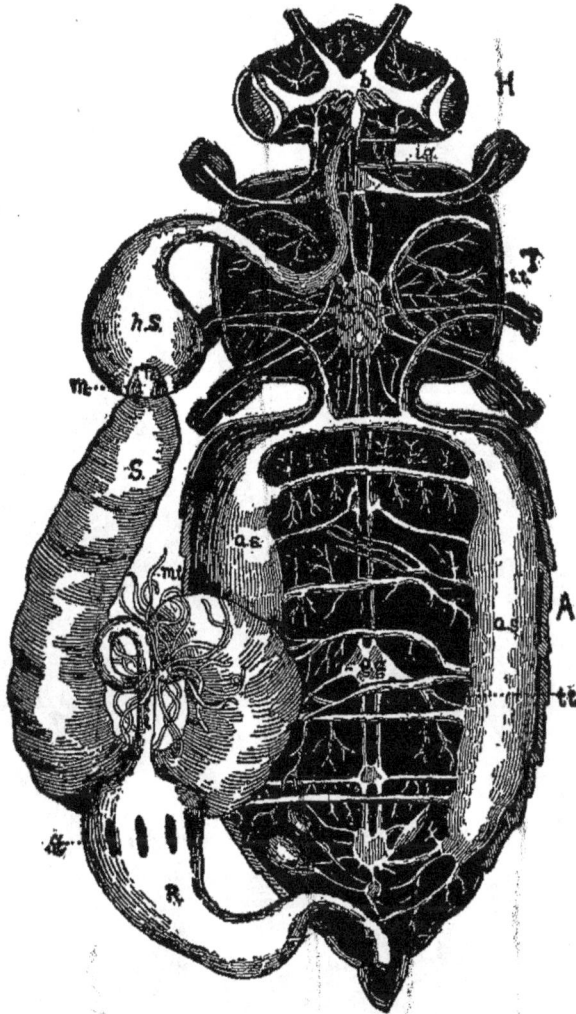

Diagram showing Internal Organs of Bee, (modified), from Cowan.

H Head.	*ig* Supra œsophageal ganglia.	*hs* Honey-stomach.
T Thorax.	*gg* Ganglia on nerve cord.	*s* Stomach.
A Abdomen.	*as* Air sacs.	*m* Stomach-mouth.
b Brain.	*tt* Tracheæ.	*i* Ileum.
R Rectum.	*mt* Malpighian tubules.	*r,g* Rectal glands.

our bodies, some are knotted, or have ganglia, and some are not. The main nervous cord is double, and has several enlargements (Fig. 27, 28) or ganglia. It runs along the under or ven-

tral side of the body, separates near the head, and after passing around the œsophagus, enlarges to form the largest of the
ganglia, which serves as a brain (Fig. 27, 28).   The uncovered

FIG. 28.

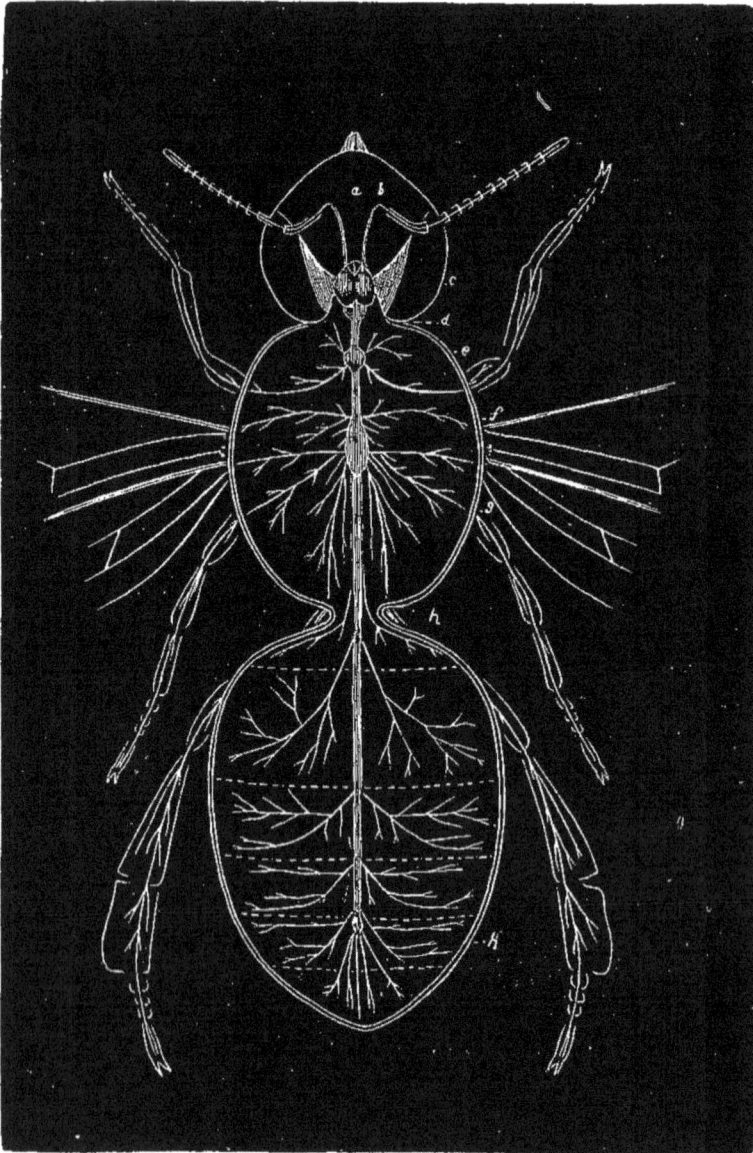

Nervous System of Drone, after Duncan.

brain shows marked convolutions (Fig. 30).   Dujardin states
that the brain of the worker-bee is 1–174 of the body ; *in the*
drone it is relatively much smaller ; the ant, 1–286 ; the ich-

neumon, 1–400; water beetle, 1–4200. In man it is 1–40. So we see that the bee is at the summit of insect intelligence, as man is of the vertebrate. The convolutions (Fig. 30) add to the argument.

From the brain many fibers extend on each side to the compound eyes. The minute nerves extend everywhere, and in squeezing out the viscera of an insect, are easily visible.

In the larva the nerve cord is much as in the adult insect, except the ganglia are more numerous. Girard says, that at first in the larva of the bee there are seventeen ganglia. The supra-œsophageal of the brain, three sub-œsophageal, three thoracic—one for each ring—and ten abdominal. Soon the three sub-œsophageal merge into one, as do also the last three abdominal, when there are in all thirteen (Fig. 31). In the

FIG. 29.

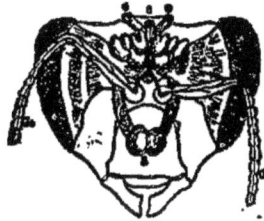

Brain of Insect, after Dujardin.

a a Antennæ. *o o o* Ocelli.

pupa, the last two of the thorax, and the first two abdominal, unite into the twin-like post-thoracic (Fig. 31), which supplies the meso, and meta-thoracic legs, and both pairs of wings with nerves. The fourth and fifth ganglia also unite, so that the adult worker-bee has nine ganglia in all. The brain or supra-œsophageal (Fig. 27), supplies nerves to the compound eyes, ocelli, antennæ and labrum; the sub-œsophageal gives off nerves to the mandibles, maxillæ, and labium; the first ganglion of the thorax sends nerves to the anterior legs. There are only four abdominal ganglia in the drone. The brain (Fig. 29, 30), like our own, is enclosed in membranes, is composed of white and gray matter, and is undoubtedly the seat of intelligence. Hence, as we should suppose, the brain of the worker is much larger than that of either the drone or queen. The ganglia along the cord are the seat of reflex acts the same as is the gray matter of our own spinal cord. Indeed, the beheaded bee uses its members much more naturally than do the higher animals after they have lost their heads. This may arise from their more simple organism, or from a higher development of the ganglia in question.

The organs of circulation in insects are quite insignificant. The heart (Fig. 32, *H*) is a long tube situated along the back, to which it is held by large muscles (Fig. 32, *m*), and receives

FIG. 31.

FIG. 30.

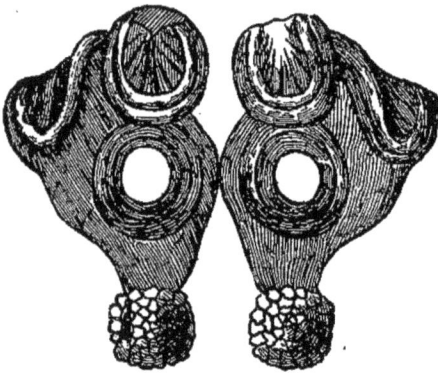

Brain of Bee, from Cowan.

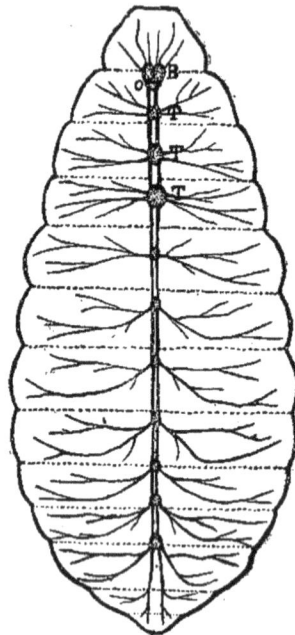

Nervous System of Worker Larva, after Duncan.

the blood at valvular openings (Fig. 32 *o*, 33 *a*,) along its sides which only permit the fluid to pass in, when, by contraction, it is forced toward the head and emptied into the general cavity. Valves prevent the blood from flowing back (Fig. 33, *b*.) Thus the heart only serves to keep the blood in motion.

There are no vessels to carry the blood to the various organs, nor is this necessary, for the nutritive fluid everywhere bathes the digestive canal, and thus easily receives nutriment, or gives waste by osmosis ; everywhere surrounds the tracheæ or air-tubes—the insect's lungs—and thus receives that most needful of all food, oxygen, and gives the baneful carbonic acid ; everywhere touches the various organs, and gives and takes as the vital operations of the animal require.

The heart, like animal vessels, generally, consists of an outer serous membrane, an inner, epithelial coat, and a middle muscular layer. Owing to the opaque crust, the pulsations of the heart can not generally be seen ; but in some transparent larvæ, like many maggots, some parasites—those of our common cabbage butterfly show this admirably—and especially in aquatic larvæ, the pulsations are plainly visible, and are most interesting objects of study.

The heart, as shown by Lyonet, is held to the dorsal wall by muscles (Fig. 32, *m*). Beneath the heart are muscles which,

FIG. 32.

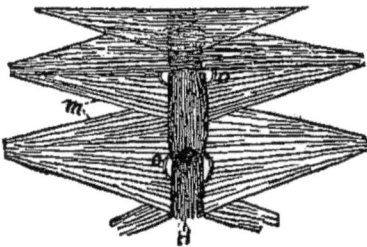

Portion of Heart of an Insect, after Packard.

H Heart.   *m* Muscles.   *o* Openings.

FIG. 33.

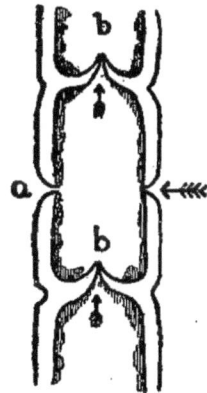

Diagram of Heart, from Cowan.

to quote from Girard, form a sort of horizontal diaphragm (Fig. 34, *d*), which as Graber shows contract, and thus aid circulation.

The blood is light colored, and entirely destitute of red discs or corpuscles, which are so numerous in the blood of higher animals, and which give our blood its red color. The function of these red discs is to carry oxygen, and as oxygen is carried everywhere through the body by the ubiquitous air-tubes of insects, we see the red discs are not needed. Except for these semi-fluid discs, which are real organs, and nourished as are other organs, the blood of higher animals is entirely fluid, in all normal conditions, and contains not the organs themselves, or any part of them, but only the elements, which are absorbed by the tissue and converted into the organs, or, to be scientific, are assimilated. The blood of insects is nearly destitute of discs, having only white corpuscles. The white corpuscles are called leucocytes. They are now known to act as so many animals, and are powerful for good in destroying microbes. We thus call them phagocytes. These phagocytes, in insect transformations, remove, we may say eat up, the no longer useful organs. It is this way that a tadpole's tail is

FIG. 34.

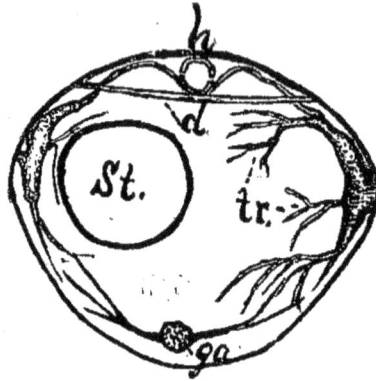

Cross Section of Bee, after Cheshire.

h Heart.	Tr. Tracheæ.
St. Stomach.	ga Ganglion.
d Diaphragm.	

removed. This process is known as phagocytosis. The leucocytes are also found in the digested food, and like the same in higher animals, are amœboid. Schonfeld has shown that the blood, chyle, the digested food, and larval food, are much the same.

The respiratory or breathing system of insects consists of a very complicated system of air-tubes (Fig. 1, 27). These tubes (Fig. 35), which are constantly branching, and almost infinite in number, are very peculiar in their structure. They are composed of a spiral thread, and thus resemble a hollow cylinder formed by closely winding a fine wire spirally about a rod, so as to cover it, and then withdrawing the latter, leaving the wire unmoved. This spiral elastic thread, like the rings of cartilage in our own trachea, serves to make the tubes rigid ; and like our trachea—wind pipe—so these tracheæ or air-tubes in insects are lined within and covered without by a thin membrane. Nothing is more surprising and interesting than this labyrinth of beautiful tubes, as seen in dissecting a bee under the microscope. I have frequently detected myself taking long pauses, in making dissections of the honey-bee, as my attention would be fixed in admiration of this beautiful breathing apparatus. In the bee these tubes expand in large lung-like sacs (Fig. 1, f), one on each side of the body. Doubtless some of my readers have associated the quick movements and surprising activity of birds and most mammals with their well developed lungs, so in such animals as the bees, we see the relation between this intricate system of air-tubes—their

FIG. 35.

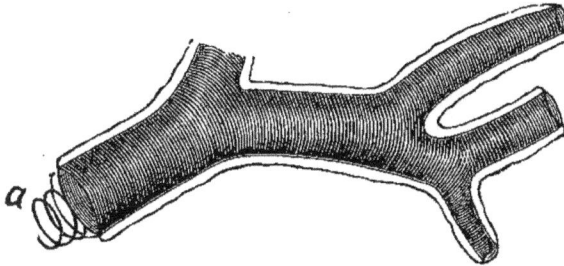

A Trachea, magnified.—Original.

lungs—and the quick, busy life which has been proverbial of them since the earliest time. Along the sides of the body are the spiracles or breathing-mouths, which vary in number. The full-grown larva has twenty, while the imago has seven pairs ; two on the thorax—one on the prothorax, and one on the metathorax—and five on the abdomen. The drone has one more on each side of the abdomen. We see, then, that to strangle an insect we would not close the mouth, but these spiracles along the sides of the body. We now understand why the bee so soon dies when the body is daubed with honey. These spiracles are armed with a complex valvular arrangement which excludes dust or other noxious particles. From these extends the labyrinth of air-tubes (Fig. 1, f, f, 27 t, t), which carries vitalizing oxygen into every part of the insect organism. As shown long ago by Leydig and Weismann, these air-tubes are but an invagination of the derm of the insect. What is more curious, these tracheæ are molted or shed with the skin of the larvæ. In the more active insects—as in bees—the main tracheæ, one on each side of the abdomen, are expanded into large air-sacs (Fig. 1, *f*). Insects often show a respiratory motion, which in bees is often very marked. Newport has shown that in bees the rapidity of the respiration, which varies from twenty to sixty per minute, gauges the heat in the hive, and thus we see why bees in times of severe cold, which they essay to keep at bay by forced respiration, consume much food, exhale much foul air and moisture, and are liable to disease. Newport found that in cases of severe cold there would be quite a rise of mercury in a thermometer which he suspended in the hive amidst the cluster.

In the larval state, many insects breathe by fringe-like gills. The larval mosquito has gills in the form of hairy tufts, while in the larval dragon-fly the gills are inside the rectum, or last part of the intestine. The insect, by a muscular effort, draws the water slowly in at the anus, where it bathes these singularly placed branchiæ, and then makes it serve a further turn by forcibly expelling it, when the insect is sent darting ahead. Thus, this curious apparatus not only furnishes oxygen, but also aids in locomotion. In the pupæ of insects there is little or no motion, yet important organic changes are taking place—the worm-like, ignoble, creeping, often repulsive, larva, is soon to appear as the airy, beautiful, active, almost ethereal imago. So oxygen, the most essential—the *sine qua non*—of all animal food is still needed. The bees are too wise to seal the brood-cell with impervious wax, but rather add the porous capping, made of wax from old comb and pollen. The pupæ, no less than the larvæ of some two-winged flies which live in water, have long tubes which reach far out for the vivifying air, and are thus called rat-tailed. Even the pupæ of the mosquito, awaiting in its liquid home the glad time when it shall unfold its tiny wings and pipe its war-note, has a similar arrangement to secure the gaseous pabulum.

The digestive apparatus of insects is very interesting, and, as in our own class of animals, varies very much in length and complexity, as the hosts of insects vary in their habits. As in mammals and birds, the length, with some striking exceptions, varies with the food. Carnivorous or flesh-eating insects have a short alimentary canal, while in those that feed on vegetable food it is much longer.

The mouth I have already described. Following this (Fig. 27) is the throat or pharynx, then the œsophagus or gullet (Fig. 36, *o*), which may expand, as in the bee, to form the honey-stomach (Fig. 36, *hs*), may have an attached crop like the

FIG. 36.

A

Cross Section of Ileum, after Schiemenz.

Alimentary Canal of Honey-Bee, modified, from Wolff.

o Œsophagus.	*sm* Stomach-mouth.
hs Honey-stomach.	*s* True stomach.
c Urinary tubes.	*i* Small intestine or ileum.
rg Rectal glands.	*r* Large intestine or rectum.

chicken, or may run as a uniform tube, as in the human body, to the true stomach (Fig. 36, *s*). Following this is the intestine—separated by some authors into an ileum (Fig. 36, *i*), and a rectum which ends in the vent or anus.

The entire alimentary canal, except the stomach, is developed from the ectoderm, or skin derm, and all is shed in molting. The stomach, often called the mid-stomach, to distinguish it from the fore and hind, is derived from the endoderm, and is not molted. Connected with the mouth are salivary glands (Fig. 58, 59), which are structurally much like those in higher animals. There is an inner and an outer chitinous layer, and the intervening cellular or epithelial, where secretion takes place.

In those larvæ that form cocoons these are the source of silk. In the glands this is a viscid fluid, but as it leaves the duct it changes instantly into the gossamer thread. Bees and wasps use this saliva in forming their structures. With it and mud some wasps make mortar ; with it and wood, others form their paper cells ; with it and wax, the bee fashions the ribbons that are to form the beautiful comb. As will be seen later, these glands are very complex in bees, and the function of the secretion very varied in both composition and function.

Lining the entire alimentary canal are mucus glands which secrete a viscid fluid that keeps the tube soft and promotes the passage of food. These lining cells also absorb, and may secrete a digestive fluid.

The true stomach (Fig. 36, s; 27 S), is very muscular; and often a gizzard, as in the crickets, where its interior is lined with teeth. The interior of the stomach is glandular, for secreting the gastric juice which is to liquefy the food, that it may be absorbed, or pass through the walls of the canal into the blood.

Appended to the anterior end of the stomach are the from two to eight coeca, or, as in some beetles, very numerous villi or tubules. These are believed by Plateau and others to be pancreatic in function. These are not found in bees. Attached to the lower portion of the stomach are the urinary or Malpighian tubules (Fig. 27, m, t), so named from their discoverer, Malpighi. There may be two to eight long tubes, or many short ones as in the bees, where we find 150. The finding in these of urea, uric acid and the urates settles the matter of their function. Cuvier and others thought these bile-tubules. Siebold thinks that some of the mucous glands secrete bile, and others act as a pancreas.

The intestine, when short, as in larvæ and most carnivora, is straight, and but little, if any, longer than the abdomen, while in most plant-eaters it is long, and thus zigzag in its course. It is a very interesting fact that the alimentary canal in the larva may be partly shed at the time of molting.

Strange as it may seem, the fecal pellets of some insects are beautiful in form, and of others pleasant to the taste. These fecal masses under trees or bushes often reveal the presence of caterpillars. I find my children use them to excellent purpose in finding rare specimens. In some caterpillars they are barrel-shaped, artistically fluted, of brilliant hue, and, if fossilized, would be greatly admired, as have been the coprolites—fossil feces of higher animals—if set as gems in jewelry. As it is, they would form no mean parlor ornament. In other insects, as the Aphides, or plant-lice, the excrement, as well as the fluid that escapes from the general surface of the body, the anus, or in some species from special tubes called the nectaries, is very sweet, and in absence of floral nectar will often be appropriated by bees and conveyed to the hives. In those insects that suck their food, as bees, butterflies, moths, two-winged flies and bugs, the feces are liquid, while in case of solid food the excrement is nearly solid. It is doubtless this liquid excreta falling from bees that has been referred to often as a fine mist.

SECRETORY ORGANS OF INSECTS.

I have already spoken of the salivary glands, which Kirby describes as distinct from the true silk-secreting tubes, though Newport thinks them one and the same. In many insects these seem absent. I have also spoken of the mucus glands, the urinary tubules, etc. Besides these, there are other secretions which serve for purposes of defense. In the queen and workers of bees, and in ants and wasps, the poison intruded with the sting is an example. This is secreted by glands at the posterior of the abdomen, stored in sacs (Fig. 38, *pg*), and extruded through the sting as occasion requires. I know of no insects that poison while they bite, except mosquitoes, gnats, and some bugs. Mosquitoes and some flies, in biting, convey, as do ticks, germs of malaria or noxious protozoans, and so induce disease.

A few exceedingly beautiful caterpillars are covered with branching spines, which sting about like a nettle. We have three such species. They are green, and of rare attraction, so that to capture them is worth the slight inconvenience arising from their irritating punctures. Some insects, like many bugs, flies, beetles, and even butterflies, secrete a disgusting fluid, or gas, which affords protection, as by its stench it renders these filthy bugs so offensive that even a hungry bird or half-famished insect passes them by on the other side. Some insects secrete a gas which is stored in a sac at the posterior end of the body, and shot forth with an explosion in case danger threatens; thus by noise and smoke it startles its enemy, which beats a retreat. I have heard the little bombardier beetle at such times, even at considerable distances. The frightful reports about the terrible horn of the tomato-worm larva are mere nonsense; a more harmless animal does not exist. My little boy of four years, and girl of only two, used to bring them to me in the summer, and regard them as admiringly as would their father upon receiving them from the delighted children.

If we except bees and wasps, there are no true insects that need be feared; nor need we except them, for with fair usage even they are seldom provoked to use their cruel weapon. The so-called "kissing bugs," which usually bite on the legs, and not on the lips, are too rare to be feared. There are two or three species of these biting bugs.

SEX-ORGANS OF INSECTS.

The male organs consist of the testes (Fig. 37, a), which are double. These are made up of tubules or vesicles, of which there may be from one, as in the drone-bee, to several, as in some beetles, on each side the abdominal cavity. In these vesicles grow the sperm cells, or spermatozoa (Fig. 50), which, when liberated, pass through a long convoluted tube, the vas deferens (Fig. 37, b, b), into the seminal sac (Fig. 37, c, c), where, in connection with mucus, they are stored. In most insects there are grandular sacs (Fig. 37, d) joined to these seminal receptacles, which, in the male bee, are very large. The sperm cells mingled with these viscid secretions, as they appear in the seminal receptacle ready for use, form the seminal fluid. Extending from these seminal receptacles is the ejaculatory duct (Fig. 37, e, f, g), which, in copulation, carries the male fluid to the penis (Fig. 37, h), through which it passes to the oviduct of the female. Beside this latter organ are the sheath, the claspers, when present, and, in the male bee, those large yellow glandular sacs (Fig. 37, *i*), which are often seen to dart forth as the drone is held in the warm hand.

FIG. 37.

Male Organs of Drone, much magnified.

a Testes.	*e* Common duct.
b b Vasa deferentia.	*f g* Ejaculatory duct.
cc Seminal sacs.	*h* Penis.
d Glandular sacs.	*i* Yellow saccules.

The female organs (Fig. 38) consist of the ovaries (Fig. 38, *o, o*), which are situated one on either side of the abdominal cavity. From these extended the two oviducts (Fig. 38, *D*), which unite into the common oviduct (Fig. 38, *D*), through which the eggs pass in deposition. In the higher Hymenoptera there is beside this oviduct, and connected with it, a sac

(Fig. 38, *s, b*) called the spermatheca, which receives the male fluid in copulation, and which, by extruding its contents, must ever after do the work of impregnation.

This sac was discovered, and its use suggested, by Mal-

FIG. 38.

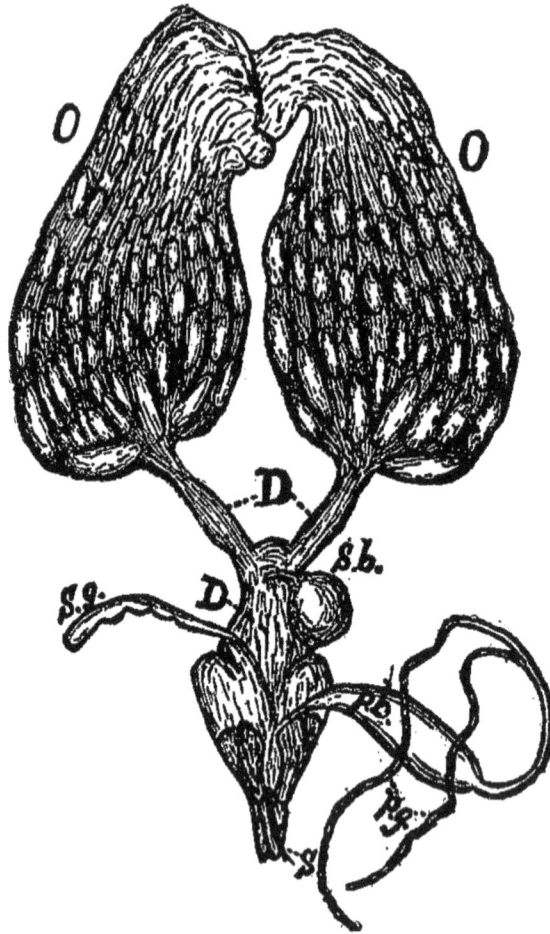

Female Organs, magnified, from Leuckart.

O Ovaries.	*Pg* Poison glands.
DD Oviducts.	*Sg* Sting glands.
S b Spermatheca.	*S* Sting.
Pb Poison sac.	

pighi as early as 1686, but its function was not fully demonstrated until 1792, when the great anatomist, John Hunter, showed that in copulation this was filled. The ovaries are multitubular organs. In some insects, as laying workers, there are but very few tubes—two or three ; while in the queen-bee there are more than one hundred. In these tubes the ova or eggs grow, as do the sperm-cells in the vesicles of the testes. The number of eggs is variable. Some insects, as the mud-wasps, produce very few, while the queen white-ant extrudes millions. The end of the oviduct, called the ovipositor, is wonderful in its variation. Sometimes it consists of concen-tric rings, like a spy-glass, which may be pushed out or drawn in ; sometimes of a long tube armed with augers or saws of wonderful finish, to prepare for eggs ; or again of a tube which may also serve as a sting. The females of all Hymenop-tera possess a very complex sting, saw, or ovipositor, which can be said of no other order.

Most authors state that insects copulate only once, or at least that the female meets the male but once. Many species like the squash-bug mate several times. In some cases, as we shall see in the sequel, the male is killed by the copulatory act. I think this curious fatality is limited to few species.

To study viscera, which of course requires very careful dissection, we need more apparatus than has been yet described. Here a good lens is indispensable. A small dis-secting-knife, a delicate pair of forceps, and some small, sharp-pointed dissecting scissors—those of the renowned Swammerdam were so fine at the point that it required a lens to sharpen them—which may also serve to clip the wings of queens, are requisite to satisfactory work. Specimens put in alcohol will be improved, as the oil will be dissolved out, and the muscles hardened. Formalin is much cheaper, and on many accounts better than alcohol. It does not evaporate as readily, and the specimens preserved in it do not smell offen-sive. Placing specimens in hot water will do nearly as well, in which case oil of turpentine will dissolve off the fat. This may be applied with a camel's-hair brush. By dissecting under water the loose portions will float off, and render effect-ive work more easy. Swammerdam, who had that most valuable requisite to a naturalist—unlimited patience—not only dissected out the parts, but with small glass tubes, fine as a hair, he injected the various vessels, as the alimentary canal and air-tubes. My reader, why may not you look in upon these wondrous beauties and marvels of God's own handiwork—Nature's grand exposition? Father, why would not a set of dissecting instruments be a most suitable gift to your son? You might thus sow the seed which would germinate into a Swammerdam, and that on your own hearth-stone. Messrs. Editors, why do not you keep boxes of these instruments for sale, and thus aid to light the torch of genius, and hasten apiarian research?

TRANSFORMATION OF INSECTS.

What in all the realm of nature is so worthy to awaken delight and admiration as the astonishing changes which insects undergo? Just think of the sluggish, repulsive caterpillar, dragging its heavy form over clod or bush, or mining in dirt and filth, changed, by the wand of Nature's great magician, first into the motionless chrysalis, decked with green and gold, and beautiful as the gem that glitters on the finger of beauty, then bursting forth as the graceful, gorgeous butterfly; which, by its brilliant tints and elegant poise, out-rivals even the birds among the life-jewels of Nature, and is made fit to revel in all her decorative wealth. The little fly, too, with wings dyed in rainbow hues, flitting like a fairy from leaf to flower, was but yesterday the repulsive maggot, reveling in the veriest filth of decaying Nature. The grub to-day drags its slimy shape through the slums of earth, on which it fattens; to-morrow it will glitter as the brilliant setting in the bracelet and ear-drops of the gay and thoughtless belle.

There are four separate stages in the development of insects: The egg, the larva, the pupa, and the imago.

THE EGG.

This is not unlike the same in higher animals. It has its yolk, the real egg, and its surrounding white or albumen, like the eggs of all mammals, and farther, the delicate shell, which is familiar in the eggs of birds and reptiles. Eggs of insects are often beautiful in form and color, and not infrequently ribbed and fluted (Fig. 41), as by a master hand. The form of eggs is very various—spherical, oval, cylindrical, oblong, straight, and curved (Fig. 39, *a*, *b*). Through the egg is an opening (Fig. 41, *A*, *B*, *m*), the micropyle, through which passes the sperm-cells. All insects seem to be guarded by a wonderful knowledge, or instinct, or intelligence, in the placing of eggs on or near the peculiar food of the larva, even though in many cases such food is no part of the aliment of the imago. The fly has the refined habits of the epicure, from whose cup it daintily sips, yet its eggs are placed in the horse-droppings of stable and pasture.

Inside the egg wonderful changes soon commence, and their consummation is a tiny larva. Somewhat similar changes can be easily and most profitably studied by breaking and examining a hen's egg each successive day of incubation. As with the eggs of our own species, and of all higher animals, the egg of insects, or the yolk, the essential part—the white is only food, so to speak—soon segments or divides into a great many cells—in the morula stage—which soon unite into three membranes, the blastoderms—blastula stage—which are the initial animal; these blastoderms soon form a single arch or sac, and not a double arch, one above the other, as in our own vertebrate branch. This sac, looking like a miniature bag of grain, grows by absorption, becomes articulated, and by budding out is soon provided with the various members. At first the sixteen or seventeen segments are much alike, and all bear appendages. From the three segments of the head come the antennæ and mouth organs, from the three thoracic rings the three pairs of legs, while the remaining abdominal joints generally soon lose all show of appendages, which are never present in the imago. The tracheæ, and fore and hind intestines, all but the stomach, are but invaginations of the ectoderm or skin membrane, and so are shed when the skin is moulted. As in higher animals, these changes are consequent upon heat, and usually, not always, upon the incorporation within the eggs of the sperm-cells from the male, which enter the egg at an opening called the micropyle. The time it takes the embryo inside the egg to develop is gauged by heat, and will, therefore, vary with the season and temperature, though in different species it varies from days to months. The number of eggs which an insect may produce is subject to wide variation.   There may be a score of them; there may be thousands.

THE LARVA OF INSECTS.

From the egg comes the larva, also called grub, maggot, caterpillar, and very erroneously worm.   These are worm-shaped (Fig. 39), usually have strong jaws, simple eyes, and the body plainly marked into ring divisions.   In some insects there are fourteen of these rings or segments, or ten besides the head and three rings of the thorax.   In bees, and nearly all other insects (Fig. 39, *f*), there is one less abdominal ring. Often, as in case of some grubs, larval bees, and maggots, there are no legs.   In most grubs there are six legs, two to each of the three rings succeeding the head.   Besides these, caterpillars have usually ten prop-legs farther back on the body, though a few—the loopers or measuring caterpillars— have only four or six, while the larvæ of the saw-flies have from twelve to sixteen of the false or prop-legs.   The alimentary canal of larval insects is usually short, direct, and quite simple, while the sex-organs are slightly if at all developed. The larvæ of insects are voracious eaters—indeed, their only work seems to be to eat and grow fat.   This rapid growth is well shown in the larva of the bee, which increases during its brief period from egg to full-grown larva—less than five days— from 1200 to 1500 times its weight.   As the entire growth occurs at this stage, their gormandizing habits are the more excusable.   I have often been astonished at the amount of food that the insects in my breeding cases would consume.   The skin or crust of insects is unyielding, hence growth requires that it shall be cast.   This shedding of the skin is called moulting.   It is a strange fact, already mentioned, that the treacheæ and a part of the alimentary canal are cast off with the skin.   Most insects moult from four to six times.   That bees moult was even known to Swammerdam.   Vogel speaks of the thickening of the cells because of these cast-skins.   Dr. Packard observed many years since, that in the thin-skinned larvæ, such as those of bees, wasps, and gall-flies, the moults are not apparent; as these larvæ increase in size, they outgrow the old skin which comes off in shreds.   The length of time which insects remain as larvæ is very variable. The maggot revels in decaying meat but two or three days; the larval bee eats its rich pabulum for nearly a week; the apple-tree borer gnaws away for three years; while the seventeen-year cicada remains a larva for more than sixteen years, groping in darkness and feeding on roots, only to come forth for a few days of hilarity, sunshine, and courtship. Surely, here is patience exceeding even that of Swammerdam. The name larva, meaning masked, was given to this stage by Linnæus, as the mature form of the insect is hidden, and can not be even divined by the unlearned.

THE PUPA OF INSECTS.

In this stage the insect is in profound repose, as if resting after its meal, the better to enjoy its active, sportive days—the joyous honeymoon—soon to come. At this time the insect may look like a seed, as in the coarctate pupa of diptera, so familiar in the "flaxseed" state of the Hessian-fly, or in the pupa of the cheese-maggot, or the meat-fly. The form of the adult insect is very obscurely shown in butterfly pupæ, called, because of their golden spots, chrysalids, and in the pupæ of moths. Other pupæ, as in the case of bees (Fig. 39, g) and beetles, look not unlike the mature insect with its antennæ, legs, and wings closely bound to the body by a thin membrane, hence the name pupa which Linne gave—referring to this condition—as the insect looks as if wrapped in swaddling clothes, the old cruel way of torturing the infant, as if it needed holding together. The pupa, and so of course the imago, has less segments than has the larva. In the bee, the first ring of the abdomen becomes the petiole, and the last three are merged into one, and thus the number of segments in the adult are only six. The drone has one more. The spiracles and ganglia are also reduced in number. Aristotle called pupæ "nymphs" —a name still in use. The word nymph is now used to designate the immature stages, both larval and pupal, of insects with incomplete transformations like locusts. Inside the pupa skin great changes are in progress, for either by modifying the larval organs or developing parts entirely new by use of the accumulated material stored by the larva during its pro-

longed banquet, the wonderful transformation from the sluggish, worm-like larva, to the active, bird-like imago is accomplished. Sometimes the pupa is surrounded by a silken
cocoon, either thick, as the cocoon of some moths, or thin and
incomplete as the cocoon of bees. The cocoon is spun by moving the head back and forth. The liquid thread quickly dries,
and is drawn forth as the head moves. These cocoons are

FIG. 39.

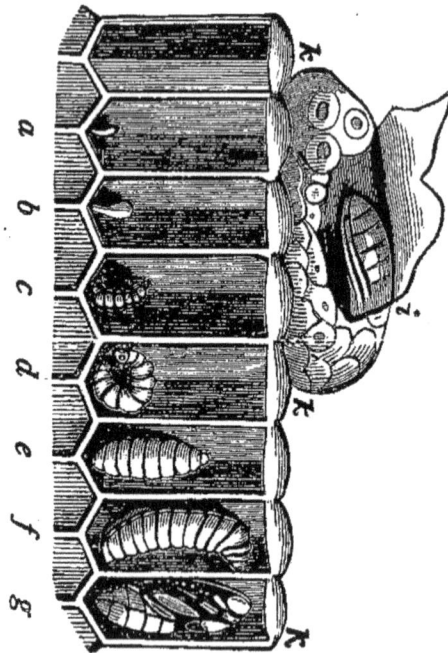

Development of the Bee, after Duncan.

a b Eggs.	g Pupa.
c d e f Larvæ.	k Caps.
i Queen-cell.	

spun by the larvæ as their last toil before assuming the restful
pupa state. The length of time in the pupa stage varies from
a few days to as many months. Sometimes insects which are
two-brooded remain as pupæ but a few days in summer, while
in winter they are moths passing the quiescent period. Our
cabbage-butterfly illustrates this peculiarity. Others, like the
Hessian-fly and coddling-moth, remain through the long, cold
months as larvæ. How wonderful is this! The first brood of

larvæ change to pupæ at once, the last brood, though the weather be just as hot, wait over inside the cocoon till the warm days of coming spring.

THE IMAGO STAGE.

This term refers to the last or winged form (Fig. 40), and was given by Linnæus because the image of the insect is now

FIG. 41.

FIG. 40.

Queen-Bee, magnified.—Original.

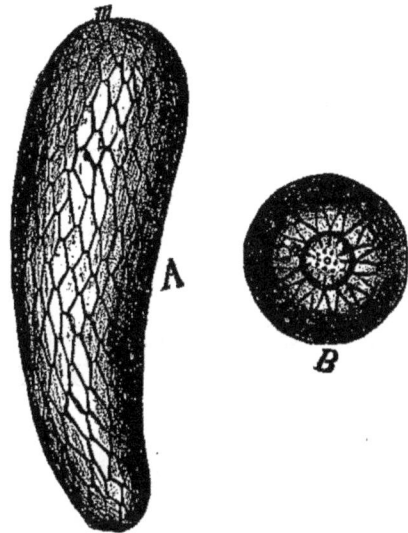

Bee-Egg.—Original.

A Egg.   B Large end.
m Micropyle.

real and not masked as when in the larva state. Now the insect has its full-formed legs and wings, its compound eyes, often complex mouth-parts—a few insects, like the bot-flies, have no mouth organs—and the fully developed sex-organs. In fact, the whole purpose of the insect now seems to be to reproduce itself. Many insects do not even eat, only flit in merry marriage mood for a brief space, when the male flees this life to be quickly followed by the female, she only waiting to place her eggs where the prospective infants may find suitable food. Some insects not only place their eggs, but feed and care for their young, as do ants, wasps and bees. Again, as in case of some species of ants and bees, abortive females perform all, or most, of the labor in caring for the young. The life of the imago also varies much as to duration. Some imagos live but for a day, others make merry for several days, while a few species live for months. Very few imagos survive the whole year. The queen-bee may live for five years, and Lubbock has queen-ants which are fifteen or more years old.

INCOMPLETE TRANSFORMATION.

Some insects like the bugs, lice, grasshoppers, and locusts, are quite alike at all stages of growth, after leaving the egg. The only apparent difference is the smaller size and the absence or incomplete development of the wings in the larvæ and pupæ. The larva and pupa are known as nymphs. The habits and structure from first to last seem to be much the same. Here, as before, the full development of the sex-organs occurs only in the imago.

ANATOMY AND PHYSIOLOGY OF THE HONEY-BEE.

With a knowledge of the anatomy and some glimpses of the physiology of insects in general, we shall now find it easy to learn the special anatomy and physiology of the highest insects of the order.

THREE KINDS OF BEES IN EACH FAMILY.

As we have already seen, a very remarkable feature in the economy of the honey-bee, described even by Aristotle, which is true of some other bees, and of ants, is the presence in each family of three distinct kinds, which differ in form, color, structure, size, habits and function. Thus, we have the queen (Lubbock has shown that there are several queens in an ant colony), a number of drones, and a far greater number of workers. Huber, Bevan, Munn, and Kirby, also speak of a fourth kind, blacker than the usual workers. These are accidental, and are, as conclusively shown by Baron von Berlepsch, ordinary workers, more deeply colored by age, loss of hair, dampness, or some other atmospheric condition. American apiarists are too familiar with these black bees, for after our severe winters they prevail in the colony, and, as remarked by the noted Baron, "They quickly disappear." Munn also tells of a fifth kind, with a top-knot, which appears at swarming seasons. I am at a great loss to know what he refers to, unless it be the pollen-masses of the asclepias, or milk-weed, which sometimes fasten to our bees and become a severe burden.

THE QUEEN-BEE.

The queen (Fig. 40), although referred to as the mother-bee, was called the king by Virgil, Pliny, and by writers as late as the last century, though in the "Ancient Bee-Master's Farewell," by John Keyes, published in London in 1796, I find an admirable description of the queen-bee, with her function correctly stated. Reaumur, as quoted in "Wildman on Bees," published in London in 1770, says, "This third sort has a grave and sedate walk, is armed with a sting, and is mother of all others."

Huber, to whom every apiarist owes so much, and who, though blind, through the aid of his devoted wife and intelligent servant, Francis Burnens, developed so many interesting truths, demonstrated the fact of the queen's maternity. This author's work, second edition, published in Edinburgh in 1808, gives a full history of his wonderful observations and experiments, and must ever rank with the work of Langstroth as a classic, worthy of study by all.

The queen, then, is the mother-bee; in other words, a fully developed female. Her ovaries (Fig. 38, o, o) are very large, nearly filling her long abdomen. The tubes, already described as composing them, are very numerous, there being more than one hundred, while the spermatheca (Fig. 38, s b) is plainly visible. This is a membranous sac, hardly 1–20 of an inch in diameter. It is fairly covered with interlacing nerves, which give to it its light, glistening appearance. The spermatheca has a short duct, joined to which is the duct of the double appendicular glands which closely embrace the spermatheca. These are described by Siebold and Leuckart, who suppose that they furnish mucus to render the sperm-cells more mobile, so that they will move more freely. Leuckart also describes muscles, which connect with the duct of the spermatheca (Fig. 38), which he thinks act as sphincters or dilators of this duct, to restrain or permit the passage of the spermatozoa. When the duct is opened the ever-active sperm-cells rush out, aided in their course by the secretion from the appended glands.

The spermatheca, according to Leuckart, may contain 25,000,000 spermatozoa. We see, then, why it does not run

FIG. 43.

FIG. 42.

Labium of Queen.—Original.

a Ligula.
d d Paraglossæ.
b Labial palpi.

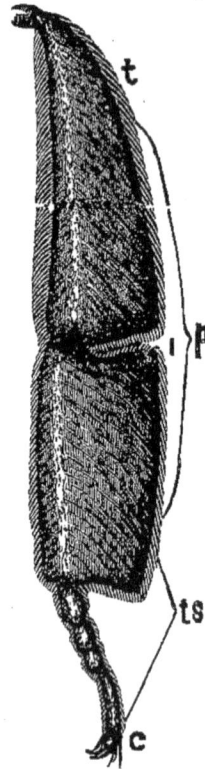

Part of Leg of Queen, magnified, after Duncan.

t Tibia.   p Broadened tibia and basal tarsus.
t s Tarsal joints.

empty, even though Siebold thought that each of the one and one-half million of eggs that a queen may lay, receives two or three sperm-cells. I think it is now proved that but one sperm-cell enters each egg. The eggs, which, as Girard states, do not form as early in the ovaries as do the sperm-cells in the organs of the drone, which are matured while the drone is yet a pupa, are a little more than 1-16 of an inch long, slightly curved, and rather smaller at the end of attachment to the comb. The outer membrane (Fig. 41) appears cellular when magnified, and shows the micropyle at the larger end (Fig. 41, B, m). The possession of the ovaries and attendant organs is the chief structural peculiarity which marks the queen, as these are the characteristic marks of females among all animals. But she has other peculiarities worthy of mention : She is longer than either drones or workers, being more than

FIG. 44.

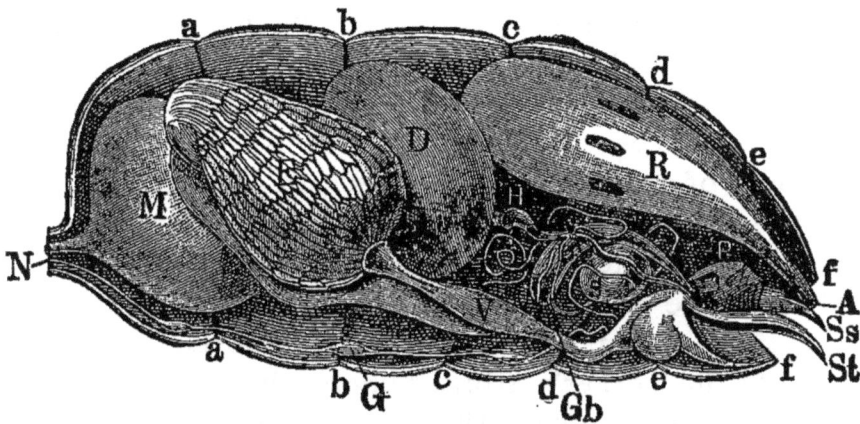

Diagram of Abdomen of Queen, from Cowan.

F Ovaries.	R Rectum.	S t Sting.
M Honey stomach.	N Œsophagus.	A Anus.
D Stomach.	S s Sheath.	V Oviduct.

seven-eighths of an inch in length, and with her long, tapering abdomen, is not without real grace and beauty. The queen's mouth organs are developed to a less degree than are those of the worker-bees. Her jaws (Fig. 65, b) or mandibles are weaker, with a rudimentary tooth, and her tongue or ligula (Fig. 42, a, and 49), as also the labial palpi (Fig. 42, b, and 49) and maxillæ, are considerably shorter. Of the four pairs of glands (Fig. 59) so elegantly figured, and so well described by Schiemenz, the queen has the first pair very rudimentary, and the others well developed. The fourth pair, or Wolff's glands, are much larger than in the worker-bees. Her eyes, though like, yet hardly as large as the same in the worker-bee (Fig. 4), are smaller than those of the drones, and do not meet above. So the three ocelli are situated above and between the compound eyes. The queen's wings (Fig. 40) are relatively shorter than those of either the workers or drones, for instead of attaining to the end of the body, they reach but little beyond the third joint of the abdomen. The queen, though she has the characteristic posterior tibia and basal tarsus (Fig. 43, *p*) in respect to breadth, has not the cavity and surrounding hairs which form the pollen-baskets of the workers. The legs of the queen (Fig. 43) are large and strong, but, like her body, they have not the pollen-gathering hairs which are so well marked in the worker. The queen possesses a sting which is longer than that of the worker, and resembles that of the bumble-bee in being curved (Fig. 44, *Ss*), and that of the bumble-bees and wasps in having few and short barbs—the little projections which point back like the barb of a fish-hook, and which, in case of the workers, prevent the withdrawing of the instrument, when once fairly inserted. While there are seven quite prominent barbs on each shaft of the worker's sting (Fig. 74), there are only three on those of the queen, and these are very short. As in case of the barbs of the worker's sting, so here, they are successively shorter as we recede from the point of the weapon. Even Aristotle discovered that a queen will rarely use her sting. I have often tried to get a queen to sting me, but without success. Neighbour gives three cases where queens used their stings, in one of which she was disabled from farther egg-laying. She stings with slight effect. The use of the queen's sting is to dispatch a rival queen. The brain of the queen is relatively small. We should expect this, as the queen's functions are vegetative. So the worker, possessed of more intricate functions, is much more highly organized. Figure 44 gives the relation of the viscera of the queen.

Schiemenz and Schonfeld are unquestionably correct in the belief that the queen, and the drones as well, are *fed by* the workers, the same food that the larvæ are fed. Thus, the digestion is performed for both queen and drones.

I have known queens to lay over 3,000 eggs a day. These I find weigh .3900 grams, while the queen only weighs .2299 grams. Thus, the queen may lay daily nearly double her own weight of eggs. This, of course, could only be possible as she was fed highly nutritious food, which was wholly digested for her. The larval bee fed the same food doubles in weight in a single day—a further proof of the excellence of this diet. Schonfeld finds that the queen, like the drones, will soon die if she be shut away from the workers by a double wire-cage, even though in the hive and surrounded with honey. The fact that pollen-husks—cuticula—are never found in the queen's stomach, gives added proof of the above fact. The contents are grayish. I never saw a queen void her feces. Vogel reports having seen it, and Mr. Cowan reports to me that he has seen a queen pass a yellowish gray liquid. We also find the queen's alimentary canal comparatively small, though the renal tubules are large and numerous. The queen, like the worker-bees, is developed from an impregnated egg, which, of course, could only come from a queen that had previously mated. These eggs are not placed in a horizontal cell, but in one specially prepared for their reception (Fig. 39, *i*). The queen-cells (Fig. 45) are usually built on the edge of the comb, or around an opening in it, which is necessitated from their size and form, as usually the combs are too close together to permit their location elsewhere. These cells extend either vertically or diagonally downward, are very rough (Fig. 45, *c*), and are composed of wax cut from the old combs, mixed with pollen (Mr. Cheshire says all kinds of refuse is used in constructing queen-cells), and in size and form much resemble a peanut. The eggs must be placed in these cells, either by the queen or workers. Huber, who, though blind, had wondrous eyes, witnessed this act of the queen. I have frequently seen eggs in these cells, and without exception in the exact position in which the queen always places her eggs in the other cells. John Keyes, in the old work already referred to, whose descriptions, though penned so long ago, are wonderfully accurate, and indicate great care, candor, and conscientious truthfulness, asserts that the queen is five times as long laying a royal egg as she is the others. From the character of his work, and its early publication, I can but think that he had witnessed this rare sight. Some candid apiarists of our own time and country—E. Gallup among the rest—claim to have witnessed the same. The eggs are so well glued, and are so delicate, that, with Neighbour, I should doubt the possibility of a removal except that some persons assert that they have positive proof that it is sometimes done. Possibly the young larvæ may at times be removed from one cell to another.. The opponents to the view that the queen lays eggs in the queen-cells, base their belief on a supposed discord between the queen and neuters.

The conditions which lead to the building of queen-cells, and the peopling of the same are : Loss of queen ; when a worker-larva from one to four days old will be surrounded by a cell ; inability of a queen to lay impregnated eggs, her spermatheca having become emptied ; any disability of the queen ; great number of worker-bees in the hive ; restricted quarters, the queen not having place to deposit eggs, or the workers little or no room to store honey ; or lack of ventilation, so that the hive becomes too close. These last three conditions are most likely to occur at times of great nectar-secretion.

A queen may be developed from an egg, or, as first shown by Schirach, from a worker-larva less than three days old. (Mr. Doolittle has known queens to be reared from worker-larvæ taken at four-and a-half days from hatching.) In such cases the cells adjacent to the one containing the selected larva are removed, and the larva surrounded by a royal cell. The development of the queen-larva is much like that of the worker, soon to be detailed, except that it is more rapid, and the queen-larva is fed richer and more plenteous food, called royal jelly. This is an excellent name for this substance, as Dr. A. de Planta has shown (B. B. J., 1887, p. 185) that this royal jelly is different from the food both of the worker and drone larva. It is doubtless digested pollen, as first suggested by Dufour, and so ably proved by Schonfeld. I have fed bees honey with finely pulverized charcoal in it, and found the same in the royal jelly. This could not be true if the latter were a secretion, as the carbon is not osmotic. Dr. Planta's researches show that the royal jelly is richer in fatty elements and proteids than the larval food either of the drones or workers ; but not as rich in sugar. It contains more albuminous material, and much more fatty matter than the food of the drone-larvæ. Quite likely evaporation may change the nature of this royal jelly. There is never undigested food fed to queen or worker larvæ, but the drone-larva is thus fed, as the microscope shows the pollen. This peculiar food, as also its use and abundance in the cell, was first described by Schirach, a Saxon clergyman, who wrote a work on bees in 1771. It is thick, like rich cream ; slightly yellow, and so abundant that the queen-larva not only floats in it during all its period of growth, but quite a large amount remains after her queenship vacates the cell. We sometimes find this royal jelly in incomplete queen-cells, without larvæ.

What a mysterious circumstance is this : These royal scions simply receive a more abundant and nutritious diet, and occupy a more ample habitation—for I have more than once confirmed the statement of Mr. Quinby, that the direction of the cell is immaterial—and yet what a marvelous transformation. Not only are the ovaries developed and filled with eggs, but the mouth organs, the wings, the legs, the sting, aye, even the size, form, and habits, are all wondrously changed. The food stimulates extra development of the ovaries, and, through the law of compensation, other parts are less developed. That the development of parts should be accelerated, and the size increased, is not so surprising—as in breeding other insects I have frequently found that kind and amount of food would hasten or retard growth, and might even cause a dwarfed imago—but that food should so essentially modify the structure, is certainly a rare and unique circumstance, hardly to be found except here and in related animals. Bevan has suggested that laying workers, while larvæ, have received some of this royal jelly from their position near a developing queen. As the workers vary the food for the several larvæ, as Dr. Planta has shown, may they not sometimes make a mistake and feed royal jelly to workers ? Surely, in caring for so many young, this would be very pardonable. Langstroth supposes that they receive some royal jelly, purposely given by the workers, and I have previously thought this reasonable and probably true. But these pests of the apiarist, and especially of the breeder, almost always, so far as I have observed, make their appearance in colonies long queenless, and I have noticed a case similar to that given

by Quinby, where these occurred in a nucleus where no queen
had been developed. May it not be true that a desire for
eggs or unrest stimulates in some worker, which was perhaps
over-fed as a larva, the growth of the ovaries, growth of eggs
in the ovarian tubes, and consequent ability to deposit? The
common high-holder, Colaptes auratus—a bird belonging to the
wood-pecker order, usually lays five eggs, and only five; but
let cruel hands rob her of these promises of future loved ones,
and, wondrous to relate, she continues to lay more than a
score. One thus treated, on the College campus, actually laid
more than thirty eggs. So we see that animal desires may
influence and move organs that are generally independent of
the will. It may be that in queenless colonies the workers
commence to feed some worker or workers, the rich nitrogenous food, and thus their ovaries are stimulated to activity.

The larval queen is longer, and more rapid of development
than the other larvæ. When developed from the egg—as in
case of normal swarming—the larva feeds for five days, when
the cell is capped by the workers. At any time during this
period the larva can be removed, as first shown by Mr. J. L.
Davis, of Michigan, in 1874, and a newly hatched larva placed
in it instead. This is easily done by use of a quill toothpick. The infant queen then spins her cocoon, which occupies
about one day. The fibrous part of the cocoon, which is also
true of both drone and worker larvæ, is confined to the outer
end, as is easily seen by microscopic examination. Yet a thin
varnish continues this over the whole interior of the cell. This
latter becomes very thick in worker-cells, as many bees are
reared in each cell, while in the queen-cell it is thin, as but
one bee is reared in each cell. A similar varnish coats the
cocoons of all silk-moths. This may be the contents of the
alimentary canal simply, which, of course, is moulted with the
last larval skin, very likely a special secretion is added. These
cocoons are shown nicely when we melt old comb in the solar
wax-extractor. The queen now spends nearly three days in
absolute repose. Such rest is common to all cocoon-spinning
larvæ. The spinning, which is done by a rapid motion to-and-
fro of the head, always carrying the delicate thread, much like
the moving shuttle of the weaver, seems to bring exhaustion

and need of repose. She now assumes the pupa state (Fig. 39, *i*). At the end of the sixteenth day she comes forth a queen. A short time before the queen emerges the workers thin off the wax from the end of the cell (Fig. 45, *D*). The reason for this is obscure, as the queen could easily come forth without it. The queen cuts her way out by use of her jaws, and leaves the cap hanging as a lid to the cell (Fig. 45, *C*).

FIG. 45.

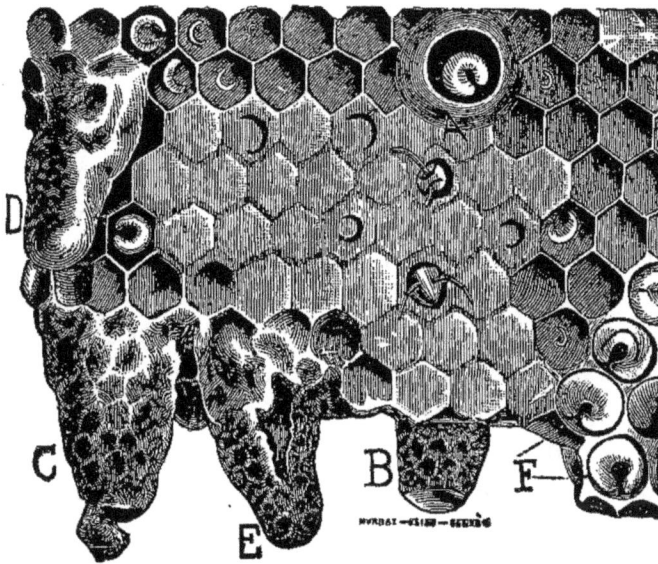

Queen-Cells, from A. I. Root Co.

A Queen-cell from modified worker-cell just started.
B Imcomplete cell.
C Cell, after queen has emerged, showing cap hanging.
D Thinned cell.   *E* Cell cut into from side.

While a queen usually comes forth in sixteen days, there may be a delay. Cold will delay hatching of the egg, and retard development. Sometimes queens are kept for a time in the cell, after they are really ready to come forth. Thus, there may be rarely a delay of even two days. Huber states that when a queen emerges the bees are thrown into a joyous excitement, so that he noticed a rise in temperature in the hive from 92 degrees F. to 104 degrees F. I have never tested this matter accurately, but I have failed to notice any marked demonstration on the natal day of her ladyship the queen, or extra respect paid her as a virgin. When queens are started from worker-larvæ they will issue as imagoes in ten or twelve days from the date of their new prospects. Mr. Doolittle writes me that he has known them to issue in eight and one-half days. My own observations sustain the assertion of Mr. P. L. Viallon, that the minimum time is nine and one-half days.

As the queen's development is probably due to superior quality and increased quantity of food, it would stand to reason that queens started from eggs, or larvæ just hatched, are preferable ; the more so as, under normal circumstances, I believe they are almost always thus started. The best experience sustains this position. As the proper food and temperature can best be secured in a full colony—and here again the natural economy of the hive adds to our argument— we should infer that the best queens would be reared in strong colonies, or at least kept in such colonies till the cells were capped. Experience also confirms this view. As the quantity and quality of food and the general activity of the bees are directly connected with the full nourishment of the queen-larva, and as these are only at the maximum in times of active gathering—the time when queen-rearing is naturally started by the bees—we should also conclude that queens reared at such seasons are superior. My experience—and I have carefully observed in this connection—most emphatically sustains this view.

Five or six days after issuing from the cell—Neighbour says the third day—if the day is pleasant the queen goes forth on her " marriage flight ;" otherwise she will improve the first pleasant day thereafter for this purpose. Mr. Doolittle says that mid-summer queens fly out in from four to nine days, while early spring and fall queens may not mate for from two to four weeks. Rev. Mr. Mahin has noticed, as have many of us, that the young queens fly out several times simply to exercise, and then he thinks they often go from two to five miles to mate ; while Mr. Alley thinks the mating is performed within one-half mile of the hive. I have known queens to be out on their mating tour for thirty-five minutes, in which case it would seem certain that they must have gone more than one-half mile. It has been reported by reliable persons that the queens are out from ten minutes to two hours. Sometimes queens will meet drones, as shown by the white thread tipping the body, and yet not be impregnated. The spermatozoa did not reach the spermatheca. In such cases, a second, and perhaps a third, mating is required. Huber was the first to prove that impregnation always takes place on the wing. Bonnet also proved that the same is true of ants, though in this case millions of queens and drones often swarm out at once. I have myself witnessed several of these wholesale matrimonial excursions among ants. I have also taken bumble-bees that were copulating while on the wing. I have also seen both ants and bumble-bees fall while united, probably borne down by the expiring males. That butterflies, moths, dragon-flies, etc., mate on the wing is a matter of common observation. It has generally been thought impossible for queens in confinement to be impregnated. Prof. Leuckart believes that successful mating demands that the large air-sacs (Fig. 1, f) of the drones shall be filled, which he thinks is only possible during flight. The demeanor of the drones suggests that the excitement of flight, like the warmth of the hand, is necessary to induce the sexual impulse.

Many others, with myself, have followed Huber in clipping the virgin queen's wing, only to produce a sterile, or drone-laying queen. One queen, however, whose wing was clipped just as she came from the cell, and the entrance to whose hive was guarded by perforated zinc so the queen could not get out, was impregnated, and proved an excellent queen. I should doubt this if I could see any other way to explain it. Yet, from a great number of experiments, I feel sure that mating in confinement can never be made practical, even if desirable. And if Leuckart is correct in the above suggestion, which is very probable, it is not desirable. Some bee-keepers claim to have mated queens by hand. I have tried this thoroughly, as also mating in boxes, green-houses, etc., and from entire lack of success I believe such mating is impossible, at least with most bee-keepers. J. S. Davitte, of Georgia, claims to have mated many queens in a large circular tent.   The drones are permitted to fly only in the tent, and so are at home.

If the queen fails to find an admirer the first day, she will go forth again and again till she succeeds.   Huber states that after twenty-one days the case is hopeless.   Bevan states that if impregnated from the fifteenth to the twenty-first she will be largely a drone-laying queen.   That such absolute dates can be fixed in either of the above cases is very questionable.   Yet all experienced breeders know that queens kept through the winter as virgins are sure to remain so.   It is quite likely that the long inactivity of the reproductive apparatus, especially of the oviduct and spermatheca, wholly or in part paralyzes it, so that queens that are late in mating can not impregnate the eggs as they desire.   This would accord with what we know of other muscular organs.   Berlepsch believed that a queen that commenced laying as a virgin could never lay impregnated eggs, even though she afterwards mated.   Langstroth thought that he had observed to the contrary.

If the queen be observed after a successful " wedding tour," she will be seen as first pointed out by Huber, to bear the marks of success in the pendant drone appendages, which are still held in the vulva of the queen.

It is not at all likely that a queen, after she has met a drone, ever leaves the hive again except when she leaves with a swarm.   It has been stated that an old queen may be impregnated.   I feel very certain that this is an error.

If the queen lays eggs before meeting the drone, or if for any reason she fail to mate, her eggs will only produce male bees.   This strange anomaly—development of the eggs without impregnation—was discovered and proved by Dzierzon, in 1845.   Dr. Dzierzon, who, as a student of practical and scientific apiculture, ranks very high, is a Roman Catholic priest of Carlsmarkt, Germany.   This doctrine—called parthenogenesis, which means produced from a virgin—is still doubted by some quite able bee-keepers, though the proofs are irrefragable :

1st. Unmated queens will lay eggs that will develop, but drones always result.

2d. Old queens often become drone-layers, but examination shows that the spermatheca is void of seminal fluid.   Such an examination was first made by Prof. Siebold, the great German anatomist, in 1843, and later by Leuckart and Leidy. I have myself made several such examinations. The spermatheca can easily be seen by the unaided vision, and by crushing it on a glass slide, by compressing with a thin glass cover, the difference between the contained fluid in the virgin and in the impregnated queen is very patent, even with a low power. In the latter it is more viscid and yellow, and the vesicle more distended. By use of a high power, the active spermatozoa or sperm-cells (Fig. 50) become visible.

3d. Eggs in drone-cells are found by the microscopist to be void of the sperm-cells, which are always found in all other fresh-laid eggs. This most convincing and interesting observation was first made by Von Siebold, at the suggestion of Berlepsch. It is quite difficult to show this. Leuckart tried before Von Siebold, at Berlepsch's apiary, but failed. I have also tried to discover these sperm-cells in worker-eggs, but as yet have been unsuccessful. Siebold has noted the same facts in eggs of wasps.

4th. Dr. Donhoff, of Germany, reports that, in 1855, he took an egg from a drone-cell, and by artificial impregnation produced a worker-bee.

Late investigation by Mr. Weismann, of Germany, leaves no doubt of this fact of parthenogenesis in the production of drone-bees.

Parthenogenesis, in the production of males, has also been found by Siebold to be true of other bees and wasps, and of some of the lower moths in the production of both males and females. Adler has shown that this agamic reproduction prevails among the Chalcididæ, a family of parasitic Hymenoptera, and it has long been known to characterize the cynips or gall-flies ; while the great Bonnet first discovered what may be noticed on any summer day all about us, even on the house-plants at our very windows, that parthenogenesis is best illustrated by the aphides, or plant lice. In the fall males and females appear which mate, when the females lay eggs which in the spring produce only females ; these again produce only females, and thus on for several generations, sometimes fifteen or twenty, till with the cold of autumn come again the males and females. Any person can easily demonstrate this fact for himself. The summer plant-lice are hatched within the mother-louse, or are ovoviviparous. It is easy to capture a young louse just as it is born, and isolate it on a plant, when soon we shall find it giving birth to young lice, though it has never even seen any louse, male or female, since birth. Bonnet observed seven successive generations of productive virgins. Duval noted nine generations in seven months, while Kyber observed production exclusively by parthenogenesis in a heated room for four years. So we see that this strange and almost incredible method of increase is not rare in the great insect world.

In two or three days after she is impregnated, the queen, under normal circumstances, commences to lay, usually worker-eggs. It is rare not to find eggs by the tenth day from the birth of the queen. The queens rarely go three weeks before laying. Such tardiness does not recommend them. It is reported that giving unhatched brood will start the queen to laying. If this be true, it is doubtless explained by her receiving different food from the workers. If the condition of the hive impels to no further swarming that season, no drones will be required, and so only worker-eggs will be laid. In many localities, and in certain favorable years in all localities, however, further swarming will occur.

It is frequently noticed that the young queen at first lays quite a number of drone-eggs. Queen-breeders often observe this in their nuclei. This continues for only a few days. This does not seem strange. The act of freeing the sperm-cells from the spermatheca is muscular and voluntary, and that these muscles should not always act promptly at first, is not strange, nor is it unprecedented. Mr. Wagner suggested that the size of the cell determined the sex, as in the small cells the pressure on the abdomen forced the fluid from the spermatheca. Mr. Quinby also favored this view. I greatly question this theory. All observing apiarists have known eggs to be laid in worker-cells ere they were hardly commenced, when there could be no pressure. In case of queen-cells, too, if the queen does lay the eggs—as I believe—these would be unimpregnated, as the cell is very large. I know the queen sometimes passes from drone to worker cells very abruptly while laying, as I have witnessed such a procedure—the same that so greatly rejoiced the late Baron of Berlepsch, after weary hours of watching—but that she can thus control at the instant this process of adding or withholding the sperm-cells certainly seems not so strange as that the spermatheca, hardly bigger than a pin-head, could supply these cells for months, yes, and for years. Who that has seen the bot-fly dart against the horse's legs, and as surely leave the tiny yellow egg, can doubt but that insects possess very sensitive oviducts, and can extrude the minute eggs just at pleasure. That a queen may force single eggs at will, past the mouth of the spermatheca, and at the same time add or withhold the sperm-cells, is, I think, without question true. What gives added force to this view is the fact that other bees, wasps and ants exercise the same volition, and can have no aid from cell-pressure, as all the eggs are laid in receptacles of the same size. As already remarked, the males and workers of Apis dorsata are developed in the same sized cells, while the males of A. indica are smaller than the workers. The Baron of Berlepsch, worthy to be a friend of Dzierzon, has fully decided the matter. He has shown that old drone-cells are as small as new worker-cells, and each harbors its own brood. Very small queens, too, make no mistakes. With no drone-cells, the queen will sometimes lay drone-eggs in worker-cells, in which drones will then be reared, and she will, if she must, though with great reluctance, lay worker-eggs in drone-cells.

Before laying an egg the queen takes a look into the cell, probably to see if all is right. If the cell contains any honey, pollen, or an egg, she usually passes it by, though, when crowded, a queen will sometimes, especially if young, insert two or three eggs in a cell, and sometimes, when in such cases she drops them, the bees show their dislike of waste, and appreciation of good living, by making a breakfast of them. If the queen find the cell to her liking, she turns about, inserts her abdomen, and in an instant the tiny egg is glued in position (Fig. 39, *b*) to the bottom of the cell.

The queen, when considered in relation to the other bees of the colony, possesses a surprising longevity. It is not uncommon for her to attain the age of three years in the full possession of her powers, while queens have been known to do good work for five years. Lubbock has queen ants in his nests that are fifteen or more years old, and still they are vigorous layers. Queens, often at the expiration of one, two, three or four years, depending on their vigor and excellence, either cease to be fertile, or else become impotent to lay impregnated eggs—the spermatheca having become emptied of its sperm-cells. In such cases the workers usually supersede the queen, that is, they rear a new queen before all the worker-eggs are gone, and then destroy the old one.

It sometimes happens, though rarely, that a fine looking queen, with the full-formed ovaries and large spermatheca well filled with male fluid, will deposit freely, but none of the eggs will hatch. Readers of bee-papers know that I have frequently received such for dissection. I received one Aug. 12, 1900, from Mr. E. R. Root. The first one I ever got was a remarkably fine looking Italian, received from the late Dr. Hamlin, of Tennessee. All such queens that I have examined seem perfect, even though scrutinized with a high-power objective. We can only say that the egg is at fault, as frequently transpires with higher animals, even to the highest. These females are barren; through some fault with the ovaries, the eggs grown therein are sterile. To detect just what is the trouble with the egg is a very difficult problem, if it is capable of solution at all. I have tried to determine the ultimate cause, but without success. Cases have also been observed where mated and impregnated queens fail to lay impregnated eggs. Here the delicate organism of the spermatheca and its duct is at fault. Queens that have been chilled, as shown by Siebold, Leuckart, and our own Langstroth, are often made drone-layers—that is, they lay only unimpregnated eggs. I have also had one queen that produced many hermaphroditic bees. These hermaphrodites are not really hermaphrodites; as, so far as I have examined, they have only ovaries or testes, but externally they have drone-organs in part, as, for instance, the appendages of the head and thorax; and worker-organs in part, as the abdomen, will be like that of a drone. Indeed, I now have a very strange hermaphrodite, where one side is worker, the other drone. It is very probable that these peculiarities arise from a diseased condition of the queen, or else from diseased spermatozoa. I have known one queen, many of whose bees were thus abnormal. If a queen is not impregnated for three or four weeks, she often commences to lay without impregnation, and then is a "drone-layer," and, of course, worthless. She may lay as regularly as if impregnated, though this is not usual. She is, of course, betrayed by the higher cappings, and exclusive drone-brood.

The function of the queen is simply to lay eggs, and thus keep the colony populous, and this she does with an energy that is fairly startling. A good queen in her best estate will lay two or three thousand eggs a day. I have seen a queen in my observing hive lay for some time at the rate of four eggs per minute, and have proved by actual computation of brood-cells that a queen may lay over three thousand eggs in a day. Both Langstroth and Berlepsch saw queens lay at the rate of six eggs a minute. The latter had a queen that laid three thousand and twenty-one eggs in twenty-four hours, by actual count, and in twenty days she laid fifty-seven thousand. This queen continued prolific for five years, and must have laid, says the Baron, at a low estimate, more than 1,300,000 eggs. Dzierzon says queens may lay 1,000,000 eggs, and I think these authors have not exaggerated. As already stated, a queen may lay nearly double her weight of eggs daily. Yet, with even these figures as an advertisement, the queen-bee can not boast of superlative fecundity, as the queen white-ant—an insect closely related to the bees in habits, though not in structure, as the white-ants are lace-wings, and belong to the order Neuroptera (Isoptera), which includes our day-flies, dragon-flies, etc.—is known to lay over 80,000 eggs daily. Yet this poor, helpless thing, whose abdomen is the size of a man's thumb, and composed almost wholly of eggs, while the rest of her body is no larger than the same in our common ants, has no other amusement; she can not walk; she can not even feed herself, or care for her eggs. What wonder then that she should attempt big things in the way of egg-laying? She has nothing else to do, or to feel proud of.

Different queens vary as much in fecundity as do different

kinds of life. Some queens are so prolific that they fairly demand hives of India rubber to accommodate them, keeping their hives gushing with bees and profitable activity; while others are so inferior that the colonies make a poor, sickly effort to survive at all, and usually succumb early, before the adverse circumstances which are ever waiting to confront all life on the globe. This lack of fecundity may be due to disease, improper development, or to special race or strain. This fact promises rich fruit to the careful, persistent breeder. The activity of the queen is governed largely by the activity of the workers. The queen will either lay sparingly, or stop altogether, in the interims of storing honey, while, on the other hand, she is stimulated to lay to her utmost capacity when all is life and activity in the hive. As the worker-bees feed the laying queen, it is more than probable that with no nectar to gather, the food is withheld, and so the queen is unable to produce the eggs which demand a great amount of nutritious food all ready to be absorbed. Thus, the whole matter is doubtless controlled by the workers. This refusal to lay when nectar is wanting does not hold true, apparently, with the Cyprian and Syrian bees.

The old poetical notion that the queen is the revered and admired sovereign of the colony, whose pathway is ever lined by obsequious courtiers, whose person is ever the recipient of loving caresses, and whose will is law in this bee-hive kingdom, controlling all the activities inside the hive, and leading the colony whithersoever it may go, is unquestionably mere fiction. In the hive, as in the world, individuals are valued for what they are worth. The queen, as the most important individual, is regarded with solicitude, and her removal or loss is noted with consternation, as the welfare of the colony is threatened; yet, let the queen become useless, and she is dispatched with the same absence of emotion that characterizes the destruction of the drones when they have become supernumeraries. It is very doubtful if emotion and sentimentality are ever moving forces among the lower animals. There are probably certain natural principles that govern in the economy of the hive, and anything that conspires against, or tends to intercept, the action of these principles, becomes an enemy to the bees. All are interested, and doubtless more united than is generally believed, in a desire to promote the free action of these principles. No doubt the principle of antagonism among the various bees has been overrated. Even the drones, when they are being killed off in the autumn, make a sickly show of defense, as much as to say, the welfare of the colony demands that such worthless vagrants should be exterminated. How relentlessly the bees drag out even the worker-bees that have become loaded with the pollen-masses of milkweed, or otherwise disabled. Such bees are of no more use, and useless members are not tolerated in the bee-community. It is most probable that what tends most for the prosperity of the colony is well understood by all, and without doubt there is harmonious action among all the denizens of the hive to foster that which will advance the general welfare, or to make war on whatever may tend to interfere with it. If the course of any of the bees seems wavering and inconsistent, we may rest assured that circumstances have changed, and that could we perceive the bearing of all the surrounding conditions, all would appear consistent and harmonious. The holding of young queens in the cells, and guarding them, seems an exception.

THE DRONES.

These are the male bees, and are generally found in the hive only from May to November, though they may remain all winter, and are not infrequently absent during the summer. Their presence or absence depends upon the present and prospective condition of the colony. If they are needed, or likely to be needed, then they are present. There are in nature several hundred, and often thousands, in each colony. The number may and should be greatly reduced by the apiarist. The drones (Figs. 46, 47) are shorter than the queen, being less than three-fourths of an inch in length, and are more robust and bulky than either the queen or workers. The drones weigh about 1-2000 of a pound, while the workers only weigh 1-5000. They are easily recognized, when flying, by their loud, startling hum. As in other societies, the least useful make the most noise. This loud hum would seem to be caused by the less rapid vibration of their large, heavy wings. Landois showed many years since, that the hum of bees and other insects, was due first to vibrations of wings, secondly to vibrations of the abdominal rings, and, thirdly, to what he styled true voice in the thoracic spiracles, where there are cavities which he thought were voice cavities. He thought the humming tone of bees and other insects came from the spiracles. The drone's flight is more heavy and lumbering than that of the workers. Their ligula (Fig. 49), labial palpi and maxillæ—like the same

FIG. 46.

FIG. 48.

(Original.)

FIG. 47.

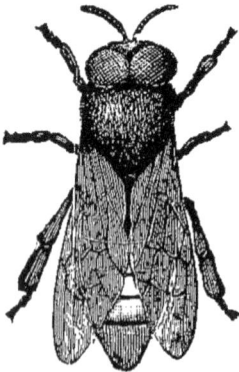

Drone-Bees, magnified, from
Newman.

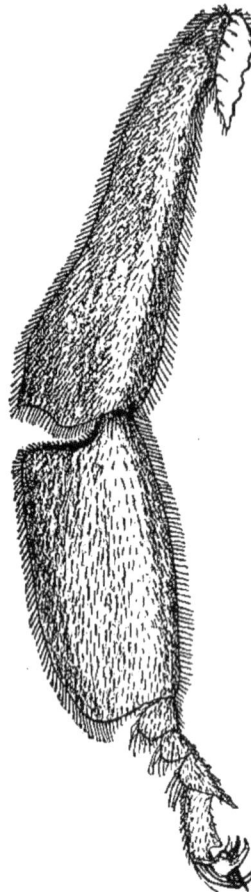

Part of Leg of Drone, magnified, after Duncan.

in the queen-bee—are short, while their jaws (Fig. 65, a) possess the rudimentary tooth, and are much the same in form as those of the queen, but are heavier, though not so strong as those of the workers. Their eyes (Figs. 3, 47) are very prominent, meet above, and thus the simple eyes are thrown forward. The ommatidia, or simple eyes which form the compound eyes of the drone (Figs. 3, 47), are, as shown by Lacodaire, more than twice as numerous as those of either queen or worker. The drones also have longer and broader antennæ, with far more of the olfactory cavities, though not so many tactile hairs as are found in the antennæ of the workers. Entomologists now believe that the better sight and smell, as also the large wings, are very useful to the drone. They make success more probable, as the drone flies forth with hundreds of other drones in quest of a mate. We can also see how, through the law of natural selection, all these peculiarities are con-

FIG. 49.

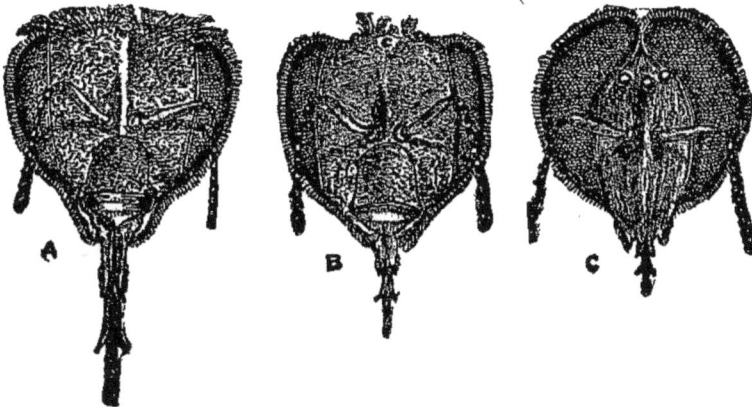

Heads of Worker, Queen and Drone, showing comparative length of Tongues, from Cowan.

A Worker.   B Queen.   C Drone.

stantly strengthened. Their posterior legs are convex on the outside (Fig. 48), so, like the queens, they have no pollen-baskets. As we should expect, the branching hairs, both on the body and legs, are almost absent in drones ; what there are are coarse, and probably aid in mating. The drones are without the defensive organ, having no sting, while their special sex-organs (Fig. 37) are very interesting. These have been fully described and illustrated by Leuckart. The testes are situated in the abdomen, in an analogous position to that of the ovaries in the queen. Like these organs in higher animals, there are in each testis hundreds of tubes in which are developed the sperm-cells in bundles. As Leuckart shows, the

testes are larger in the pupa than in the imago, for even then
the spermatozoa have begun to descend to the versiculæ semi-
nales (Fig. 37, c, c).   Thus, in old drones, the testes have
shrunken.   The spermatozoa are very long, with a marked
head (Fig. 50), which, as Mr. Cowan remarked to me, look
like cat-tail flags, as there is a short, small projection beyond
the head.   These sperm-cells are so very small, and so long
and slender, that it is difficult to isolate or trace them ; hence,
in microscopic preparations they look like one hopeless tangle
(Fig. 50).   It is incomprehensible how they can be separated

FIG. 50.

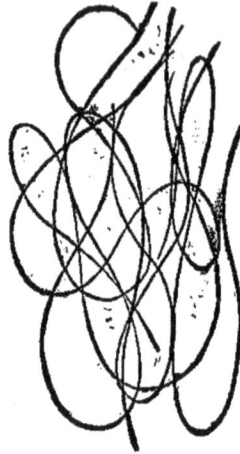

Spermatozoa.—Original.

and passed, one, two, or more at a time, by the queen as the
eggs are to be impregnated.   Appended to the versiculæ semi-
nales (Fig. 37, c,c) just where they pass to the ejaculatory duct
are two large glandular sacs (Fig. 37, d), which add mucus to
the seminal fluid.   The ejaculatory duct (Fig. 37, e) is rather
long and very muscular.   This passes to a pouch (Fig. 37, f),
where the sperm-cells are massed, preparatory to coition.
Leuckart called this mass of spermatozoa the spermatophore.
This is what is passed to the spermatheca of the queen during
coition.   Below this is the organ proper.   It has, as may be
seen by pressing a drone, three pairs of appendages, somewhat
horn-like, and certain roughness or pleats (Fig. 37, h, i), which
serve to make connection more close during coition.   These
little barb-like teeth, rough projections and horns, as they are

grasped and firmly pressed by the vulva or enlargement just at the end of the oviduct of the queen, are held as in a vice; and so we see why they are torn from the drone during coition. As Leuckart has so admirably described, the external organs of the drone are drawn up into the so-called bean or sac (Fig. 37, *f*), as the finger of a glove, to use the words of Girard, often turns in as we draw the glove off the hand. As we press a drone, or hold it in our warm hand as it has just returned from a long flight, when its air sacs are distended; or when it meets the queen, the sexual act is accomplished wholly or in part, and the external organ is everted or turned out as we turn the glove-finger out. In case of coitus, this eversion is very complete, so that the bean or sac (Fig. 37, *f*) turns out, and the spermatophore is passed into the oviduct of the queen, and by her muscular oviduct pushed into the spermatheca. This seems a wonderful operation, almost beyond the possible. Yet the passage of the egg from the ovaries in higher animals is almost as surprising. Leuckart is undoubtedly correct in suggesting that for full and complete impregnation the drone needs tense muscles, full air sacs, and thus the vehement exercise on the wing is very important in the sexual act. If this be true, then impregnation of the queen in confinement is as undesirable as it seems to be exceptional. While it may not be absolutely necessary to have these conditions for impregnation, as I think I have positive proof, it doubtless is better, and usually necessary, that they exist. At this time the queen's ovaries are small, and thus her smaller size before impregnation. Hence, there is lack of high tension within the abdomen of the queen, which also tends to aid in the sexual act.

The drone has not the wax-glands beneath the abdomen. On the ventral plates are scattering compound hairs, which doubtless have importance in the sexual act. The drone, like the queen, is without the lower head or pollen-digesting glands, and so is largely fed by the workers. Schonfeld has proved this by caging drones in full colonies. If caged in a single-walled cage, so as to be accessible to the workers, they live; if in a double-walled cage they all soon die, though all have abundant honey. While honey is necessary it is not enough.

It was discovered by Dzierzon in 1845, that the drones hatch from unimpregnated eggs. This strange phenomenon, seemingly so incredible, is, as has been shown in speaking of the queen, easily proved and beyond question. These eggs may come from an unimpregnated queen, a laying worker— which will soon be described—or an impregnated queen which may voluntarily prevent impregnation. It is asserted by some that the workers can change a worker-egg to a drone egg at will. When the workers are able to abstract the sperm-cells, which are so small that we can see them only by using a high-power microscope, then we may expect to see wheat turn to chess. Such eggs will usually be placed in the larger horizontal cells (Fig. 78, a), in manner already described.

The drone-cells are one-fourth of an inch in diameter, and project beyond the worker-cells, so they are a little more than one-half an inch long. Very rarely drones are produced in worker-cells. Such drones are diminutive, and undesirable in the apiary. As stated by Bevan, the drone feeds six and a half days as a larva before the cell is capped. As the microscope shows, undigested pollen is given to the drone-larvæ after the fourth day, which is not true of either the queen or worker. The capping of the drone-cells is very convex, and projects beyond the plane of the same in worker-cells, so that the drone-brood is easily distinguished from worker, and from the darker color—the wax being thicker and less pure—the capping of both drone and worker brood-cells enables us easily to distinguish them from honey-cells. In twenty-four days from the laying of the eggs, the drones come forth from the cells. Of course, variation of temperature and other conditions, as variable amount of diet, may slightly retard or advance the development of any brood, in the different stages. The drones—in fact all bees—when they first emerge from the cells, are gray, and are easily distinguished from the mature bee.

Just what the longevity of the male bee is, I am unable to state. It is probable, judging from analogy, that they live till accident, the worker-bees, or the performance of their natural function, cause their death. The worker-bees may kill off the drones at any time, which they do by constantly biting and worrying them; though principally, I think, by withholding their albuminous food. They may also destroy the drone-brood. It is not very rare to see workers carrying out immature drones even in midsummer. At the same time they may destroy inchoate queens. Such action is prompted by a sudden check in the yield of honey, and in case of drones is common only at the close of the season. The bees seem very cautious and far-sighted. If the signs of the times presage a famine, they stay all proceedings looking to the increase of colonies. On the other hand, nectar secretion by the flowers, rapid increase of brood, crowded quarters—whatever the age of the queen—are sure to bring many of the male bees, while any circumstances that indicate a need of drones in the near future, like loss or impotency of the queen, will prevent their destruction even in late autumn.

The function of the drones is solely to impregnate the queen, though when present they add to the heat of the hive. Yet for this they were far better replaced by worker-bees. That their nutrition is active, is suggested by the fact that, upon dissection, we usually find their capacious honey-stomachs filled with honey.

Impregnation of the queen always takes place, as before stated, while on the wing, outside the hive, usually during the heat of a warm, sunshiny day. After mating, as before suggested, the drone-organs adhere to the queen, and may be seen hanging to her for some hours. The copulatory act is fatal to the drone. By holding a drone just returned from a long flight in the hand, the ejection of the sex-organs is quickly produced, and is always followed by immediate death. As the queen meets only a single drone, and that only once, it might be asked why nature was so improvident as to decree hundreds of drones to an apiary or colony, whereas a score would suffice as well. Nature takes cognizance of the importance of the queen, and as she goes forth amidst the myriad dangers of the outer world, it is safest and best that her stay abroad be not protracted, that the experience be not repeated, and, especially, that her meeting a drone be *not delayed*. Hence, the superabundance of drones—especially under natural conditions, isolated in forest homes, where ravenous birds are ever on the alert for insect game—is most wise and provident. Nature is never "penny wise and pound foolish." In our apiaries the need is wanting, and the condition, as it exists in nature, is not enforced. Again, close impregnation or in-breeding, which is not conducive to animal vigor, is thus prevented, where otherwise it would be necessary and always the practice.

The fact that parthenogenesis prevails in the production of drones, has led to the theory that from a pure queen, however mated, must ever come a pure drone. My own experience and observation, which have been very extended, and under circumstances most favorable for a correct judgment, have fully and completely confirmed this theory. Yet, if telegony or the impure mating of our cows, horses, and fowls renders the females of mixed blood ever afterward, as is believed and taught by many who would seem most competent to judge—though I must say I am very skeptical in the matter—then we must look closely as to our bees, for certainly, if a mammal, and especially if a fowl, is tainted by impure mating, then we may expect the same of insects. In fowls such influence, if it exist, must come simply from the presence in the female generative organs of the sperm-cells, or spermatozoa, and in mammals, too, there is little more than this, for though they are viviporous, so that the union and contact of the offspring and mother seem very intimate during the fetal development, yet there is no intermingling of blood, for a membrane ever separates that of the mother from that of the fetus, and only the nutritious and waste elements pass from one to the other. To claim that the mother is tainted through the circulation, is like claiming that the same result would follow her inhaling the breath of her progeny after birth. If such taint be produced, it probably comes through the power of a cell to change those cells contiguous to it. That cells have such power is proved every day in case of wounds, and the spread of any disease. I can only say that I believe this whole matter is still involved in doubt, and still needs more careful, scientific and prolonged observation. I have tried very extensive experiments with both chickens and bees, and all the evidence was against telegony. My brown Leghorn hens ran with light Brahma roosters all winter, then were removed for three weeks, after which they were purely mated, and every one of the two hundred chickens were without any Brahma marks. Even the legs were absolutely clean. Likewise, thousands of drones, reared from pure Syrian queens, but mated to Italian drones, showed not the slightest Italian taint. I believe telegony is a very doubtful hypothesis.

THE NEUTERS, OR WORKER-BEES.

These, called "the bees" by Aristotle, and even by Wildman and Bevan, are by far the most numerous individuals of

FIG. 51.

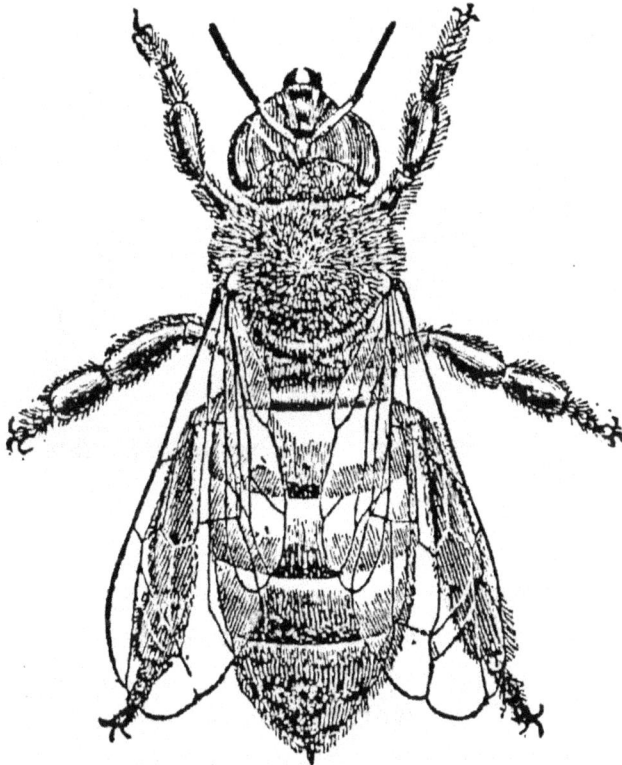

Worker-Bee much magnified, from Newman.

the hive—there being from 15,000 to 40,000 in every good colony. It is possible for a colony to be even much more populous than this. (Lubbock says that there are often 50,000 worker-ants in a nest.) These are also the smallest bees of the colony, as they measure but little more than one-half of an inch in length (Fig. 51.) As already stated, it takes about

5,000 worker-bees to weigh a pound. Prof. W. R. Lazenby found the weight of a worker to be .0799 grams, a load of honey weighed .043 grams. This is maximum. The average is .022 grams ; a load of pollen weighs .006 grams. Prof. Lazenby is probably correct in the assertion that usually only honey or pollen is carried by the bees ; but I have repeatedly known of bees carrying both honey and pollen at the same time.

The workers—as taught by Schirach, and proved by Mlle. Jurine, of Geneva, Switzerland, who, at the request of Huber, sought for and found, by aid of her microscope, the abortive

FIG. 52.

FIG. 53.

Ovaries of Worker-Bee, from
Leuckart.

Ovaries of Laying-Worker, from
Leuckart.

ovaries (Fig. 52) are undeveloped females. Rarely, and probably very rarely except when a colony is long or often queen-less, as is frequently true of our nuclei, these bees are so far developed as to produce eggs, which, of course, would always be drone-eggs. Such workers—known as "fertile"—were first noticed by Riem, while Huber saw one in the act of egg-aying. Paul L. Viallon and others have seen the same thing often. Several laying workers, sent me by Mr. Viallon, were examined, and the eggs and ovaries (Fig. 53) were plainly visible. Leuckart found, as seen in the figure, the rudiment of the spermatheca in both the common and the laying worker. Except in the power to produce eggs, they seem not unlike the other workers. Huber supposed that these were reared in cells contiguous to royal cells, and thus received royal food by accident. The fact, as stated by Mr. Quinby, that these occur in colonies where queen-larvæ were never reared, is fatal to the above theory. Langstroth and Berlepsch thought that these bees, while larvæ, were fed, though too sparingly, with the royal aliment, by bees in need of a queen, and hence the accelerated development. As already stated, the queen-larva is fed different and more abundant food than is the worker, and hence her accelerated and varied development. Is it not possible that these laying workers receive an excess of food as larvæ? Again, we have seen that laying workers occur in hopelessly queenless colonies ; and that queens are fed by the workers. May it not be that colonies hopelessly queenless take to feeding some special workers the chyle, and thus arise the laying workers? These are interesting inquiries that await solution. The generative organs are very sensitive, and exceedingly susceptible to impressions, and we may yet have much to learn as to the delicate forces which will move them to growth and activity. Though these laying workers are a poor substitute for a queen, as they are incapable of producing any bees but drones, and are surely the harbingers of death and extinction to the colony, yet they seem to satisfy the workers, for often the latter will not brook the presence of a queen when a laying worker is in the hive, frequently will not suffer the existence in the hive of a queen-cell, even though capped. They seem to be satisfied, though they have very slight reason to be so. These laying workers lay indifferently in large or small cells—often place several eggs in a single cell, and show their incapacity in various ways. Laying workers seem to appear more quickly and in greater abundance in colonies of Cyprian and Syrian bees, after they become hopelessly queenless, than in Italian colonies.

The maxillæ and labium of the worker-bee (Fig. 56) are much elongated (Fig. 54). The maxillæ (Fig. 54, A, mx, mx) are deeply grooved, and are hinged to the head by strong chitinous rods (Fig. 54, A, c, c, St, St), to which are attached the muscles which move these parts. The gutter-like extremities (Fig. 54, A, l, l) are stiffened with chitine, and, when approximated, form a tube which is continued by a membrane to the mouth-opening of the pharynx, just between the bases

FIG. 54.

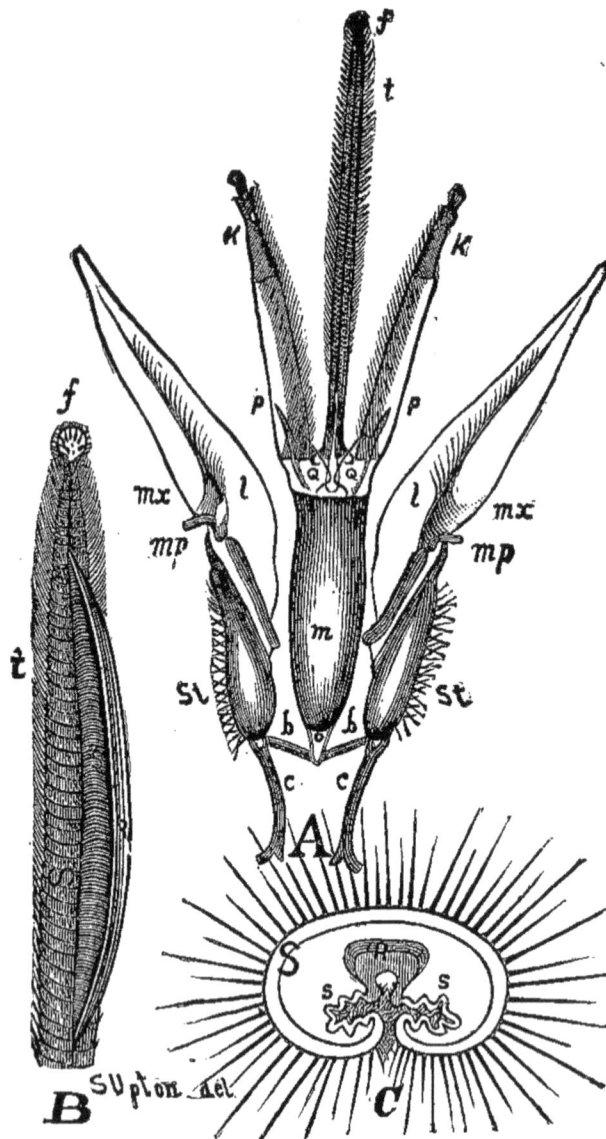

Tongue of Worker-Bee, much magnified.—Original.

mx mx Maxillæ.	*mp, mp* Max. palpi.	*k k* Labial palpi.
b b Lora.	*o* Sub-mentum.	*t* Tongue.
c c Cardines.	*m* Mentum.	*f* Funnel.
St, St Stipes.	*p, p,* Paraglossæ.	*R* Tubular rod.
l, l Laciniæ.	*B* Ligula, with sac	*s s* Colorless membrane.
s Colorless membrane.	distended.	*f* Funnel.
S Sheath.	*A* Maxillæ and labium.	*C* Cross-section of
R Tubular rod.		ligula.

(The above figure is drawn to the same scale as Fig. 27.)

of the jaws. This tube forms the largest channel through which nectar passes to the pharynx. The labium varies in length from .23 to .27 of an inch. By the sub-mentum (Fig. 54, *A, o*) and two chitinous rods, the lora (Fig. 54, *A, b, b*), it is hinged to the maxillæ. The base or mentum is chitinous beneath and membranous above. From the mentum extends the tongue or ligula (Fig. 54, *A, t*), the paraglossæ (Fig. 54, *A, p, p*), sac-like organs which connect with the cavity of the mentum, and so are distended with blood when the mentum is pressed. They also stand out like leaves or plates, and aid in directing the nectar which is drawn through the ligula into the mouth (Fig. 16). The labial palpi (Fig. 54, *A, k, k*) are four jointed, and in arrangement, form and function resemble the maxillæ. The tongue or ligula consists of an annulated sheath (Fig. 54, *C, S*) which is slitted along its under side to near the end. This is very hairy. Within this is a tubular rod (Fig. 54, *B* and *C, R*) which is also slitted along its under side to near the end, and opens above at its base between the paraglossæ (Fig. 54, *C*). Each margin of this slitted rod is united by a thin pubescent membrane to the corresponding margin of the surrounding sheath (Fig. 54, *C, s*). (So far as I know I was the first to discover this membrane.) Hence any pressure within the annular sheath may throw the central rod out (Fig. 54, *B, R*). This results when we press on the mentum ; as the blood pushes into the sheath and straightens the folded membrane (Fig. 54, *C, s*). The bee then can take nectar in three ways, first rapidly when sipping from flowers containing much nectar (Figs. 54, *A*, 57, *o, o*) by the large channels formed by approximating its maxillæ and labial palpi (Fig. 54, *A*, Fig. 57, *o, o*) ; secondly, slowly from deep tubular flowers, when it sips through the central rod ; and, thirdly, it may lap from a smeared surface because of the slitted ligula. By use of colored liquids I have demonstrated that the bee does actually sip in all these ways. At the end of the ligula there is a funnel (Fig. 54, *A, f*, 56, *b*).

Strange to say the structure and physiology of the tongue of the honey-bee were more correctly explained by old Swammerdam, than by most modern writers. Both he and Reaumur were quite accurate in their descriptions. Wolff, in his elegant monograph from which I have taken several figures, described with beautiful illustrations the mouth organs of the honey-bee, but was in doubt as to their physiology. Dr. Hyatt, of New York, did much to explain the anatomy of the bee's tongue; but so far as I know I was the first to explain accurately the anatomy and physiology of this organ. Within the mentum (Fig. 55, *C, m*) are strong muscles for retracting

FIG. 55.

B

Tongue bent under Head.

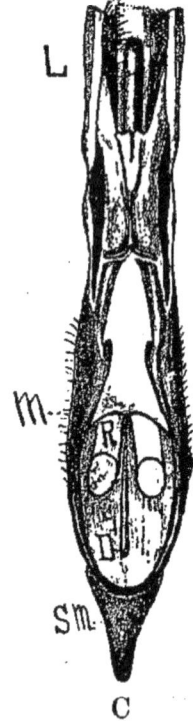

A

Tongue extended for sucking.

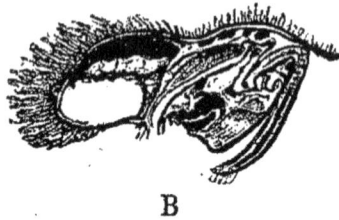

m Maxillæ.
  L Ligula.
s m Sub-mentum.
  D Duct from upper head and thoracic glands.

R Retractor muscles. The opening opposite L. at upper base of tongue between paraglossæ. All from Wolff.

C

Base of Labium.

the organ. The force of suction is doubtless analogous to the act of drinking on our own part. The rhythmical motion of the ligula in sipping honey is thus explained. By the muscles of the mouth the cavity is enlarged, producing suction, when by pressure swallowing is accomplished.

When not in use, the tongue with the attendant mouth organs, are bent back under the head (Fig. 55, *B*).

GLANDULAR ORGANS.

These important organs, which have been so fully described by Siebold, Wolff, and especially by Schiemenz, are

FIG. 56.

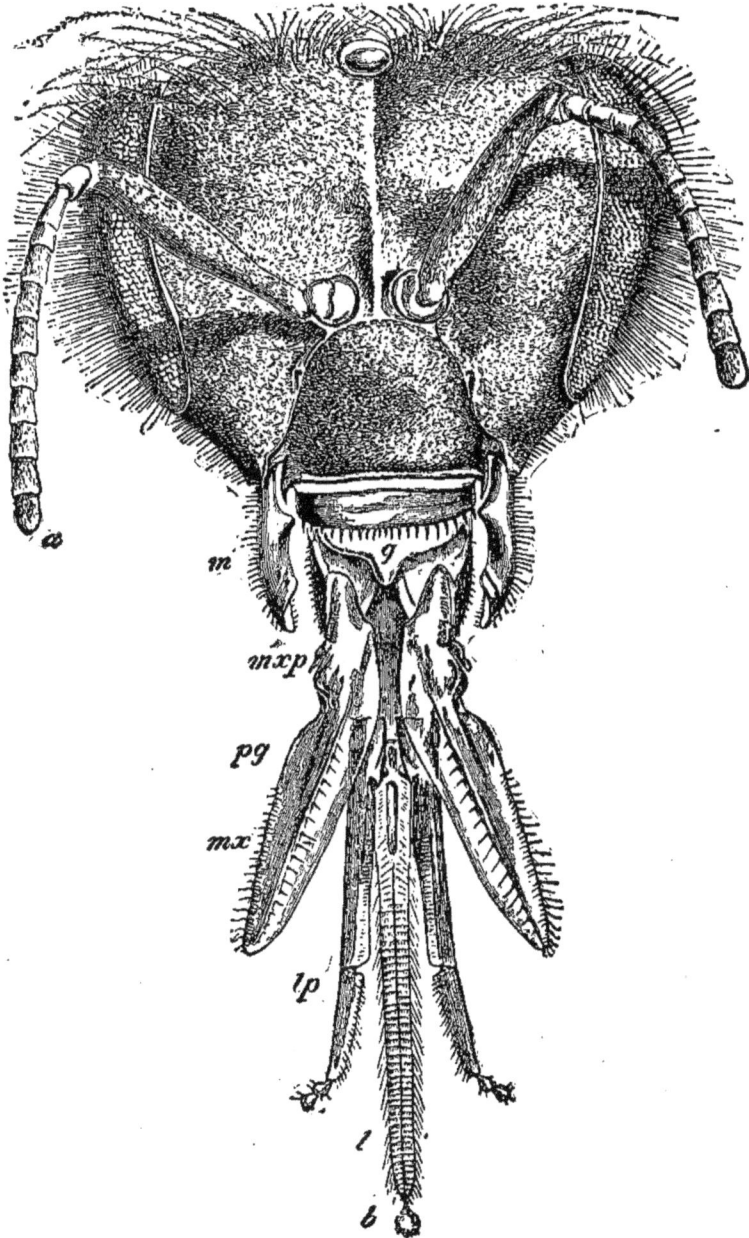

Head and Tongue of Bee, magnified twelve times.

(From Department of Agriculture.)

a Antenna.	*m x p* Maxillary palpus.	*l p* Labial palpus.
m Mandibles.	*p g* Paraglossa.	*l* Ligula.
g Epipharynx.	*m x* Maxilla.	*b* Funnel of tongue.

so intimately connected with the mouth organs, are so evidently useful in digestion, and are so well developed in the worker-bees, that they deserve full consideration. All the glands have a chitinous inner intivna and outer propria, and a middle epithelial membrane.

The spinning gland of the larval bee is a simple tubular gland, and is well illustrated by Schiemenz (Fig. 58). On each side within the head of the worker-bee (Fig. 59, *u h g*) are large glands, discovered by Meckel in 1846, and fully described by Siebold in 1870, which are very rudimentary in the queen and entirely absent in the drone. They are often called the lower head-glands. These are in form of the meibomian

FIG. 57.

Cross section of Tongue in use, after Cowan.

l l Labial palpi.	*o o* Tube for sucking the nectar.
m m Maxillæ.	*p* Overlapping maxillæ.

glands in our own eyelids; that is, a long duct bears many follicles rich with secreting cells, the whole looking like a compound leaf with small leaflets. Dr. Packard says each follicle is unicellular. While all the others are acinose. The ducts empty on the floor of the mouth. These glands are very marked in nurse-bees, but smaller in aged bees. Schiemenz believes that these glands secrete the food for the larval bees and also for the laying queen. Their large size, their full development only in the nurse-bees, and their entire absence in queen and drones, surely seem to give great force to this view. As already stated, the queen-larva is fed very liberally, and almost exclusively, of this so-called bee-milk. Berlepsch says that the little pollen sometimes (?) found in the digestive tube of the queen-larva is accidental. The worker-larva receives less of this secretion, and to that fed to the drone is

added, just at the last, some partially digested pollen which is
shed when the alimentary canal is moulted with the last larval
skin. The fact that undigested pollen is found in the larval
food shows that this food is from the stomach, and is not a
secretion. It has been suggested that the difference which Dr.
Planta and others find in the composition of the larval food of
worker, queen and drone larvæ is wholly due to this partially
digested pollen which is withheld from the inchoate queen and
workers.

There are also large compound racemose or acinose glands
(Fig. 59, *l h g*) in the head, and also a similar pair (Fig. 58, *l g*)

FIG. 58.

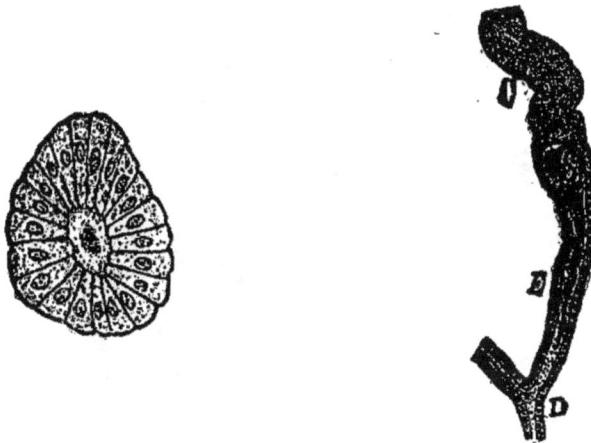

Spinning Gland of Larva, and cross section of same, after Schiemenz.

C Gland.	*S* Sinus.
I Duct.	*D* Common duct.

in the thorax, which are by some thought to be the modified
spinning glands of the larva. These four glands unite into a
common duct, which passes through the mentum and opens
just at the base of tongue on top in the groove between the
paraglossæ (Fig. 55, *C, L*, and 56). The thoracic glands were
discovered by Ramdohr in 1811, while Meckel also discovered
the second pair of cephalic glands, these are the upper head-
glands; Schiemenz is probably correct in thinking that these
glands, which are present in all bees, are for digesting the

nectar.   The cane-sugar of nectar is certainly digested or
changed into the more osmotic and assimilable glucose-like
sugar of honey.   Very likely these compound racemose glands
supply the digestive ferment which accomplishes this part of

FIG. 59.

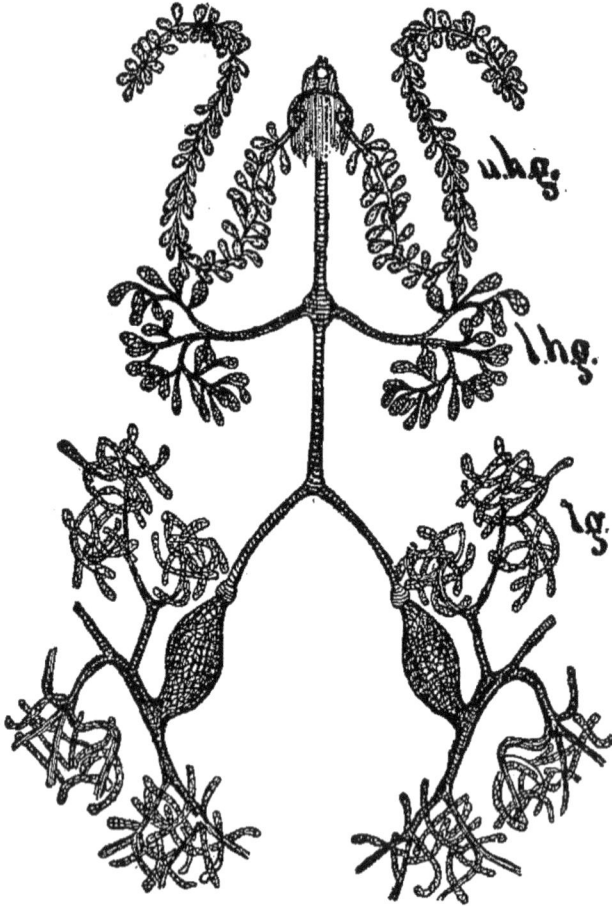

Gland System of Bee, after Girard.

digestion.   We similarly digest all the cane-sugar that we eat.
As honey is not always fully digested, the drones and queens,
as well as the workers, possess these glands.

Wolff's glands are large follicular glands (Fig. 60), situated
at the base of the mandibles.   From their position we might
suppose that their secretion was useful in forming wax into
comb, but their large size in the queens, and the fact that the

secretion from them is acid, would rather argue that they, like the racemose glands, were also digestive in their function. I would suggest that we call the thoracic glands, the glands of Ramdohr; the racemose glands of the head, the glands of Meckel, and the other glands of the head-glands of Siebold, in compliment to the excellent work which has been done in their study and elucidation; while the glands at the base of the mandibles may well be called, from their discoverer, Wolff's glands. In studying the digestive organism we are greatly

FIG. 60.

Jaw of Worker showing Wolff's gland, after Wolff.

M Muscles.   *J* Jaws.   *G* Gland.

indebted to Schiemenz and Schonfeld, who have not only explained by use of beautiful illustrations the detailed anatomy of the alimentary canal, but have been equally happy in describing the wonderful physiology of digestion in bees. Schonfeld, from a very elaborate series of experiments, concludes that the theory of Schiemenz and v. Siebold is not correct. He thinks the lower head-glands secrete saliva which moistens the pollen, and aids in digesting it. The fact that it is acid adds force to the theory. They empty on the floor of the mouth just where they should pour out the saliva. As the queen and drones never eat pollen, but are fed by the workers, they do not need these glands. Schonfeld thinks the larval food is digested pollen, and he claims to have found this in the true stomach of nurse-bees. Partially digested pollen he terms chyme, which, just before the drone-larvæ are to be

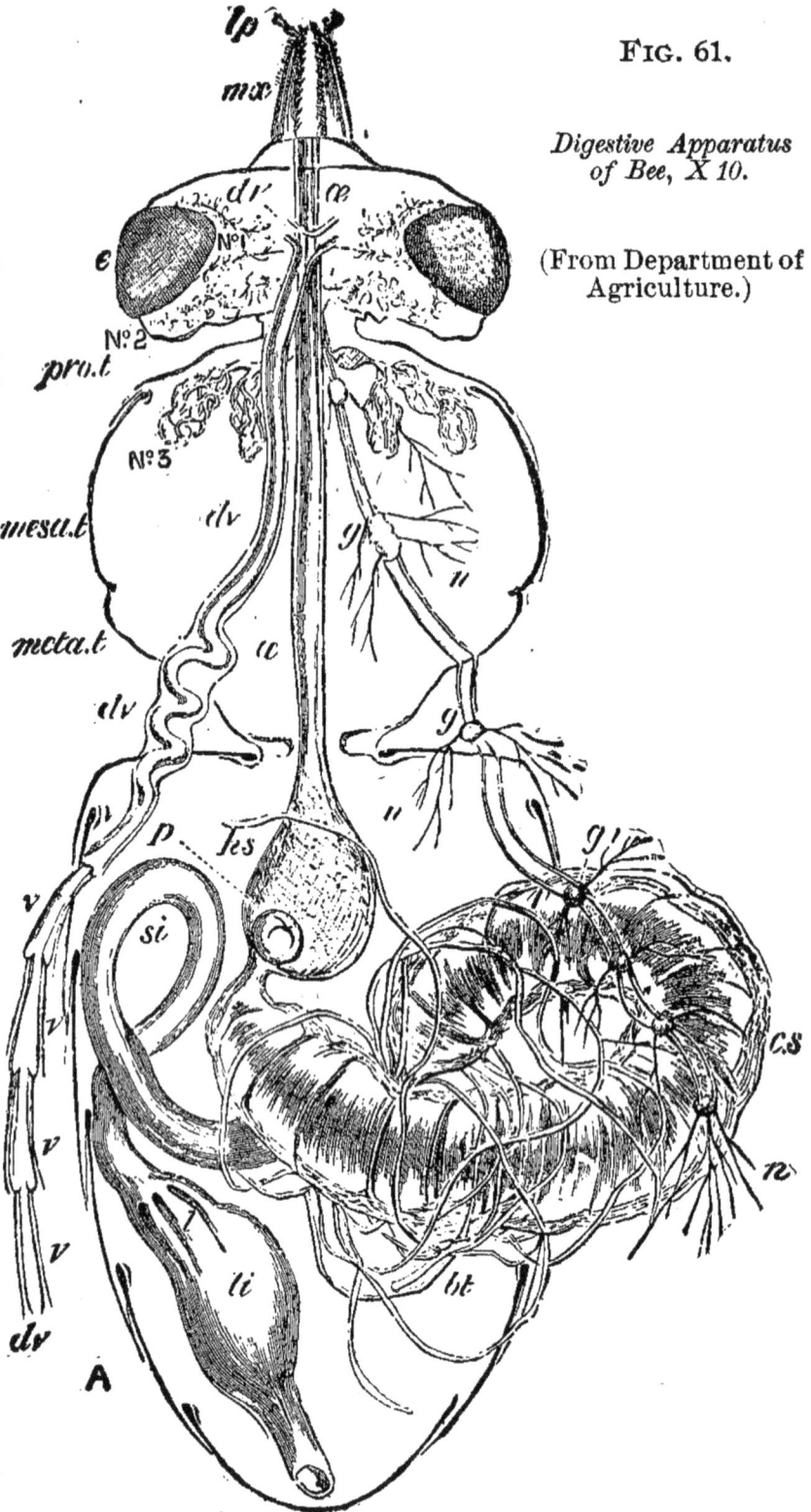

FIG. 61.

*Digestive Apparatus
of Bee,* X 10.

(From Department of
Agriculture.)

l p Labial palpi.
m x Maxilla.
e Compound eyes.
d v Dorsal vessel or heart.
v v Segments of heart.

p Stomach-mouth.
c.s. True stomach.
b t Malpighian tubules.
l.i. Large intestine.
g.g. Ganglia.

pro.t Prothorax.
mesa Mesathorax.
meta Metathorax.
œ Œsophagus.
h s Honey-stomach.

No. 1 Lower head glands.
No. 2 Upper head glands.
No. 3 Thoracic glands.
S i Small intestine.
l Rectal glands.

sealed up, is fed to them. The chyle and larval food he finds to contain blood corpuscles, and he thinks them identical with the same in the blood of the bee. Schonfeld fed indigestible material like iron particles to starving bees that had brood. The chyle, the larval food, but not the blood of the nurse-bees contained this iron. This food of the larvæ then must be chyle and not a secretion. I confirmed this by feeding bees sugar syrup in which I mixed finely pulverized charcoal. The charcoal appeared in the royal jelly in the queen-cells. As the charcoal is utterly non-osmotic, it could not pass to the blood, and so could not appear in any secretion, but could and would be in any regurgitated food. This secretion then appears to answer to the gastric juice in our own digestion. Again, the fact that it is acid, makes this conclusion more than warranted. This experiment certainly settles the matter.

Again, these same lower head-glands are found in some insects that do not feed their larvæ at all, as species of Eristalis—wasp-colored two-winged flies—and of Nepa, a genus of water-bugs.

Dr. Planta and others have shown that the chyle fed to queen-larvæ is not the same as that fed to drone-larvæ, nor yet like that fed to worker-larvæ. If this is chyle the difference could be explained, as it would arise from variation of food. If a secretion, it could not be easily explained. This view is adopted by Mr. Cowan, the ablest and most learned British authority on bees. Bordas has found two other pairs of glands in both worker and drone bees, which he terms, from their position, the internal mandibular and sublingual. It would be interesting but difficult to determine what secretion, if not all the secretions, aided in kneading the wax.

As in our own development, so in the embryo bee, the mid-intestine arises from the endoderm or inner layer of the initial animal. As the ectoderm or outer layer is around this, not only the mouth and vent, but the fore and hind intestine—all but the true stomach—arise by absorption at these points, or from invagination (a turning in) of the outer layer. Infants are not infrequently born with an imperforate anus. In such cases there is an arrest, the absorption does not take place, and the surgeon's knife comes to Nature's relief. Strangely

enough in the bee—this is also true of ants and some wasps—
this condition persists all through the larval period. Thus
bee-larvæ have no anus or vent, and so void no excreta. But
as known both to Swammerdam and Newport, when the last
larval skin is moulted the whole canal, with its contents, is

FIG. 62.

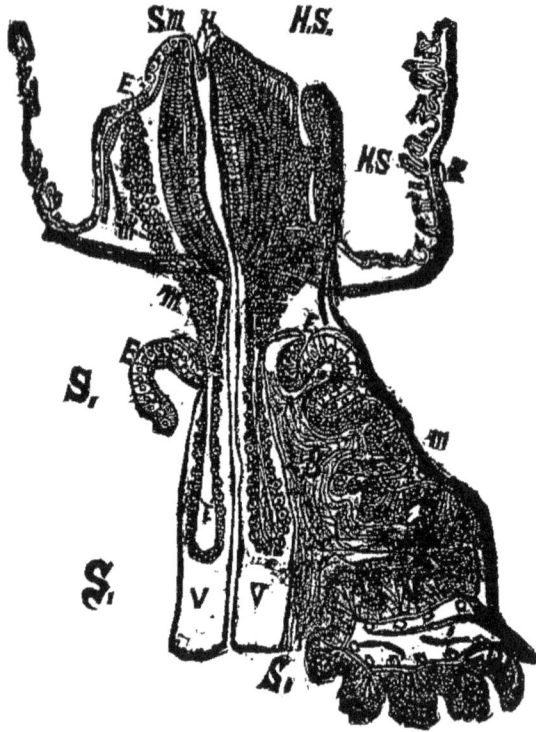

Section showing structure of Honey-stomach, Stomach-mouth and Stomach,
after Schiemenz.

H S Honey-stomach.	*E* Epithelial cells.
S Stomach.	*V* Stomach valve.
m Muscles.	*h* Hairs to hold pollen.
S m Stomach-mouth.	

moulted with the skin. As already stated, the spinning
glands in the larva become the thoracic, or glands of Ramdohr,
in the adult bee.

The œsophagus or gullet, the fine thread which is pulled
out as we behead a bee, passes from the mouth through the
muscular thorax (Figs. 25 and 27) to the honey-stomach, which
is situated in the abdomen. Often, as every bee-keeper knows,
this honey-stomach (Fig. 36, *hs*, 61 *hs*) comes along with the

œsophagus as we pull the bee's head from the body. The œsophagus (Fig. 61, $æ$) is about .2 of an inch long and .02 of an inch in diameter. In form and function the œsophagus is not different from the same organ in other animals. It is simply a passageway for the food (Fig. 27, 61 oe).

The honey-stomach (Fig. 62, h, s) or honey-sac is a sort of a crop or proventriculus. This sac is oval about .1 of an inch in diameter. While this organ is lined with a cellular layer (Fig. 62, HS, E), the cells are not large and numerous as in

FIG. 63.

Four pieces forming Stomach-Mouth, after Schiemenz.

c Cells. $T\,m$ Transverse muscles. $H\,s$ Longitudinal muscles.

the true stomach (Fig. 62, S, E). The muscular layers of this sac are quite pronounced (Fig. 62, m), as we should expect, as the honey has to be regurgitated from it to the honey-cells. This is truly a digestive chamber, as the nectar—cane-sugar— is here changed to honey—glucose-like sugar—but this is probably through the ferment received from the glands of Meckel and Ramdohr, and not from any secretion from the organ itself. The pollen is also very slightly digested here, as Schonfeld has shown, through the action of the saliva from the glands of Siebold, or lower head-glands. At the posterior end of this honey-stomach is the stomach-mouth (Fig. 36, 62, s, m, and 61, p) of Burmeister, which is admirably described by Schiemenz. It is really a stomach-mouth. Spherical in form, .02 of an inch in diameter, and, as Schonfeld well says, reminds one of a flower-bud. It (Fig. 61 p) can be seen by the

unaided eye, and as Schonfeld suggests, is easily studied with
a low-power microscope.   There are four jaw-like plates which
guard this stomach-mouth (Fig. 63), and as Schimenz shows,
open to let food pass to the true stomach.   This same author
tells us how by pressing with a needle, while viewing the
stomach-mouth under a microscope, we can see the jaws open
and shut.   These plates have fine hairs, pointing down (Fig.
62, *h*), which would, if a portion of the honey-containing pollen
were taken by this very muscular stomach-mouth, retain the
pollen-grains, while the honey could be passed back into the
honey-stomach.   Hence, Schiemenz very naturally concludes
that this is a sort of strainer, constantly separating the pollen
and honey as the bee is sipping nectar from flower to flower.

FIG. 64.

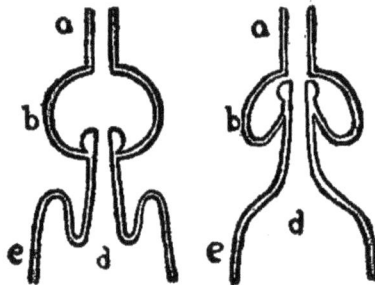

Stomach-mouth in Honey-Stomach, after Cowan.

A Normal.	*a* Œsophagus.	*d* Vales.
B Raised in regurgitation.	*b* Honey-stomach.	*e* True stomach.

As will be seen, this stomach-mouth has not only great longitudinal muscles (Fig. 62, *m*), but also circular muscles as well
(Fig. 62, *m*).   If Schiemenz is correct, then this stomach-mouth
is to separate the honey and pollen.   Even with this interesting apparatus, much of our honey has not a few pollen-grains,
as every observing bee-keeper knows.   The fact that nectar
has much more pollen in it than does honey, makes Schiemenz's view all the more probable.

There is also a long prolongation (Fig. 62, *v*) from the
stomach-mouth into the true stomach.   This is .04 of an inch
long, and is rich in cells, which are held by a very delicate

membrane which extends on still further. Schiemenz believes that this is a valve, and certainly unless drawn by the strong muscles in the walls out of the stomach as Schonfeld believes, it would act as a most efficient valve. If this does act as a perfect valve, then of course the nurse-bees can never feed the larvæ or queen any digested food from the true stomach. This is Schiemenz's view. Pastor Schonfeld, however, still holds, and seems to have proved, that while this may serve as a valve it is under the control of the bee, and may be so drawn up by the very muscular honey-stomach as to permit regurgitation (Fig. 64). In this regurgitation of chyle, the stomach-mouth closely approximates the stomach end of the œsophagus (Fig. 64, *B*); and so the chyle does not pass into the honey-stomach. This prolongation then is a valve under the control of the bee, and is another wonderful structure in this highly organized insect.

The true stomach (Fig. 61, *c*, *s*) is curved upon itself, and is .4 of an inch long and .1 of an inch in diameter. It is rugose, and the circular wrinkles or constrictions are quite regular. It is richly covered within by secreting cells (Fig. 62, *s*, *c*). The mucous membrane is folded, and hence there are very numerous gastric cells. Undoubtedly the function of the gastric juice is the same as in our own stomachs, it aids to liquify or render osmotic—capable of being absorbed—the albuminous food, in this case the pollen. This view is confirmed by the fact that we almost always find pollen in all stages of digestion in the true stomach of the bee. We may not wonder at the varied source of this digestive secretion; these gastric cells, the lower head-glands, and possibly Wolff's glands. Where among animals is such thorough digestive work accomplished? Emptying into the pyloric or posterior end of the stomach (Fig. 61, *bt*) are numerous tubules, the Malpighian tubules. These are the urinary organs, and remove waste elements from the blood. They are really the bee's kidneys. Like our own kidneys, they are nothing more than tubules lined with excreting cells. The small intestine is often called ileum (Fig. 61, *l i*). This portion of the digestive tube is lined with very minute, sharp chitinous teeth, which Schiemenz believes are used to further masticate the pollen-grains, that have not yielded to the digestive action of the stomach. This opinion is sustained by the strongly muscular nature of the tube (Fig. 36, A). The diameter of the ileum is hardly .02 of an inch. The rectum, or last portion of the intestine (Fig. 61, $l\,i$), is much larger than the ileum, and carries on its mucous or inner surface six glands (Fig. 36, r, g), which Schiemenz calls rectal glands. It is quite likely that these may be excretory in function. Their position would make this view seem probable at the least. Minot claims that these are not glands nor absorbant organs. Fernald thinks them valvular, and believes they restrain the injesta.

Before leaving the subject it seems well to remark that it now seems certain that the old view of Dufour, so ably advocated by Pastor Schonfeld is, despite the arguments and researches of Schiemenz, the correct one. Our experiments with charcoal prove this absolutely. The queen, drone and larvæ do not get their food as a secretion—a sort of milk—but it is rather the digested pollen modified, as the bees desire by varying their own food. In addition to this albuminous food

FIG. 65.

a Jaw of Drone.   b Jaw of Queen.   c Jaw of Worker.
(*Original.*)

the queen and drones also take much honey. Thus they need the glands which furnish the ferment that changes cane to reducible sugar, and they have them. If all honey were fully digested, then the drones and queen would not need any glands at all. The fact that the pollen that the larvæ do get is partially digested is further proof that this is chyme, or partially digested pollen.

The jaws (Fig. 65, c) are very strong, without the rudimentary tooth, while the cutting edge is semi-conical, so that when the jaws are closed they form an imperfect cone. Thus these organs are well formed to cut comb, knead wax, and perform their various functions. As we should expect, the muscles of the jaw are very large and powerful (Fig. 60). Wolff's glands empty at the base of these, and are doubtless excited by their action—a proof that their secretion is gastric in nature. The worker's eyes (Fig. 4) are like those of the queen, while their wings, like those of the drones (Fig. 46), attain the end of the body. These organs (Fig. 2), as in all insects with rapid flight, are slim and strong, and, by their more or less rapid vibrations, give the variety of tone which characterizes their hum. Thus we have the rapid movements and high pitch of anger, and the slow motion and mellow note of content and joy.

Landois proved many years since, that aside from the noise made by the wings, bees have a true voice. Thus he showed that a bumble-bee without wings, or with wings glued fast, would still hum. This voice is produced in the spiracles. Who has not noticed that a bumble-bee imprisoned closely in a flower still hums? I have also heard a carpenter-bee in a tunnel hardly larger than its body, hum loudly. Landois found this hum ceased when the spiracles were closed with wax. He describes quite an intricate voice-box, with a complex folded membrane, the tension of which is controlled through the action of a muscle and tendon. Thus we see that bees have a vocal organization not very unlike our own in the method of its action. The piping of the queen is probably this true voice. Landois also states that bees and other insects also make noise by the movement of the abdominal segments, the one on the other. From the enormous muscles in the thorax (Fig. 25) we should expect rapid flight in bees. Marked bees have been known to fly one-half mile, unload and return in six minutes, and double that distance in eleven minutes. In thirty minutes they went two and one-half miles, unloaded and returned. Thus they fly slower when foraging at a distance. These experiments were tried by my students, and the time was in the afternoon. I think they are reliable. Possibly, early in the day the rapidity would be greater. Sometimes swarms go so slowly that one can keep up with them. At other times they fly so rapidly that one needs a good horse to follow them closely. Here the rate doubtless depends upon the queen.

The legs of worker-bees are very strangely modified. As they are exceedingly useful in the bee economy, this is not strange. We find in the progressive development of all animals, that such organs as are most used are most modified, and thus we see why the legs and mouth organs of the worker-bees are so wonderfully developed.

The abundant compound hairs on the first joints of all the legs are very marked in the worker-bees. These are the pollen-gathering hairs, and from their branching, fluffy nature are well suited to gather the pollen-grains.

On the anterior legs the antenna cleaner (Fig. 66) is well marked, as it is in all Hymenoptera except the lowest families where it is nearly or quite absent. In the honey-bee, this is found in the queen and drone as well as in the worker. It is situated at the base of the first tarsus, and consists of a nearly semi-cylindrical concavity (Fig. 66, c), armed on the outer side with from seventy-eight to ninety projecting hairs. These teeth-like hairs projecting as fringe form a very delicate brush. Extending from the tibia is a blade-like organ—really

FIG. 66.

Antenna-Cleaner of Worker-Bee.—Original.

C Cavity.   S Spur.

the modified tibial spur (Fig. 66, S)—which when the leg is bent at this joint, comes squarely over the notch in the tarsus. Near the base on the inside a projecting knob is seen which perhaps acts as a strengthener. The part of this blade or spur that opposes the notch when in use consists of a delicate membrane. In other Hymenoptera this spur is greatly varied. Often, as in the ants and mud-wasps, it is also delicately fringed. Sometimes it has a long projecting point, and is thickly set with spinous hairs.

That this organ is an antenna-cleaner is quickly seen by watching a bee—preferably a bumble-bee—come from a tubu-

FIG. 67.

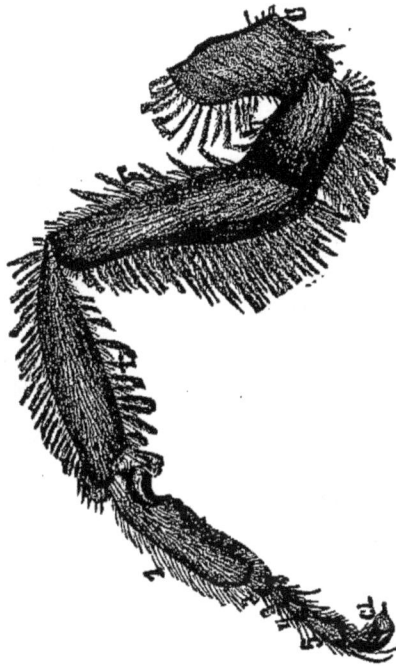

Anterior Leg of Worker-Bee.—Original.

C Coxa.	*T* Trochanter.
F Femur.	*Ti* Tibia.
1 2 3 4 5 Tarsal joints in order.	*Cl* Claws.

lar flower, like that of the malva, or by placing a honey-bee, bumble-bee or wasp on the inside of a window-pane and dusting its antennæ with flour or pulverized chalk. The insect at once draws its antennæ, one and then the other, through these admirable dusters, till the organs are entirely free from the dust. The bee in turn cleans its antenna-cleaners by scraping them between the inner brush-like faces of the basal tarsi of its middle legs, which is done each time after they are used to clean an antenna. The paper-making wasps, and I presume all wasps clean these organs by passing them between their jaws, much as a child cleans his fingers after eating candy, except here lips take the place of jaws. We can hardly conceive of a better arrangement for this purpose, a delicate brush and a soft membrane; even better than the housewife armed with soft brush and a silk kerchief, for this antenna cleaner just fits the organs to be dusted. We have seen the important function of the antennæ, as most delicate touch-

FIG. 69.

FIG. 68.

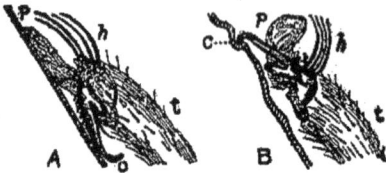

Tip of Foot of Bee.—Original.

A Pulvilli in use.
B Claws in use.
c.c. Claws.   *h.h.* Hairs.
p.p. Pulvilli.
t.t. Last joint of Tarsus.

End of Middle Leg of Worker-Bee.—Original.

organs, and as organs of smell, two senses of marvelous development in the bee. It is as imperative that the bee keeps its antennæ dust-free as that the microscopist keeps his glasses immaculate. A delicate brush (Figs. 66 and 67) on the end of the tibia opposite the spur and also the brush of rather spinous hairs on the tarsus (Fig. 66) are of use to brush the hairs, eyes and face, as may be seen by careful observation.

The claws and pulvilli—the delicate gland between the claws—are well marked on all the feet of bees. The claws (Fig. 67, *cl*) are toothed, and are very useful in walking up wooden or other rough surfaces (Fig. 68, *B*), as they are used just as a squirrel uses its claws in climbing a tree. These claws are also used in holding the bees to some object, or together while clustering. What a grip they must have. It is as if we were to grasp a limb or branch and then hold hundreds, yes thousands, of other persons as heavy as ourselves who had in turn grasped hold of us. When walking up a vertical wall of glass or other smooth metal, the claws are of no use, and so are turned back (Fig. 68, *A*), and the pulvilli—glandular organs—are spread out and serve to hold the bee. These secrete a viscid or adhesive substance which so sticks that the bee can even walk up a window-pane. This is why bees soon cloud or befoul glass over which they constantly walk. We thus understand why a bee finds it laborious and difficult to walk up a moist or dust-covered glass or metal surface.

The middle legs of the worker-bee are only peculiar in the prominent tibial spur (Fig. 69), and the brushes or pollen-combs on the inside of the first tarsus. It has been said that the spur is useful in prying off the pollen-masses from the posterior legs, as the bee enters the hive to deposit the pollen in the cells. This is doubtless an error. The queen and drone have this spur even longer than does the worker; the pollen comes off easily, and needs no crow-bar to loosen it. It is common among insects, and there are often two. The coarse, projecting hairs on all the feet are doubtless the agents that push off the loads of pollen.

We have already seen how the brushes or combs on the inner face of the first tarsus of the middle legs serve to remove the dust from the antenna cleaner. These also serve as combs, like similar but more perfect organs on the posterior legs, to remove the pollen from the pollen-hairs, and pack it in the pollen-baskets on the hind legs. Mr. Root speaks of the tongue as the organ for collecting pollen. Are not these hairs really the important agents in this important work?

But the posterior legs are the most interesting, as it is rare to find organs more varied in their uses, and so as we should expect, these are strangely modified. The branching or pollen-gathering hairs (Fig. 71) are very abundant on the coxa trochanter and femur, and not absent, though much fewer (Fig. 70) on the broad triangular tibia. The basal tarsus (Fig. 70) is quadurate, and it and the tibia on the outside (Fig. 70)

FIG. 70.

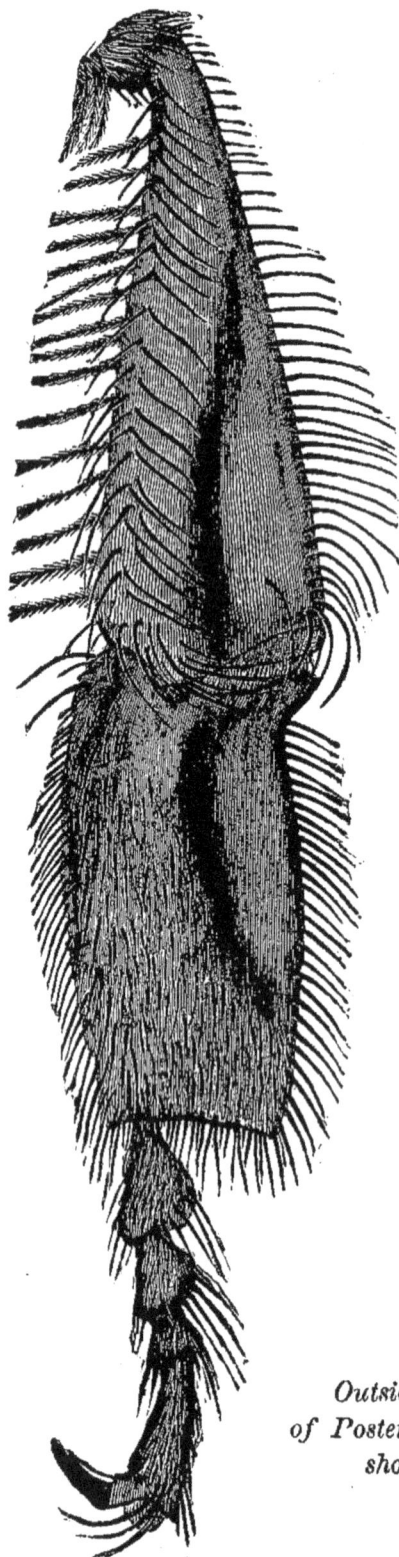

*Outside of Tibia and Tarsi
of Posterior Leg of Worker-Bee,
showing Corbicula.—
Original.*

FIG. 71.

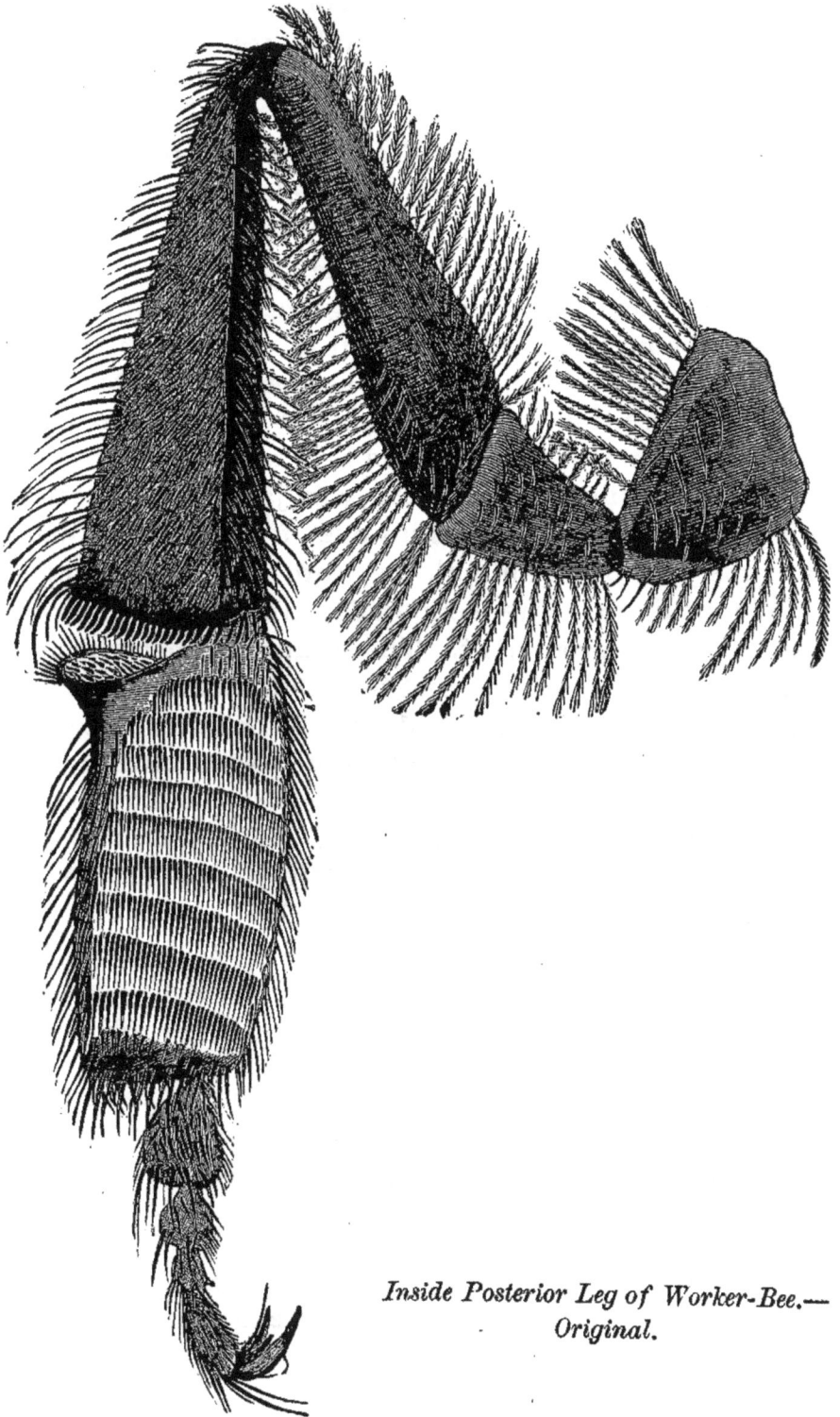

Inside Posterior Leg of Worker-Bee.—
Original.

are smooth and concave, especially on the posterior portion, which shallow cavity forms the corbicula or "pollen-basket." This is deepened by stiff marginal hairs, which stand up like stakes in a sled. These spinous hairs not only hold the pollen-mass, as do stakes, but often pierce it, and so bind the soft pollen to the leg. Opposite the pollen cavity of the first tarsus, or on the inside (Fig. 71), are about eleven rows of stiff hairs. They are of golden color, and very beautiful. These may be called the pollen-combs, for it is they that gather, for the most part, the pollen from the pollen-gathering hairs of legs and body, and convey it to and pack it in the pollen-baskets. As we have seen (Fig. 69), there are less perfect combs—similar in character, position and function—on the middle legs. The contiguous ends of the tibia and first tarsus or planta are most curiously modified to form the wax-jaws. The back part of this joint (Figs. 70, 71) reminds one of a steel trap with teeth, or of the jaws of an animal, the teeth in this case consisting of spinous hairs. The teeth on the tibia, the pecten or comb, are strong and prominent. These shut against the upper ear-like auricle of the planta, and thus the function of these wax-jaws is doubtless to grasp and remove the wax-scales from the wax-pockets, and carry them to the jaws of the bees. These wax-jaws are not found in queens or drones, nor in other than wax-producing bees. They are well developed in Trigona and Melipona, and less, though plainly marked, in bombus. Girard gives this explanation in his admirable work *Les Abeilles;* and as he is no plagiarist, as he gives fullest credit to others, he may be the discoverer of these wax-jaws. If he is not, I know not who is. The genus Apis is peculiar among our bees, and really exceptional among insects in having no posterior tibial spurs. They would, of course, be in the way of action of the wax-jaws. As before stated, there are six segments to the abdomen, in the queen and worker-bee (Fig. 9), and seven in the male. Each of these abdominal rings consists of a dorsal piece or plate—tergite or notum and pleurites united—which bears the spiracle, and which overlaps the ventral plate or sternite. These plates are strengthened with chitine. These rings are connected with a membrane, so that they can push in and out, something as the sections of a spy-glass are worked.

The ventral or sternal abdominal plates of the second, third, fourth and fifth segments of the worker (Fig. 72) are

FIG. 72.

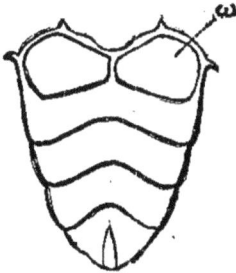 

Underside of Abdomen of Worker-Bee.—Original.

w Wax Scales.   w.w. Wax Scales.

modified to form the "wax-pockets;" though wax-plate would be a more appropriate name. These wax-plates (Fig. 73) are smooth, and form the anterior portion of each of these ventral plates. Each is margined with a rim of chitine, which gives it strength, and makes "pocket" a more appropriate name,

FIG. 73.

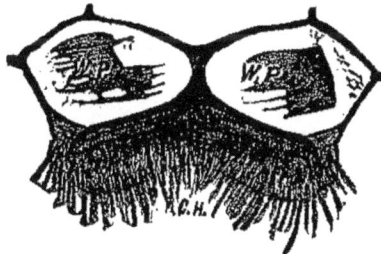

Original.

w p Wax-Plates.   c h Compound Hairs.

especially as the preceding segment shuts over these wax-plates. The posterior portion—less than half the sternite (Fig. 73)—bears compound hairs, and shuts over the succeeding wax-pocket. These wax-pockets are absent, of course, in queen and drones.

Inside the wax-plates are the glands that secrete the wax. When the wax leaves these glands it is liquid, and passes by osmosis through the wax-plate and is molded on its outer face.

The worker-bees possess at the end of the abdomen an organ of defense, which they are quick to use if occasion demands. Female wasps, the females of the family Mutillidæ, and worker and queen ants, also possess a sting. In all other Hymenoptera, like Chalcid and Ichneumon flies, gall-flies, saw-flies, horn-tails, etc., while there is no sting, the females have a long, exserted ovipositor, which, in these families, replaces the sting, and is useful, not as an organ of defense, but as an auger or saw, to prepare for egg-laying, or else, as in case of the gall-flies, to wound and poison the vegetable tissue, and thus by irritation to cause the galls.

This organ in the worker-bee is straight, and not curved as is the sting of the queen. The poison which is emitted in stinging, and which causes the severe pain, is both an acid and an alkaline liquid, which Carlet shows are both necessary for maximum results. These are secreted by a double tubular gland (Fig. 38, *Pg.*) and stored in a sac (Fig. 74, *c*, and 38, *Pb.*) which is about the size of a flax-seed. This sac is connected by a tube (Fig. 74, *M*) with the reservoir of the sting. The sting is a triple organ consisting of three sharp hollow spears, which are very smooth and of exquisite polish. If we magnify the most beautifully wrought steel instrument, it looks rough and unfinished; while the parts of the sting, however highly magnified, are smooth and perfect. The true relation of the three parts of the sting was accurately described by Mr. J. R. Bledsoe, in the American Bee Journal, Vol. VI, page 29. The action in stinging, and the method of extruding the poison, are well described in a beautifully illustrated article by Mr. J. D. Hyatt, in Vol. I, No. 1, of American Quarterly Microscopical Journal. The larger of the three awls (Fig. 74, *A*) usually, though incorrectly, styled a sheath, has a large cylindrical reservoir at its base (Fig. 74, *S*), which is entirely shut off from the hollow (Fig. 74, *H*) in the more slender part of the awl, which latter serves no purpose whatever, except to give strength and lightness. Three pairs of minute barbs (Fig. 74) project like the barbs on a fish-hook, from the end of this awl.

The reservoir connects at its base with the poison-sac and below, by a slit, with the opening (Fig. 74, N) made by the

FIG. 74.

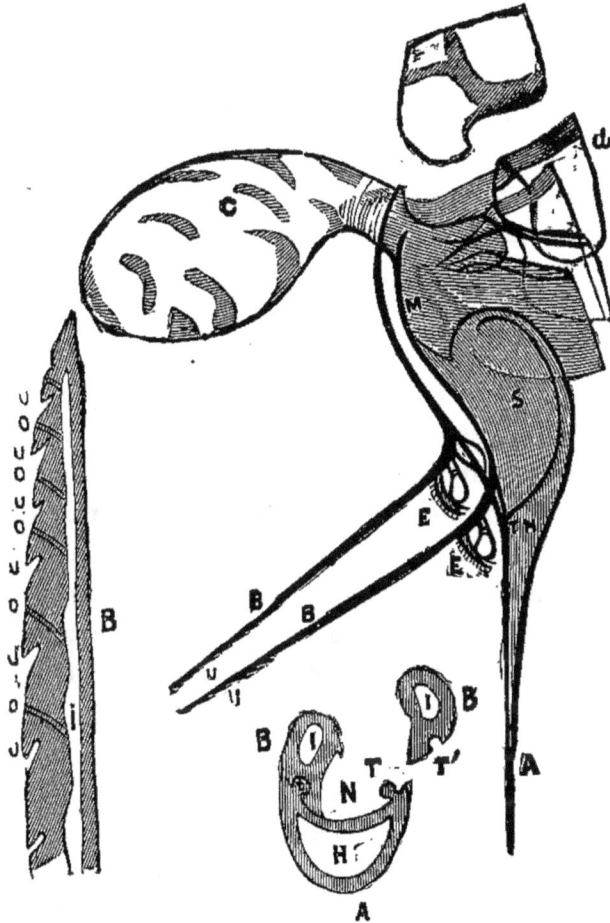

Sting with Lancets drawn one side, cross-section of Sting, and a Lancet, much magnified.—Original.

C Poison sac.	*M* Tube from sac to	*S* Reservoir.
A Awl.	reservoir.	*E, E* Valves.
U, U Barbs.	*B, B* Lancets.	*H* Hollow in awl.
I, I Hollows in lancets.	*O, O* Openings from hollow	*T, T* Ridges in awl.
T' Groove in lancet.	in lancets.	

approximation of the three awls. The other two awls (Fig. 74, *B, B*), which we call lancets, are also hollow (Fig. 74, *I, I*). They are barbed (Fig. 74, *U, U*) much like a fish-hook, except

that there are eight or ten barbs instead of one. Five of the
barbs are large and strong. These barbs catch hold and cause
the extraction of the sting when the organ is used. Near the
base of each lancet is a beautiful valvular organ (Fig. 74, E, E).
Mr. Hyatt thought these acted like a hydraulic ram, and by
suddenly stopping the current forced the poison through the
hollow lancets. It seems more probable that the view of Mr.
T. G. Bryant (Hardwick's Science Gossip, 1875) is the more
correct one. He suggests that these are really suction-valves—
pistons, so to speak—which, as the piston-rods—the lancets—
push out, suck the poison from the sacs. Carlet shows that the
poison-sac is not muscular, so the pumping is necessary. The
hollow inside each lancet (Fig. 74, I, I), unlike that of the awl,
is useful. It opens anteriorly in front of the first six barbs
(Fig. 74, o, o), as shown by Mr. Hyatt, and posteriorly just
back of the valves into the central tube (Fig. 74, N), and
through it into the reservoir (Fig. 74, S). The poison then can
pass either through the hollow lancets (Fig. 74, I, I) or through
the central tubes (Fig. 74, N), between the three spears.

The lancets are held to the central piece by projections
(Fig. 74, T, T) from the latter, which fit into corresponding
grooves (Fig. 74, T) of the lancets. In the figure the lancets
are moved one side to show the barbs and valves ; normally
they are held close together, and thus form the tube (Fig.
74, N, Fig. 44, St.)

At the base of the central awl two flexible arms (Fig. 75, b,b)
run out and up, where they articulate with strong levers (Fig. 75,
D, D). The two lancets are singularly curved and closely
joined to the flexible arms by the same kind of dovetailed
groove and projection already described. These lancets connect at their ends (Fig. 75, c, c) with heavy triangular levers
(Fig. 75, B, B), and these in turn with both C and D at j and s.
All of these levers, which also serve as fulcra (Fig. 75, B, C
and D), are very broad, and so give great space for muscular
attachment (Fig. 75, m). These muscles, by action, serve to
compress the poison-sac, also cause the lever (Fig. 75, B) to
rotate about S as a center, and thus the whole sting is thrown
out something as a knee-joint works, and later the lancets are
pushed alternately further into the wound, till stopped by the

valves striking against the farther end of the reservoir, in the central awl (Fig. 74, *S*). As Hyatt correctly states in his excellent article, the so-called sheath first cuts or pierces, then the lancets deepen the wound. Beside the sting are two feeler-like organs (Fig. 75, *E*, *E*), which doubtless determine where best to insert the sting, though usually there would seem little time for consideration. Leuckart discovered a second smaller gland (Fig. 38, *Sg*,) mentioned also by Girard and Vogel, which also has a sac or reservoir where its secretion is stored. This secretion, as first suggested by Leuckart, is supposed to act as a lubricant to keep the sting in good condition. The fact that muscles connect the various parts (Fig. 75) explains

FIG. 75.

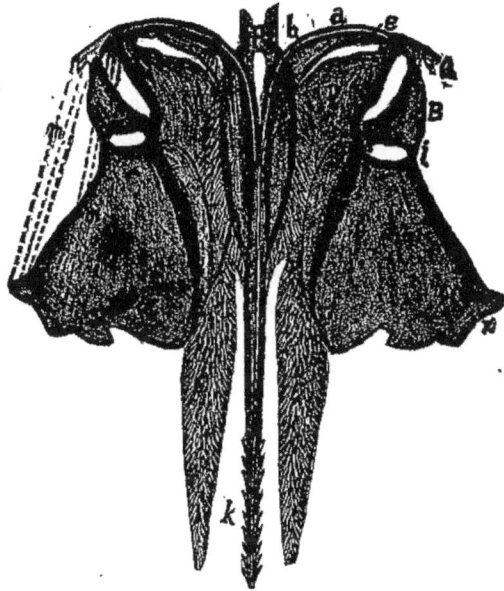

Sting of Worker-Bee, modified from Hyatt and Bryant.

how a sting may act, even after the bee is apparently lifeless, or, what is even more wonderful, after it has been extracted from the bee. Dr. Miller thinks a sting extracted months before may still act. The barbs hold one lancet as a fulcrum for the other, and so long as the muscles are excitable, so long is a thrust possible. Thus I have known a bee, dead for hours, to sting. A wasp, dead more than a day, with the abdomen cut off, made a painful thrust, and stings extracted for several minutes could still bring tears by their entering the flesh. In stinging, the awl first pierces, then the lancets follow. As the lancets push in, the valves force the poison already crowded into the reservoir forward, close the central tube, when the poison is driven through the lancets themselves, and comes out by the openings near the barbs (Fig. 74, *o, o*). The drop of poison which we see on the sting when the bee is slightly irritated, as by jarring the hive on a cold day, is pushed through the central opening by muscular contraction attendant upon the elevation of the abdomen and extrusion of the sting. The young microscopists will find it difficult to see the barbs, especially of the central awl, as it is not easy to turn the parts so that they will show. Patience and persistence, however, will bring success. Owing to the barbs the sting is often sacrificed by use. As the sting is pulled out, the body is so lacerated that the bee dies. Sometimes it will live several hours, and even days, but the loss of the sting is surely fatal, as my students have often shown by careful experiment. It is hardly necessary to say that there is no truth in the statement that the sting is used to polish the comb ; nor do I think there is any shadow of foundation for the statement that poison from the sting is dropped into the honey-cells to preserve the honey. The formic acid of honey doubtless comes from the honey-stomach. Each is an animal secretion.

The workers hatch from impregnated eggs, which can only come from a queen that has met a drone, and are always laid in the small, horizontal cells (Fig. 78, *c*). It is true that workers are very rarely reared in drone-cells when the rim is constricted. Mr. Root found that larger cells of foundation were likewise narrowed. These eggs are in no wise different, so far as we can see, from those which are laid in the drone or queen cells. All are cylindrical and slightly curved (Fig. 39, *a, b*), and are fastened by one end to the bottom of the cell, and a little to one side of the center. The eggs will not hatch unless a little food is added. Is this absorbed, or does it soften the shell so as to make exit possible ? Girard says that the egg on the first day stands oblique to the bottom of the cell, is more inclined the second day, and is horizontal the third day. As in other animals, the eggs from different queens vary perceptibly in size. As already shown, these are voluntarily fertilized by the queen as she extrudes them, preparatory to fastening them in the cells. These eggs, though small—one-sixteenth of an inch long—may be easily seen by holding the comb so that the light will shine into the cells. With experience they are detected almost at once, but I have often found it quite difficult to make the novice see them, though very plainly visible to my experienced eye.

The egg hatches in three days. The larva (Fig. 39, *d*, *e*, *f*), incorrectly called grub, maggot—and even caterpillar, by Hunter—is white, footless, and lies coiled up in the cell till near maturity. It is fed a whitish fluid, the chyle already described, though this seems to be given grudgingly, as the larva never seems to have more than it wishes to eat, so it is fed quite frequently by the mature workers. It would seem that the workers fear an excessive development, which, as we have seen, is most mischievous and ruinous, and work to prevent the same by a mean and meager diet. Not only do the worker-larvæ receive the chyle grudgingly, but just at the last, before the cell is sealed, a different diet is given. There are more albuminoids and fats, and less carbohydrates, as shown by Dr. de Planta. It is probable that honey is also given them, and so Dufour was wholly right in urging that digested food was fed to the larvæ, for honey is digested nectar. This added honey is what probably changes the food. He was also correct in supposing the food of the larva to be a sort of chyle. M. Quinby, Doolittle, and others, say water is also an element of this food. But bees often breed very rapidly when they do not leave the hive at all, and so water, other than that contained in the honey, etc., can not be added. The time when bees seem to need water, and so repair to the rill and the pond, is during the heat of spring and summer, when they are the most busy. May this not be quaffed for the most part to slake their own thirst? If water is carried to the hives it is doubtless given to the nurse-bees. They may need water when the weather is hot and brood-rearing at its very height. There is no reason to doubt that bees, like all other active animals, need water as they do salt, to aid the physiological processes. They cool by evaporation, and need water to promote the process.

When they smother, is not the moisture about them in part the water of respiration rather than exclusive honey ?

At first the larvæ lie at the bottom of the cells, in the cream-like " bee-milk." Later they curl up, and, when fully grown, are straight (Fig. 39, *f*). They now turn head down and cast their skin and digestive canal, then turn with their heads towards the mouth of the cell (Fig. 39, *f*). Before this, however, the cell has been capped.

In eight days (Root says nine or ten) from the laying of the egg, the worker-cell, like the queen-cell, is capped over by the worker-bees. This cap is composed of pollen and old wax, so it is darker, more porous, and more easily broken than the caps of the honey-cells ; it is also more convex (Fig. 39, *k*). The larva, now full grown, having lapped up all the food placed before it, spins its silken cocoon, so excessively thin that it requires a great number to appreciably reduce the size of the cell. The silken part of the cocoon extends down from the cap but a short distance, but like moths and many other insects, the larval bee, just before it pupates, spreads a thin glue or varnish over the entire inner part of the cell. These cocoons, partly of silk and partly of glue, are well seen when we reduce combs to wax with the solar wax-extractor. These always remain in the cells after the bees escape, and give to old comb its dark color and great strength. Yet they are so thin that cells used even for a dozen years, seem to serve as well for brood as when first used. Indeed, I have good combs which have been in constant use nineteen years. As before stated, the larva sheds its skin, and at the last moults the alimentary canal or digestive tube with its contents as well. These, as stated by Vogel, are pushed to the bottom of the cell. In three days the insect assumes the pupa state (Fig. 39, *g*). In all insects the spinning of the cocoon seems an exhaustive process, for so far as I have observed, and that is quite at length, this act is succeeded by a variable period of repose. By cutting open cells it is easy to determine just the date of forming the cocoon, and of changing to the pupa state. The pupa looks like the mature bee with all its appendages bound close about it, though the color is still whitish.

In twenty-one days, it may be twenty with the best conditions, the bees emerge from the cells. Every bee-keeper should hold in memory these dates : Three days for the egg, six for the larva, and twelve days after the larva is sealed over. Of course, there may be slight variations, as the temperature of the colony is not always just the same.

The old writers were quite mistaken in thinking that the advent of these was an occasion of joy and excitement among the bees. All apiarists have noticed how utterly unmoved the bees are, as they push over and crowd by these new-comers in the most heedless and discourteous manner imaginable. Wildman tells of seeing the workers gathering pollen and honey the same day that they came forth from the cells. This idea is quickly disproved if we Italianize black bees. We know that for some days—usually about two weeks if the colony is in a normal condition, though if all the bees are very young it may be only one week—these young bees do not leave the hive at all, except in case of swarming, when bees even too young to fly will attempt to go with the crowd. However, the young bees do fly out for a sort of "play spell" before they commence regularly to work in the field. They doubtless wish to try their wings. These young bees, like young drones and queens, are much lighter colored when they first leave the cell.

The worker-bees never attain a great age. Those reared in autumn may live for eight or nine months, and if in queenless colonies, where little labor is performed, even longer ; while those reared in spring will wear out in three months, and when most busy will often die in from thirty to forty-five days. None of these bees survive the year through, so there is a limit to the number which may exist in a colony. As a good queen will lay, when in her best estate, three thousand eggs daily, and as the workers live from one to three months, it might seem that forty thousand was too small a figure for the number of workers. Without doubt a greater number is possible. That it is rare is not surprising, when we remember the numerous accidents and vicissitudes that must ever attend the individuals of these populous communities.

The function of the worker-bees is to do all the manual labor of the hives. They secrete the wax, which, as already

stated, forms in small scales (Fig. 72, *w*) under the over-lap-
ping rings under the abdomen. I have found these wax-
scales on both old and young. According to Fritz Muller, the
admirable German observer, so long a traveler in South
America, the bees of the genus Melipona secrete the wax on
the back.

The young bees commence work in a day from the cells.
They build the comb, ventilate the hive, feed the larvæ, queen
and drones, and cap the cells. The older bees—for, as readily
seen in Italianizing, the young bees do not usually go forth
for the first two weeks—gather the honey, collect the pollen,
or bee-bread as it is generally called, bring in the propolis or
bee-glue, which is used to close openings and as a cement,
supply the hive with water (?), defend the hive from all improper intrusion, destroy drones when their day of grace is
past, kill and arrange for replacing worthless queens, destroy
inchoate queens, drones, or even workers, if circumstances
demand it, and lead forth a portion of the bees when the conditions impel them to swarm.

When there are no young bees, the old bees will act as
housekeepers and nurses, which they otherwise refuse to do.
The young bees, on the other hand, will not go forth to glean,
at less than six days of age, even though there are no old bees
to do this necessary part of bee-duties. An indirect function
of all the bees is to supply animal heat, as the very life of the
bees requires that the temperature inside the hive be main-
tained at a rate considerably above freezing. In the chemical
processes attendant upon nutrition, much heat is generated,
which, as first shown by Newport, may be considerably augmented at the pleasure of the bees, by forced respiration. The
bees, by a rapid vibration of their wings, have the power to
ventilate their hives and reduce the temperature when the
weather is hot. Thus they are able to moderate the heat of
summer, and temper the cold of winter.

CHAPTER III.

SWARMING, OR THE NATURAL METHODS OF INCREASE.

The natural method by which an increase of colonies among bees is secured, is of great interest, and though it has been closely observed, and assiduously studied for a long period, and has given rise to theories as often absurd as sound, yet even now it is a fertile field for investigation, and will repay any who may come with the true spirit of inquiry, for there is much concerning it which is involved in mystery. Why do bees swarm at unseemly times? Why is the swarming spirit so excessive at times and so restrained at other seasons? These and other questions we are to apt to refer to erratic tendencies of the bees, when there is no question but that they follow naturally upon certain conditions, perhaps intricate and obscure, which it is the province of the investigator to discover. Who shall be first to unfold the principles which govern these, as all other actions of the bees?

In the spring or early summer, when the hive has become very populous, the queen, as if conscious that a home could be overcrowded, and foreseeing such danger, commences to deposit drone-eggs in drone-cells, which the worker-bees, perhaps moved by like consideration, begin to construct, if they are not already in existence. Drone-comb is almost sure of construction at such times. In truth, if possible the workers will always build drone-comb. No sooner is the drone-brood well under way, than the large, awkward queen-cells are commenced, often to the number of ten or fifteen, though there may be not more than three or four. The Cyprian and Syrian bees often start from fifty to one hundred queen-cells. In these, eggs are placed, and the rich royal jelly added, and soon, often before the cells are even capped, and *very rarely* before a cell is built—Mr. Doolittle says the first swarms of the season never leave until there are capped cells—if the bees are crowded, the hives unshaded, and the ventilation insufficient, some bright day, usually about eleven o'clock, after an unusual disquiet both inside and outside the hive, a large part of the worker-bees—being off duty for the day, and having previously loaded their honey-sacs—rush forth from the hive as if alarmed by the cry of fire. Crowded, unshaded and illy ventilated hives hasten swarming. Swarming rarely takes place except on bright, pleasant days, and is most common from eleven to two o'clock. The bees seem off duty for the day. They load their honey-stomachs, and amid a great commotion inside the hive and out, they push forth with the queen, though she is never leader, and is frequently late in her exit. Dr. Miller once had a swarm from a colony from which he had taken a queen an hour before. Of course, the swarm returned to the hive.

It is often asserted that bees do no gathering on the day they swarm, previous to leaving the hive. This is not true. Mr. Doolittle thinks they are just as active as on other days. The queen, however, is off duty for some time before the swarm leaves. She even lays scantily for two or three days prior to this event. This makes the queen lighter, and prepares her for her long, wearying flight. In her new home she does no laying for several hours. The assertion that bees always cluster on the outside preliminary to swarming, is not true. The crowded hive makes this common, though in a well-managed apiary it is very infrequent. The bees, once started on their quest for a new home, after many gyrations about the old one, dart forth to alight upon some bush (Fig. 76), limb, or fence, though in one case I knew the first swarm of bees to leave at once for parts unknown, without even waiting to cluster. After thus meditating for the space of from one to three hours, upon a future course, they again take wing and leave for their new home, which they have probably already sought out, and fixed up.

Some suppose the bees look up a home before leaving the hive, while others claim that scouts are in search of one while the bees are clustered. The fact that bees take a right-line to

FIG. 76.

Hiving a Swarm.—From Department of Agriculture.

their new home, and fly too rapidly to look as they go, would argue that a home is pre-empted, at least, before the cluster is dissolved. The fact that the cluster remains sometimes for hours—even over night—and at other times for a brief period, hardly more than fifteen minutes, would lead us to infer that the bees cluster while waiting for a new home to be found. Yet, why do bees sometimes alight after flying a long distance, as did a first swarm one season upon our College grounds? Was their journey long, so that they must needs stop to rest, or were they flying at random, not knowing whither they were going? This matter is no longer a matter of question. I now know of several cases where bees have been seen to clean out their new home the day previous to swarming. In each case the swarm came and took possession of the new home the day after the house-cleaning. The reason of clustering is no doubt to give the queen a rest before her long flight. Her muscles of flight are all "soft," as the horsemen would say. She must find this a severe ordeal, even after the rest.

If for any reason the queen should fail to join the bees, and rarely when she is among them, possibly because she finds she is unfit for the journey, they will, after having clustered, return to their old home. They may unite with another swarm, and enter another hive. Many writers speak of clustering as rare unless the queen is with the swarm. A large experience convinces me that the reverse is quite the case.

The youngest bees will remain in the old hive, to which those bees which are abroad in quest of stores will return. Most of these, however, may be in time to join the emigrants.

The presence of young bees on the ground immediately after a swarm has issued—those with flight too feeble to join the rovers—will often mark the previous home of the swarm. Mr. Doolittle confines a teacupful, or less, of the bees when he hives the swarm, and after the colony is hived he throws the confined bees up in the air, when he says they will at once go to the hive from which the swarm issued.

Soon, in seven days, often later if Italians—Mr. E. E. Hasty says in from six to seventeen days—the first queen will come forth from her cell, and in two or three days she will, or may, lead a new swarm forth; but before she does this, the peculiar note, known as the piping of the queen, may be heard.

This piping sounds like "peep," "peep," is shrill and clear, and can be plainly heard by placing the ear to the hive, nor would it be mistaken. This sound is Landois' true voice, as it is made even in the cell, and also by a queen whose wings are cut off. Cheshire thinks this sound is made by friction of the segments, one upon the other, as the queen moves them. The newly hatched queen pipes in seven or eight hours after coming from the cell. She always pipes if a swarm is to issue, and if she pipes a second swarm will go unless weather or man interferes. The second swarm usually goes in from thirty-five to forty-five hours after the piping is heard. This piping of the liberated queen is followed by a lower, hoarser note, made by a queen still within the cell. The queen outside makes a longer note followed by several shorter ones; the enclosed queens repeat tones of equal length. This piping is best heard by placing the ear to the hive in the evening or early morning. If heard, we may surely expect a swarm the next day but one following, unless the weather be too unpleasant.

Some have supposed that the cry of the liberated queen was that of hate, while that by the queen still imprisoned was either enmity or fear. Never will an after-swarm leave, unless preceded by this peculiar note. Queens occasionally pipe at other times, even in a cage. This is probably a note of alarm, as the attendant bees are always aroused by it.

At successive periods of one or two days, though the third swarm usually goes two days after the second, one, two, or even three more swarms may issue from the old home. Mr. Langstroth knew five after-swarms to issue, and others have reported eight and ten. The cells are usually guarded by the workers in all such cases against the destruction of the queen. These last swarms, all after the first, will each be heralded by the piping of the queen. They will be less particular as to the time of day when they issue, as they have been known to leave before sunrise, and even after sunset. The well-known apiarist, Mr. A. F. Moon, once knew a second swarm to issue by moonlight. They will, as a rule, cluster further from the hive. The after-swarms are accompanied by the queen, and in case swarming is delayed, may be attended by a plurality of queens. I have counted five queens in a second swarm. Berlepsch and

Langstroth each saw eight queens issue with a swarm, while others report even more. Mr. Doolittle says the guards leave the cells when the queen goes out, and then other queens, which have been fed for days in the cells, rush out and go with the swarm. He says he had known twenty to go with third swarms. I have seen several young queens liberated in a colony. How does Mr. Doolittle explain that ? Mr. Root thinks that a plurality of queens only attends the last after-swarm, when the bees decide to swarm no more. These virgin queens fly very rapidly, so the swarm will seem more active and definite in its course than will first swarms, and are quite likely to cluster high up if tall trees are near by. When the swarming is delayed it is likely that the queens are often fed by the workers while yet imprisoned in the cells. The view is generally held that these queens are kept in the cells that the queen which has already come from the cell may not kill them.

The cutting short of swarming preparations before the second, third, or even the first swarm issues, is by no means a rare occurrence. This is effected by the bees destroying the queen-cells, and sometimes by a general extermination of the drones, and is generally to be explained by a cessation of the honey-yield. Cells thus destroyed are easily recognized, as they are torn open from the side (Fig. 45, *E*) and not cut back from the end. It is commonly observed that while a moderate yield of honey is very provocative of swarming, a heavy flow seems frequently to absorb the entire attention of the bees, and so destroy the swarming impulse entirely.

Swarming-out at other times, especially in late winter and spring, is sometimes noticed by apiarists. This is doubtless due to famine, mice, ants, or some other disturbing circumstance which makes the hive intolerable to the bees. In such cases the swarm is quite likely to join with some other colony of the apiary.

CHAPTER IV.

PRODUCTS OF BEES; THEIR ORIGIN AND FUNCTION.

Among all insects, bees stand first in the variety of the useful products which they give us, and, next to the silk-moths, in the importance of these products. They seem the more remarkable and important in that so few insects yield articles of commercial value. True, the cochineal insect, a species of bark-louse, gives us an important coloring material; the lac insect, of the same family, gives us the important element of our best glue—shellac; another scale insect forms the Chinese wax of commerce; the blister-beetles afford an article prized by the physician, while we are indebted to one of the gall-flies for a valuable element of ink; but the honey-bee affords not only a delicious article of food, but also another article of no mean commercial rank, namely, wax. We will proceed to examine the various products which come from bees.

HONEY.

Of course, the first product of bees, not only to attract attention, but also in importance, is honey. And what is honey? It is digested nectar, a sweet, neutral substance gathered from the flowers. This nectar contains much water, though the amount is very variable, a mixture of several kinds of sugar and a small amount of nitrogenous matter in the form of pollen. Nectar is peculiar in the large amount of sucrose or cane-sugar which it contains. Often there is nearly or quite as much of this as of all the other sugars. We can not, therefore, give the composition of honey. It will be as various as the flowers from which it is gathered. Again, the thoroughness of the digestion will affect the composition of honey. This digestion is doubtless accomplished through the aid of the saliva—that from the racemose glands of the head and thorax (Fig. 59, *lhg, lg*, and Fig. 61, No. 2 and No. 3).

The composition of honey is of course very varied. Thus analyses give water all the way from 15 to 30 percent. The first would be fully ripe, the last hardly the product we should like to market.

The reducing sugars—so called because they can reduce the sulphate of copper when made strongly alkaline by the addition of caustic potash or soda—include all vegetable sugars but sucrose of cane-sugar; and consist mainly of dextrose, which turns the ray of polarization to the right, and levulose, which turns the ray to the left. Dextrose and levulose are both products of various fruits, as well as honey. Dextrose and levulose are also called invert sugars ; because, when cane-sugar is heated with a mineral acid, like hydrochloric acid, it changes from cane-sugar, which revolves the polarized ray to the right, to dextrose and levulose; but the latter is most effective, so now the ray turns to the left, hence the terms inversion, or invert sugar. Glucose is a term which refers to both dextrose and levulose, and is synonymous with grape-sugar.

The amount of reducing sugars varies largely, as shown by numerous analyses, usually from 65 to 75 percent ; though a few analyses of what it would seem must have been pure honeys, have shown less than 60 percent. But in such cases there was an excess of cane-sugar. It seems not improbable that in such cases honey was gathered very rapidly, and the bees not having far to fly did not fully digest the cane-sugar of the nectar. Dr. J. Campbell Brown, in a paper before the British Association, gave as an average of several analyses 73 percent of invert or reducing sugars ; 36 and 45–100 percent was levulose, and 36 and 57–100 percent was dextrose. Almost always pure honey gives a left rotation of from two to twelve degrees. This wide variation is suggestive. Does it not show that very likely the honey from certain flowers, though pure honey, may give a right-handed rotation with a large angle because of a large amount of dextrose and little levulose? It occurs to me that these two uncertain factors, incomplete digestion and the possible variation in nectar, make determination by the analyst either by use of the polariscope or chemical reagents a matter of doubt. I speak with more confidence, as our National Chemist pronounced several specimens of what I feel sure were pure honey, to be probably adulterated. I think that now he has perfected his methods so that such mistakes would rarely occur.

While nearly or quite half of the nectar of flowers is cane-sugar, there is very little of such sugar in honey. While from one to three percent is most common it not infrequently runs to five or six percent, and occasionally to twelve or sixteen percent. Quite likely in this last case, imperfect digestion was the cause. The nectar was not long enough in the stomach to be changed ; or else for some reason there was too little of the digestive ferment present. Of course, twelve to fifteen percent of sucrose would almost surely rotate the plane to the right. There is a very interesting field for study here. What flowers yield nectar so rich in cane-sugar that even the honey is rich in the same element ? Honey often contains, we are told, as much as four percent of dextrine. This, of course, tends to make it rotate the ray to the right, and further complicates the matter. Again, it is easy to see that in case flowers secrete nectar in large quantities the bees would load quickly, and so proportionately less saliva would be mixed with it, and digestion would be less thorough.

We see now why drones and queens need salivary glands to yield the ferment to digest honey. Often the worker-bees do not thoroughly digest it. We see, too, why honey is such an excellent food. We have to digest all our cane-sugar. The honey we eat has been largely digested for us.

Albuminoids—evidently from the pollen—vary from five to seventy-five hundredths of one percent. These vary largely according to the flowers. It is quite likely that in case of bloom like basswood where the honey comes very rapidly— fifteen pounds per day sometimes for each colony—the stomach-mouth can not remove all the pollen. Here is an opportunity for close observation. If we know we have honey that was gathered very rapidly, we should have a test made for albuminous material to see if its quantity increases with the rapidity with which the honey is gathered. While there may be quite an amount of this pollen in honey, usually there will be but little.

Besides the above substances, there is a little mineral matter—fifteen hundredths of one percent—which I suppose to be mainly malate of lime; a little of the essential oils which possibly give the characteristic flavor of the different kinds of honey, and more or less coloring matter, more in buckwheat honey, less in basswood. There is also a little acid—formic acid—which probably aids to digest the nectar, and possibly with the saliva, may, like the acid gastric juice of our own stomachs, resist putrefaction, or any kind of fermentation. It has been urged that this is added to the honey by the bees dropping poison from the sting. I much doubt this theory. It is more reasonable, however, than the absurd view that the bee uses its sting to polish its cells. If the poison-glands can secrete formic acid, why can not the glands of the stomach? Analogy, no less than common sense, favors this view. The acid of honey is often recognizable to the taste, as every lover of honey knows. The acid is also shown by use of blue litmus. The specific gravity varies greatly of course, as we should expect from the great variation in the amount of water. I have found very thick honey to have a specific gravity of 1.40 to 1 50. The fact that honey is digested nectar or sucrose, shows that in eating honey our food is partially digested for us, the cane-sugar is changed to a sugar that can be readily absorbed and assimilated.

I have fed bees pure cane-sugar, and, when stored, the late Prof. R. F. Kedzie found that nearly all of this sugar was transformed in much the same way that the nectar is changed which is taken from the flowers.

It is probable that the large compound racemose glands in the head and thorax of the bees (Fig. 59, *lhg*, *lg*, and Fig. 61) secrete an abundant ferment which hastens these transformations which the sugars undergo while in the honey-stomach of the bee. I once fed several pounds of cane-sugar syrup at night to the bees. I extracted some of this the next morning, and more after it was capped. Both samples were analyzed by three able chemists—Profs. Kedzie, Scovell, and Wiley—and the sample from the capped honey was found to be much better digested. This shows that the digestion continues in the comb. Much of the water escapes after the honey is stored.

The method of collecting honey has already been described.

The principles of lapping and suction are both involved in the operation.

When the stomach is full the bee repairs to the hive and regurgitates its precious load, either giving it to the bees or storing it in the cells. This honey remains for some time uncapped that it may ripen, in which process the water is partially evaporated, and the honey rendered thicker. If the honey remains uncapped, or is removed from the cells, it will generally granulate, if the temperature be reduced below 70 degrees. Like many other substances, most honey, if heated and sealed while hot, will not crystallize till it is unsealed. In case of granulation the sucrose and glucose crystallize in the mellose. Some honey, as that from the South, and some from California, seems to remain liquid indefinitely. Some kinds of our own honey crystallize much more readily than others. I have frequently observed that thick, ripe honey granulates more slowly than thin honey. The only sure (?) test of the purity of honey, if there be any, is that of the polariscope. This, even if decisive, is not practical except in the hands of the scientist. The most practical test is that of granulation, though this is not wholly reliable. Granulated honey is almost certainly pure. Occasionally genuine honey, and of superior excellence, refuses, even in a zero atmosphere, to crystallize.

When there are no flowers, or when the flowers yield no sweets, the bees, ever desirous to add to their stores, frequently essay to rob other colonies, and often visit the refuse of cider-mills, or suck up the oozing sweets of various plant or bark lice, thus adding, may be, unwholesome food to their usually delicious and refined stores. It is a curious fact that the queen never lays her maximum number of eggs except when storing is going on. In fact, in the interims of honey-gathering, egg-laying not infrequently ceases altogether. The queen seems discreet, gauging the size of her family by the probable means of support. Or it is quite possible that the workers control affairs by withholding the chyle, and thus the queen stops per-force. Syrian bees are much more likely to continue brood-rearing when no honey is being collected than are either German or Italian bees.

. Again, in times of extraordinary yields of honey the storing is very rapid, and the hive becomes so filled that the queen is unable to lay her full quota of eggs; in fact, I have seen the brood very much reduced in this way, which, of course, greatly depletes the colony. This might be called ruinous prosperity.

The natural use of the honey is to furnish, in part, the drones and imago worker-bees with food, and also to supply, in part at least, the queen, especially when she is not laying.

WAX.

The product of the bees second in importance is wax. The older scientists thought this was a product formed from pollen. Girard says it was discovered by a peasant of Lusace. Langstroth states that Herman C. Hornbostel discovered the true source of wax in 1745. Thorley in 1774, and Wildman in 1778, understood the true source of wax. This is a solid, unctuous substance, and is, as shown by its chemical composition, a fat-like material, though not, as some authors assert, the fat of bees. This is lighter than water, as its specific gravity is .965. The melting point is never less than 144 degrees F. Thus, it is easy to detect adulteration, as mineral wax, both paraffine and ceresin, have a less specific gravity. Paraffine also has a much lower melting point. It is impossible to adulterate wax with these mineral products for use as foundation. They so destroy the ductility and tenacity that the combs are almost sure to break down. Ceresin might be used, but it is distasteful to the bees, and foundation made from wax in which ceresin is mixed would have no value. Only pure beeswax is used in manufacturing foundation in the United States. I have this on the authority of Mr. A. I. Root, whose dictum in such matters is conclusive.

As already observed, wax is a secretion from the glands just within the wax-plates, and is formed in scales, the shape of an irregular pentagon (Fig. 72, *w*) underneath the abdomen. These scales are light-colored, very thin and fragile, and are secreted by the wax-gland as a liquid, which passes through the wax-plate by osmosis, and solidifies as thin wax-scales on the outside of the plates opposite the glands. Neighbour speaks of wax oozing through pores from the stomach. This is not the case, but, like the synovial fluid about our own joints, it is formed by the secreting membrane, and does not pass through holes, as water through a sieve. There are, as already stated, four of these wax-pockets on each side (Fig. 72), and thus there may be eight wax-scales on a bee at a time. This wax can be secreted by the bees when fed on pure sugar, as shown by Huber, whose experiment I have verified. I removed all honey and comb from a strong colony, left the bees for twenty-four hours to digest all food which might be in their stomachs, and then fed pure sugar, which was better than honey, as Prof. R. F. Kedzie has shown by analysis that not only filtered honey, but even the nectar which he collected right from the flowers themselves, contains nitrogen. The bees commenced at once to build comb, and continued for several days, so long as I kept them confined. This is as we should suppose; sugar contains hydrogen and oxygen in proportion to form water, while the third element, carbon, is in the same, or about the same, proportion as the oxygen. Now, the fats usually contain little oxygen and a good deal of carbon and hydrogen. Thus the sugar, by losing some of its oxygen, would contain the requisite elements for fat. It was found true in the days of slavery in the South that the negroes of Louisiana, during the gathering of the cane, would become very fat. They ate much sugar; they gained much fat. Now, wax is a fat-like substance, not that it is the animal fat of bees, as often asserted—in fact, it contains much less hydrogen, as will be seen by the following formula from Hess:

Oxygen 7.50
Carbon..79.30
Hydrogen...13.20

—but it is a special secretion for a special purpose, and from its composition we should conclude that it might be secreted from a purely saccharine diet, and experiment confirms the conclusion. Dr. Planta has found that there is a trace of nitrogen in wax-scales, a little less than .6 of one percent, while he finds in newly made comb, nearly .9 of one percent. It has been found that bees require about twenty pounds of honey to secrete one of wax. The experiments of Mr. P. L. Viallon show this estimate of Huber to be too great. Berlepsch says sixteen to nineteen pounds when fed on sugar without pollen, and ten pounds when fed both. My own experiments would sustain Huber's statement. In these experiments the bees are confined, and so the conclusions are to be received with caution. We can not know how much the results are changed by the abnormal condition in which the bees are placed.

For a time nitrogenous food is not necessary to the secretion of wax. Probably the small amount of nitrogen in the scales and in the saliva may be furnished by the blood. This, of course, could not continue long; indeed, the general nutrition would be interfered with, and ill health can never do maximum work.

It is asserted that to secrete wax, bees need to hang in compact clusters or festoons in absolute repose. Such quiet would certainly seem conducive to most active secretion. The food could not go to form wax, and at the same time supply the waste of tissue which ever follows upon muscular activity. The cow, put to hard toil, could not give so much milk. But I find, upon examination, that the bees, even the oldest ones, while gathering in the honey season, yield up the wax-scales the same as those within the hive. During the active storing of the past season, especially when comb-building was in rapid progress, I found that nearly every bee taken from the flowers contained the wax-scales of varying sizes in the wax-pockets. By the activity of the bees, these are not infrequently loosened from their position and fall to the bottom of the hive, sometimes in astonishing quantities. This explains why wax is often mentioned as an element of honey. Its presence, however, in honey is wholly accidental. It is probable that wax-secretion is not forced upon the bees, but only takes place as required. So the bees, unless wax is demanded, may perform other duties. When we fill the sections and brood-chamber wholly with foundation, it is often difficult to find any bees bearing wax-scales. In such cases I have often looked long, but in vain, to find such scales *in situ* to show to my students. A newly-hived colony, with no combs or foundation, will show these wax-scales on nearly every bee. Whether this secretion is a matter of the bee's will, *or whether* it is excited by the surrounding conditions without any thought, are questions yet to be settled. No comb necessitates quiet. With us and all other higher animals, quiet and heavy food-taking favors fat deposits. May not the same in bees conduce to wax-production?

These wax-scales are loosened by the wax-jaws of the posterior legs, carried to their anterior claws, which in turn bear them to the mouth, where they are mixed with saliva probably from Wolff's glands (Fig. 60), or mixed saliva.

After the proper kneading by the jaws, these wax-scales are fashioned into that wonderful and exquisite structure, the comb. In this transformation to comb, the wax may become colored. This is due to a slight admixture of pollen or old wax. It is almost sure to be colored if the new comb is formed adjacent to old, dark-colored comb. In such cases chippings from the old soiled comb are used.

Honey-comb is wonderfully delicate, the base of a new cell being, according to Prof. C. P. Gillette, in worker-comb, between .0032 and .0064 of an inch, and the drone between .0048 and .008. The walls are even thinner, varying, he says, from .0018 to .0028 of an inch. The cells are so formed as to combine the greatest strength and maximum capacity with the least expense of material. It need hardly be said that queen-cells are much thicker, and contain, as before stated, much that is not wax. In the arch-like pits in queen-cells, we farther see how strength is conserved and material economized.

Honey-comb has been an object of admiration since the earliest time. Some claim that the form is a matter of necessity—the result of pressure or reciprocal resistance and not of bee-skill. The fact that the hexagonal form is sometimes assumed just as the cell is started, when pressure or resistance could not aid, has led me to doubt this view; especially as wasps form their paper nests of soft pulp, and the hexagonal cells extend to the edge, where no pressure or resistance could affect the form of the cells. Yet I am not certain that the mutual resistance of the cells, as they are fashioned from the soft wax, may not determine the form. Mullenhoff seems to have proved that mutual resistance of the cells causes the hexagonal form. The bees certainly carve out the triangular pyramid at the base. They would need to be no better geome-

tricians to form the hexagonal cells. The assertion that the
cells of honey-comb are absolutely uniform and perfect is
untrue, as a little inspection will convince any one. The late
Prof. J. Wyman demonstrated that an exact hexagonal cell
does not exist. He also showed that the size varies, so that in
a distance of ten worker-cells there may be a variation of one
cell in diameter, and this in natural, not distorted, cells. Any
one who doubts can easily prove, by a little careful examination, that Prof. Wyman was correct. This variation of one-
fifth of an inch in ten cells is extreme, but variation of one-

FIG 77.

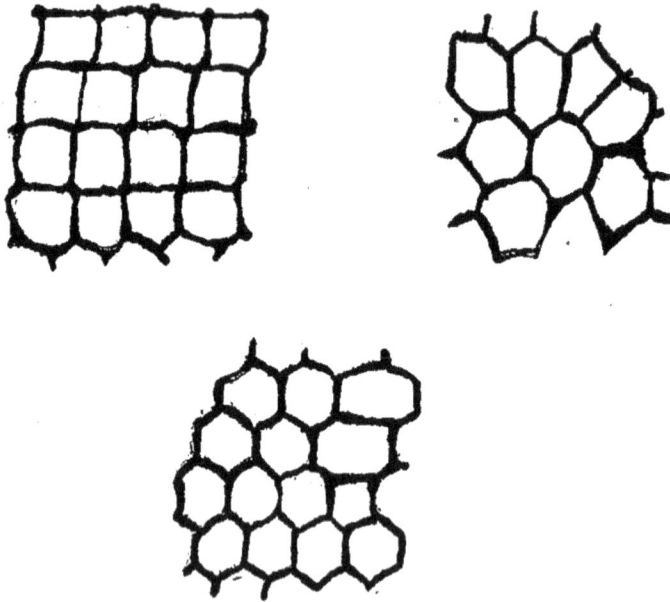

Irregular Cells, (modified) from Cowan.

tenth of an inch is common. The sides, as also the angles,
are not constant. The rhombic faces forming the bases of the
cells also vary. The idea which has come down from the past
that mathematics and measurement exactly agreed upon the
angles of the rhombs, that the two opposite obtuse angles were
each 109° 28' 16'' and the acute 70° 31' and 44'' is without foundation in fact. Mr. Cowan figures (Fig. 77) triangular, quad-
rangular, and even cells with seven sides. Of course, such
deformity is very rare.

The bees change from worker (Fig. 78, *c*) to drone cells (Fig. 78, *a*), which are one-fifth larger, and *vice versa*, not by any system (Fig. 78, *b*), but simply by enlarging or contracting. It usually takes about four rows to complete the transfor-

FIG. 78.

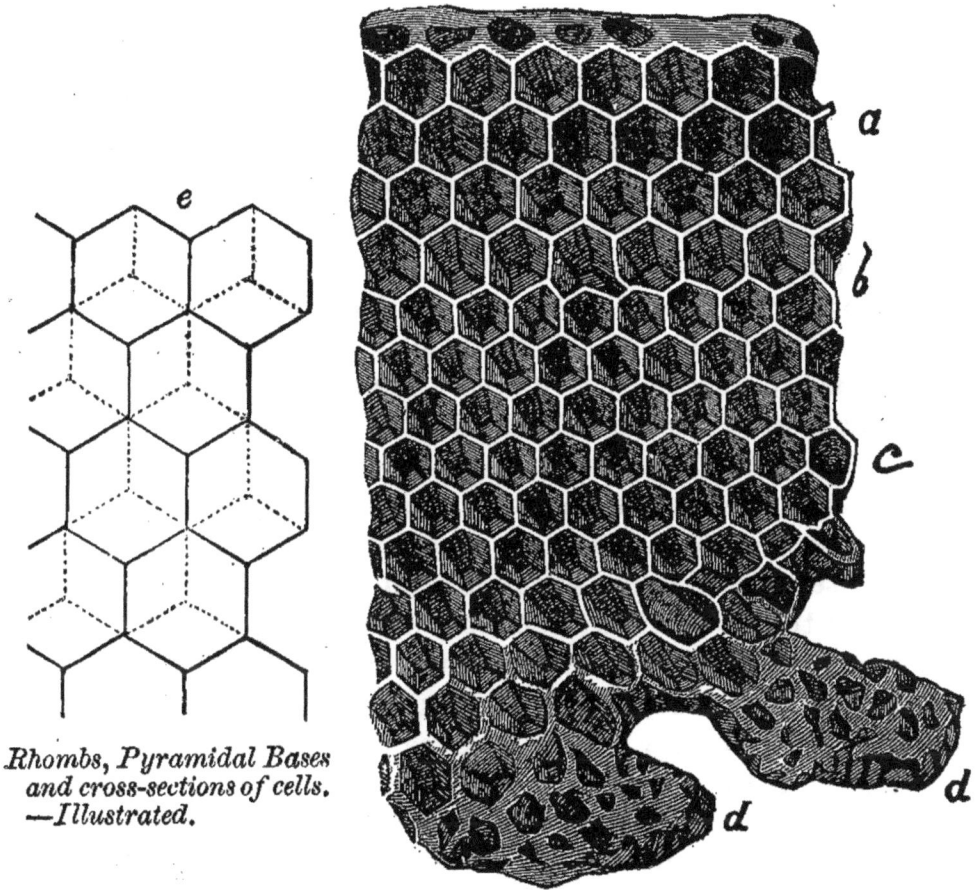

Rhombs, Pyramidal Bases
and cross-sections of cells.
—Illustrated.

Honey-Comb.—   after Duncan.

a Drone-cells.   c Worker-cells.
b Deformed cells.   dd Queen-cells.

mation, though the number of deformed cells varies from two, very rarely one, to eight. The perfect drone-cells may be, often are, contiguous to perfect worker-cells, the irregular cells being used to fill out the necessary irregularities. An English writer criticises Langstroth's representation of these irregular cells, and adds that the angles can never be less than 100 degrees. This is far from the truth, as I have found many cells where an angle was considerably less than this. Mr. Cowan, in his excellent "Honey-Bee," describes and figures cells where the angle is even acute.

The structure of each cell is quite complex, yet full of interest. The base is a triangular pyramid (Fig. 78, e), whose three faces are rhombs (Mr. Cowan has found and photographed cells with four faces), and whose apex forms the very center of the floor of the cell. From the six free or non-adjacent edges of the three rhombs extend the lateral walls or faces of the cell. The apex of this basal pyramid is a point where the contiguous faces of the three cells on the opposite side meet, and form the angles of the bases of three cells on the opposite side of the comb. Thus the base of each cell forms one-third of the base of three opposite cells. One side thus braces the other, and adds much to the strength of the comb. Each cell, then, is in the form of a hexagonal prism, terminating in a flattened triangular pyramid.

The bees usually build several combs at once, and carry forward several cells on each side of each comb, constantly adding to the number, by additions to the edge. The bees, in constructing comb, make the base or so-called mid-rib, the "fish-bone" in honey where foundation is used, thick at first, and thin this as they add to the cells in lengthening them. Prof. C. P. Gillette demonstrated this by coloring foundation black. The color reached nearly to the end of the cell, and extended an inch below the foundation. Thus we understand why bees take so kindly to foundation. To work this out is not contrary to their instincts, and gives them a lift. Huber first observed the process of comb-building, noticing the bees abstract the wax-scales, carry them to the mouth, add the frothy saliva, and then knead and draw out the yellow ribbons which were fastened to the top of the hive, or added to the comb already commenced.

The diameter of the worker-cells (Fig. 78, c) averages little more than one-fifth of an inch—Reaumur says two and three-fifths lines, or twelfths of an inch—while the drone-cells (Fig.

78, *a*) are a little more than one-fourth of an inch, or, according to Reaumur, three and one-third lines. But this distinguished author was quite wrong when he said: "These are the invariable dimensions of all cells that ever were or ever will be made." A recent English author, after stating the diameter of cells, adds: "The statement many times made that twenty-five and sixteen of these, respectively, cover a square inch, is erroneous, as they are not square." He says there are 28 13-15 and 18 178-375. I find the worker-cells per square inch vary from 25 to 29, and the drone-cells from 16 to 19 per square inch. The drone-cells, I think, vary more in size than do the worker-cells. The depth of the worker-cells is a little less than half an inch; the drone-cells are slightly extended, so as to be a little more than half an inch deep. Thus worker-comb is seven-eighths and drone-comb one and one-fourth inches thick. This depth, even of brood-cells, varies, so we can not give exact figures. The cells are often drawn out so as to be an inch long, when used solely as honey receptacles. Such cells are often very irregular at the end, and sometimes two are joined. The number of cells in a pound of comb will vary much, of course, as the thickness of the comb is not uniform. This number will vary from thirty to fifty thousand. In capping the honey the bees commence at the outside of each cell and finish at the center. The capping of the brood-cells is white and convex. The capping of honey-cells is made thicker by black bees than by the other races, and so their comb honey is more beautiful. Another reason for the whiter color comes from a small air-chamber just beneath the capping. The inner surface of the capping is, therefore, usually free from honey. This chamber is usually a little larger in the honey-comb of black bees. The cappings are strengthened by tiny braces of wax, which, as we should expect, are most pronounced in drone-comb.

The strength of comb is something marvelous. I have known a frame of comb honey eleven inches square to weigh eleven pounds, and yet to be unsupported at the bottom, and for not more than one-third of the distance from the top on the sides, and yet it held securely. The danger in cold weather, from breaking, is greater, as then the comb is very brittle.

Prof. Gillette has found that comb one inch thick will weigh only from one-twentieth to one twenty-fifth the weight of the honey which it may hold.

The character of the cells, as to size, that is, whether they are drone or worker, seems to be determined by the relative abundance of bees and honey. If the bees are abundant and honey needed, or if there is no queen to lay eggs, drone-comb (Fig. 78, a) is invariably built, while if there are few bees, and of course little honey needed, then worker-comb (Fig. 78, c) is

FIG. 79.

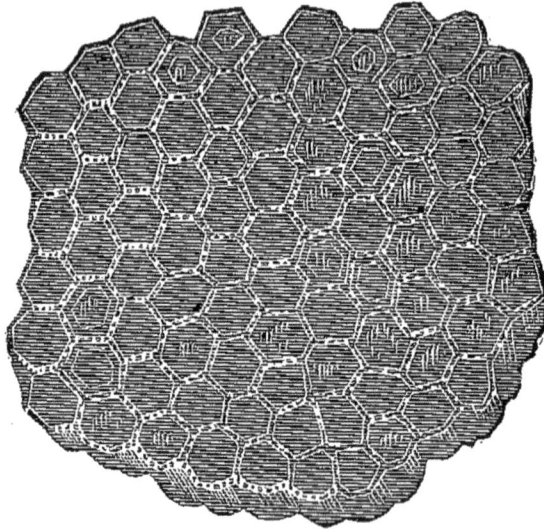

Honey-Comb Coral.—Original.

almost invariably formed. It is also a curious fact that if the queen keeps along with the comb-builders in the brood-chamber, then no drone-comb is built; but let her fail to keep cells occupied, and drone-comb is at once formed. It would seem that the workers reasoned thus : We are going to have comb for storing, for such we better fashion the large celled or drone-comb.

All comb, when first formed, is clear and translucent. The fact that it is often dark and opaque implies that it has been long used as brood-comb, and the opacity is due to the innumerable thin glue-like cocoons which line the cells. This may be separated by dissolving the wax; which may be done by putting it in boiling alcohol, or, better still, by use of the solar wax-extractor. Such comb need not be discarded, for if composed of worker-cells it is still very valuable for breeding purposes, and should not be destroyed till the cells are too small for long service, which will not occur till after many years of use. The function, then, of the wax, is to make comb and

FIG. 80.

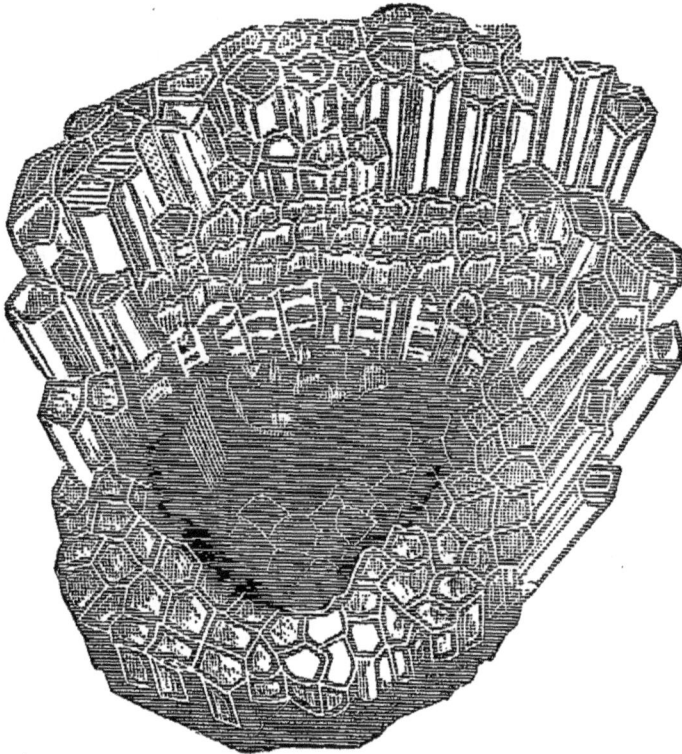

Honey-Comb Coral.—Original.

caps for the honey-cells, and, combined with pollen, to form queen-cells (Fig. 78, *d*) and caps for the brood-cells.

A very common fossil found in many parts of the Eastern and Northern United States is, from its appearance, often called petrified honey-comb. We have many such specimens in our museum. In some cases the cells are hardly larger than a pin-head; in others a quarter of an inch in diameter. These (Figs. 79, 80) are not fossil honey-comb as many are led to believe, though the resemblance is so striking that no wonder the public generally are deceived. These specimens are fossil coral, which the paleontologist places in the genus Favosites; favosus being a common species in the Northern United States. They are very abundant in the lime rock in northern Michigan, and are very properly denominated honey-comb coral. The animals of which these were once the skeletons, so to speak, are not insects at all, though often called so by men of considerable information.

The species of the genus Favosites first appeared in the Upper Silurian rocks, culminated in the Devonian, and disappeared in the early Carboniferous. No insects appeared till the Devonian age, and no Hymenoptera—bees, wasps, etc.—till after the Carboniferous. So the old-time Favositid reared its limestone columns and helped to build islands and continents untold ages—millions upon millions of years—before any flower bloomed, or any bee sipped the precious nectar. In some specimens of this honey-comb coral (Fig. 80) there are to be seen banks of cells, much resembling the paper-nests of some of our wasps. This might be called wasp-comb coral, except that both styles were wrought by the self-same animals.

POLLEN OR BEE-BREAD.

An ancient Greek author states that in Hymettus the bees tied little pebbles to their legs to hold them down. This fanciful conjecture probably arose from seeing the pollen-balls on the bees' legs.

Even such scientists as Reaumur, Bonnet, Swammerdam, and many apiarists of the last century, thought they saw in these pollen-balls the source of wax. But Huber, John Hunter, Duchet, Wildman, and others already referred to, noticed the presence and function of the wax-scales already described, and were aware that the pollen served a different purpose.

This substance, like nectar, is not secreted nor manufactured by the bees, only collected. The pollen-grains form the male element in plants. They are in plants what the spermatozoa or sperm-cells are in animals; and as the sperm-cells are much more numerous than the eggs or germ-cells, so pollen-grains are far more numerous in plants than are the ovules or seeds. In Chinese wistaria, *Wistaria sinensis*, there are, says Goodale, about 7,000 pollen-grains to each ovule. The color of pollen is usually yellow; but we often find it orange, reddish, nearly white, and in several Gilias in California it is bright blue. Pollen-grains are really single cells, and have two coats; the outer is the extine, which may be smooth, variously sculptured, or even thickly set with spines (Fig. 81). These spines, as also the color, often enable us to tell the species of plant from which the pollen came. Usually the extine is per-

FIG. 81.

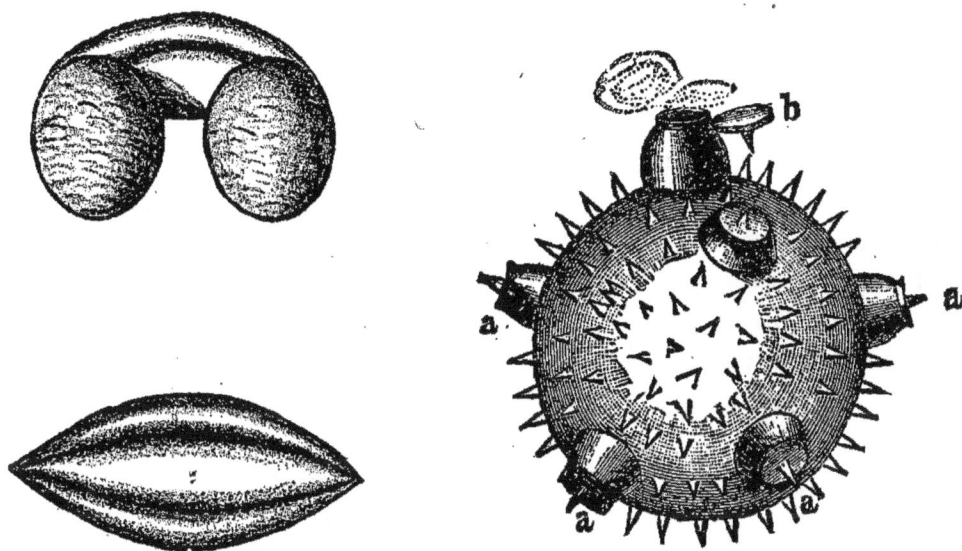

Pollen-Grains, from A. I. Root Co.

forated, though the inner wall—intine—is not. These perforations are also definite in number within the species. These holes give opportunity for the pollen-tubes (Fig. 252, *T*) to push out after the pollen-grain reaches the stigma of the flower. Where there are no perforations of the extine, the wall breaks. In some cases like orchids, pollen-grains are held together by an adhesive substance. In our milkweeds we notice a similar grouping of pollen-grains (Fig. 227) which often are very disturbing to bees and other insects.

The composition of pollen, says Goodale, is protoplasmic

matter, granular food materials, such as starch and oil and dissolved food matters, sugar and dextrine.

Dr. A. de Planta gives the following analysis of pollen of the hazel (B. B. Journal, Vol. XIV, p. 269). He finds protoplasm, oils and starch—the important food elements.

Before drying he found :

Water	9.19
Nitrogen	4.81
Ash	3.81

After drying thoroughly he found :

Nitrogenous Matter	31.63
Non-nitrogenous	64.36
Ash	4.01

He found no reducing sugar, but did find 14.70 percent of cane-sugar.

As will be seen, pollen, like our grains, is rich in the albuminoids. Like our grains, or even different specimens of the same grain, the composition of pollen will doubtless vary to quite an extent. As we note that pollen contains besides an ash, albuminoids, sugar, starch, and oils, we understand its excellence as a food ; it contains within itself all the important food elements. The bees usually obtain it from the stamens of flowers ; but if they gain access to flour when there is no bloom, they will take this in lieu of pollen, in which case the former term used above becomes a misnomer, though usually the bee-bread consists wholly of pollen. I have also known bees to gather extensively for bee-bread from the common raspberry rust. Very likely the spores of others of these fungi or low vegetables help to supply this nutritious substance. Occasionally there is a drouth of bee-bread alike in hive and flowers, then bees will seek this kind of food in meal or flour box or bin. Hence, the wisdom of feeding rye-flour which the bees will readily take if it is needed. Flour may be added to candy and fed to bees.

As already intimated, the pollen is conveyed in the pollen-baskets (Fig. 70) of the posterior legs, to which it is conveyed by the other legs, as already described, page 154, and compressed into little oval masses. The motions in this conveyance are exceedingly rapid, and are largely performed while the bee is on the wing. The bees not infrequently come to the hives not only with replete pollen-baskets, but with their whole under-surface thoroughly dusted. Dissection will also show that the same bee may have her sucking stomach distended with honey, though this is rare. Thus the bees make the most of their opportunities. It is a curious fact, noticed even by Aristotle, that the bees, during any trip, almost always gather only a single kind of pollen, or gather only from one species of bloom. Hence, while different bees may have different colors of pollen, the pellets of bee-bread on any single bee will be uniform in color throughout. It is possible that the material is more easily collected and compacted when homogeneous. It seems more probable that they prefer the pollen of certain plants, and work on such species so long as they yield the desired food, though it may be a matter of simple convenience. From this fact we see why bees cause no intercrossing of species of plants; they only intermix the pollen of different plants of the same species.

The pollen is usually deposited in the small or worker cells, and is unloaded by a scraping motion of the posterior legs, the pollen-baskets being first lowered into the cells. The bee thus freed, leaves the wheat-like masses to be packed by other bees, which is packed by pushing with the head. The cells, which may or may not have the same color of pollen throughout, are never filled quite to the top, and not infrequently the same cell may contain both pollen and honey. Such a condition is easily ascertained by holding the comb between the eye and the sun. If there is no pollen it will be wholly translucent ; otherwise there will be opaque patches. A little experiénce will make this determination easy, even if the comb is old. Combs in small sections, especially if separators are used, are not likely to receive pollen or be used for breeding. It is often stated that queenless colonies gather no pollen, but it is not true, though they gather less than they otherwise would. It is probable that pollen, at least when honey is added, contains all the essential elements of animal food. It certainly contains the very important principle which is not found in pure nectar or honey—nitrogenous material. I do not think the bee-moth larva will destroy combs that are entirely destitute of pollen, surely not unless they have been long used as brood-combs. The intruder must have proteid food.

The function of bee-bread is to furnish albuminous food to all the bees, adults no less than larvæ. As already stated, brood-rearing is impossible without it. And though it is certainly not essential to the nourishment of the adult bees when in repose, it still may be so, and unquestionably is, in time of active labor. This point is clearly proved from the fact that pollen-husks are almost always found in the intestines of bees. We may say it feeds the tissues of the imago bees, and is necessary that the workers may form the food for the queen, drones and larvæ. Schonfeld thinks the bees must have it in winter, and in case no bee-bread is in the combs, he thinks the bees scrape it from the cells and old combs. I believe bees often winter better when there is no pollen in the hive.

PROPOLIS OR BEE-GLUE.

This substance, also called bee-glue, is collected as the bees collect pollen, and is not made or secreted. It is the product of various resinous buds, and may be seen to glisten on the opening buds of the hickory and horse-chestnut, where it frequently serves the entomologist by capturing small insects. From such sources, from the oozing gum of various trees, from varnished furniture, and from old propolis about unused hives that have previously seen service, do the bees secure their glue. Probably the gathering of bees about coffins to collect the glue from the varnish, led to the custom of rapping on the hives to inform the bees, in case of a death in the family, that they might join as mourners. This custom still prevails, as I understand, in some parts of the South. Propolis has great adhesive force, and though soft and pliable when warm becomes very hard and unyielding when cold.

The use of bee-glue is to cement the combs to their supports, to fill up all rough places inside the hive, to seal up all crevices except the place of exit, which the bees often contract by aid of propolis, and even to cover any foreign substance that can not be removed. Intruding snails have thus been imprisoned inside the hive. Reaumur found a snail thus incased; Maraldi a slug similarly entombed; while I have myself observed a Bombus, which had been stripped by the bees of wings, hair, etc., in their vain attempts at removal, also encased in this unique style of a sarcophagus, fashioned by the bees. Alcohol, benzine, gasoline, ether, and chloroform are all ready solvents of bee-glue, and will quickly remove it from the hands, clothing, etc. Boiling in water with concentrated lye will remove propolis completely. Even steam and hot water used as a spray have been found to do the same.

PARTIAL BIBLIOGRAPHY.

For very full lists of books, etc., see Packard's Text-Book of Entomology.

Alley, Henry—Thirty Years Among the Bees, 1880, and Queen-Rearing, 1883.

Adair, D. L.—Annals of Bee-Keeping, 1872.

Amans, Dr.—Essai sur le vol des Insectes, 1883.

Ballantine, Rev Wm.—Bee-Culture, 1884.

"Bee-Master,"—The Times Bee-Keeping, 1864.

Benton, Frank—The Honey-Bee, 1899.

Berger, E.—Untersuchungen uber den Bau des Gehirnes und der Retina der Arthropoden, 1873.

Berlepsch, A. Baron von—Die Biene und ihre Zucht, 1873.

Bevan, Dr. E.—The Honey-Bee, 1838.

Blanchard, E.—Recherches anatomique sur le systeme nerveux les Insectes, 1846.

—— De la circulation dans les Insectes, 1848.

—— Du grand sympathique chez les Animaux articules, 1858.

Bonnet, C.—Œuvres d'histoire naturelle, 1779–1783.

Bonnier, G.—Les Nectaires, 1879.

Bordas, L.—Glandes salivaries des Apides, Apis mellifica, (Comptes rendus Acad. Sci. Paris,) 1894. Appareil glandulaire des Hymenopteres (Ann. Soc Nat. Zool. Paris,) 1894.

Brandt, E.—Comparative Anatomy of the Nerve System of Insects (in Russian,) 1878.

Briant, T. J.—Notes on the Antennæ of the Honey-Bee (Jour. Linn. Soc.,) 1883.

—— On the Anatomy and Functions of the Tongue of the Honey-Bee (Jour. Linn. Soc.,) 1884.

—— Antennæ of Honey-Bee (Jour. Linn. Soc.,) 1885.

British Bee Journal—1873 to 1889. Present Editor, T. W. Cowan, F.L.S., etc.

Brougham, Lord H.—Observations, Demonstrations, and Experiments upon the Structure of the Cells of Bees (Natural Theology,) 1856.

Buchner, L.—Mind in Animals, 1880.

Burmeister, H.—Handbuch der Entomologie, 1832.

Butschli, O.—Zur Entwicklungsgeschichte der Biene, 1870.

Cameron, P.—On Parthenogenesis in the Hymenoptera (Trans. Nat. Hist. Soc. of Glasgow,) 1888.

Chambers, V. T.—On the Tongue of some Hymenoptera (Jor. Nat. Hist. Soc. Cincin.,) 1874.

Cheshire—Bees and Bee-Keeping, two volumes, 1886.

Claparede, E.—Morphologie des zusammengesetzten Anges bei den Arthropoden (Zeit. fur Wiss. Zool.,) 1860.

Clute, Dr. O.—Blessed Bees, 1878.

Collin, Abbe—Guide du proprietaire d'Abeilles, 1878.

Comstock, H. J.—Manual for the Study of Insects, 1895. Recent and authoritative.

Cowan, T. W.—The Honey-Bee, 1890. Very accurate and full.

——— Bee-Keeper's Guide Book, 1881.

Dadant, Chas. and Son—Langstroth on the Honey-Bee, 1899.

Dahl, F.—Archiv. f. Naturg., 1884, pp. 146-193.

Darwin, C.—Origin of Species, 1859, 1872, 1878.

Debeauvoys, M.—L'Apiculteur, 1853.

Dewitz, H.—Vergleichende Untersuchungen uber Bau und Entwickelung des Stachels der Honigbiene, 1874.

Doolittle, G. M.—Scientific Queen-Rearing, 1889.

Donhoff, Dr.—Bienenzeitung, 1851-1854.

Dufour, Leon—Memo. pres. par divers savants a l'Acad. des Sci. de l'Inst. de France. Tome VII.

Dujardin, F.—Memoire sur le systeme nerveux des Insectes, 1851.

——— Observations sur les Abeilles, 1852.

Dumas et Milne Edwards—Sur la production de la cire des Abeilles, 1843-1844.

Duthiers, L.—Recherches sur l'armure genitale des Insectes (Ann des Scien. Nat.,) 1848-1852.

Dzierzon, Dr.—Bienenzeitung, 1845-1854.

———Theorie und Praxis des neuen Bienenfreundes, 1849-1852.

———Rational Bee-Keeping. English translation by Dieck and Stutterd, 1882.

Erichson—De fabrica et usu antennarum in Insectis, 1847.

Exner, S.—Ueber das Sehen von Bewegungen und die Theorie des zusammengesetzten Auges, 1875.

——— Die Frage der Functionsweise der Facettenaugen (Biol. Centralblatt,) 1880, 1882.

Figuier, L.—The Insect World, translated by P. Martin Duncan, 1872.

Fischer, G.—Bienenzeitung, 1871.

Geddes, Prof. Patrick and J. A. Thomson—The Evolution of Sex, 1889.

Girard, M.—Sur la chaleur libre degagee par les animaux invertebres et specialement les Insectes, 1869.

—————— Traite elementaire d'Entomologie, 1873.

—————— Les Abeilles, organes et fonctions, 1878.

Girdwoyn, M.—Anatomie et physiologie de l'Abeille, 1876.

Gottsche, C. M.—Beitrag zur Anat. und Physiol. des Auges der Fliegen, etc. (Mull. Arch. fur Anat.,) 1852.

Graber, Dr. V.—Ueber die Blutkorperchen der Insekten, 1871.

—————— Ueber den propulsatorischen Apparat der Insekten, 1872.

—————— Verlaufiger Bericht uber den propulsatorischen Apparat der Insekten, 1872.

—————— Ueber neue otocystenartige Sinnesorgane der Insekten, 1878.

—————— Die Chordotonalen Sinnesorgane und das Gehor der Insekten (Arch. fur. Mic. Anat.,) 1882.

Grassi, Dr. B.—Intorno allo sviluppo delle Api nell' uovo, 1883, 1884, 1886.

Grenacher, H.—Untersuchungen uber das Sehorgan der Arthropoden, 1879.

—————— Abhandlungen zur vergleichenden Anatomie des Auges, 1886.

Grimshaw, R. A. H.—Heredity in Bees (British Bee Journal,) 1889.

Gundelach, F. W.—Die Naturgeschichte der Honigbiene, 1842.

Hauser, G.—Physiologische und histologische Untersuchungen uber das Geruchsorgan der Insekten, 1880.

Haviland, J. D.—The Social Instincts of Bees, their Origin and Natural Selection, 1882.

Heddon, James—Success in Bee-Culture, 1886.

Helmholz—Sensations of Tone.

Hicks, Dr. J. Braxton—On a new structure in the Antennæ of Insects (Jour. Linn. Soc.,) 1857.

—————— On certain Sensory Organs in Insects, hitherto undescribed, 1860.

—————— The Honey-Bee (Samuelson and Hicks,) 1860.

Hickson, Dr. S. J.—The Eye and Optic Tract of Insects (Quart. Jour. Mic. Science,) 1885.

Hopkins, Isaac—Australasian Bee-Manual, 1886.

Huber, F.—Nouvelles observations sur les Abeilles, 1814, (and other editions.)

Hunter, J.—On Bees (Philosophical Trans.,) 1792.

—————— Manual of Bee-Keeping, 18—

Hutchinson, W. Z.—Advanced Bee-Culture, 1883.

—————— Comb Honey, 1897.

Hyatt, J. D.—The Structure of the Tongue of the Honey-Bee (Amer. Quart. Mic Jour.,) 1878, p. 287.

—————— The Sting of the Honey-Bee (ibid,) 1878, p. 3.

—————— The Sting of the Honey-Bee (Pop. Sc. Mon.,) 1879.

Janscha, I. A.—Hinterlassene vollstandige Lehre von der Bienenzucht, 1775.

John, Dr. Martin—Ein neu Bienen-Buchel, 1691.

Jurine, Mademoiselle—Huber's Nouvelles observations sur les Abeilles, 1792–1814.

King, H.—Bee-Keepers' Text-Book, 1883.

Kirby, W.—Monographia Apum Angliae, 1802.

Kirby and Spence—Introduction to Entomology.

Klein, Dr. E.—Handbook for the Physiological Laboratory, 1873.

—— Elements of Histology, 1884.

Kowalevsky—Embryologische Studien an Wurmern und Arthropoden, 1871.

Kraepelin, Dr. K.—Phys. und Hist. uber die Geruchsorgane der Insekten (Zeit. f. Wiss. Zool.,) 1880.

—— Ueber die Mundwerkzeuge der saugenden Insekten. (*ibid*,) 1882.

—— Ueber die Geruchsorgane der Gliederthiere, 1883.

Krancher, Dr. O.—Der Bau der Stigmen bei der Insekten, 1881.

—— Die dreierlei Bienenwesen, 1884.

Lacordaire—Introduction a l'Entomologie, 1861.

Landois, Dr. H.—Beitrage zur Entwicklungsgeschichte des Schmetterlingsflugels in der Raupe und Puppe, 1871.

—— Die ton und Stimmapparate der Insekten, 1867.

Langstroth, L. L.—The Honey-Bee, 1859–1873.

Latreille, P. A.—Eclaircissemens relatifs a l'opinion de M. Huber fils, sur l'origine et l'issue exterieure de la Cire (Acad. Roy. des Sciences,) 1821. Cours d'entomologie, 1831.

Leeuwenhoek, A.—Select works, translated by H. Hoole.

Lefebvre, A.—Note sur le sentiment olfactif des Insectes (Ann. Soc. entom. de France,) 1838.

Leuckart, Dr.—Zur Kentniss des Generationswechsels und der Parthenogenesis bei der Insekten, 1858.

Leuckart, R.—Ueber Metamorphose, ungeschlechtliche Vermehrung, Generationswechsel, 1851.

Leydig, F.—Das Auge der Gliederthiere, 1864.

—— Zur Anatomie der Insekten (Mull. Archiv. f. Anat.,) 1859.

Lhuilier, S. A. J.—Memoire sur le minimum de cire des alveoles des Abeilles, et en particulier sur un minimum minimorum relatif a cette matiere, 1781.

Lowe, J.—Trans. Ent. Soc. Vol. V. pp. 547–560, 1867.

Lowne, B. T.—On the Simple and Compound Eyes of Insects (Phil. Trans.,) 1879.

—— On the Compound Vision and the Morphol. of the Eye in Insects (Trans. Linn. Soc. Lond.,) 1884.

Lubbock, Sir J.—Ants, Bees and Wasps, 1882.

—— The Senses, Instincts and Intelligence of Animals, 1889.

Lucas, I. G.—Entwurf eines wissenschaftlichen Systems fur Bienenzucht, 1808.

Lucas, M. H.—Cas de cyclopie observe chez un insecte Hymenoptere (Apis mellifica,) 1868.

Lyonet, Pieter—Traite anatomique de la chenille qui ronge Le bois de saule, etc., 1762.

Macloskie, G.—The Endocranium and Maxillary Suspensorium of the Bee (Amer. Natural, pp. 567–573,) 1884.

Maraldi, G. F.—Observations sur les Abeilles (Mem. Acad. des Sciences,) 1712.

Marey, E. J.—Animal Mechanism: A Treatise on Terrestrial and Aerial Locomotion, 1883.

Mayer, Dr. Paolo—Sopra certi Organi di Senso nelle Antenne dei Ditteri, 1878–79.

Meckel, H.—Muller's Archiv. fur Anatomie, 1846.

Miller, Dr. C. C.—A Year Among the Bees, 1888.

Milne-Edwards—Manual of Zoology, 1863.

Moufet, T.—Insectorum sine minimorum animalium Theatrum, 1634.

Mullenhoff, Dr. K.—Formation of Honey-Comb (Pfluger's Archiv f. gesammt. Physiol., XXXII, pp. 589–618,) 1883.

—— Structure of the Honey-Bee's Cell (Arch. f. Anat. und Physiol., pp. 371–375,) 1886.

Muller, J.—Zur vergleichenden Physiologie des Gesichtsinnes, 1826.

—— Fortgesetzte anatomische Untersuchungen uber den Bau der Augen bei den Insekten und Crustaceen, 1829.

Munn, N. A.—Bevan on the Honey-Bee, 1870.

Neighbour, Alfred—The Apiary, 1878.

Newman, Thomas G.—Bees and Honey, 1892.

Newport, G.—On the Respiration of Insects, 1836.

—— Insects (Todd's Cyclopedia, Anat. and Phys.,) 1839.

—— Article 'Insecta,' in Todd's Cyclopedia of Anat. and Physiol., Vol. II, p. 980, 1839.

—— On the Uses of the Antennæ of Insects (Trans. Ent. Soc.,) 1837–40.

—— On the Structure and Development of Blood (An. of Nat. Hist., XV., pp. 281–284,) 1845.

—— On the Temperature of Insects, and its Connection with the Functions of Respiration and Circulation, 1837.

—— Extracts from Essay in Martin Duncan's Transformation of Insects.

Packard, Dr. A. S.—A Text-Book of Entomology, 1898. Very full and excellent.

—— Guide to the Study of Insects, 1869.

Pancritius, Paul.—Beitrage zur Kentniss der Flugelentwicklung bei den Insekten, 1884.

Parker & Haswell—Text-Book of Zoology, 1897.

Perez, J.—Bulletin de la Soc.d' Apicul. de la Gironde, 1878–1880

—— Les Abeilles, 1889.

Perris, Ed—Memoire sur le siege de l'odorat dans les Articules, 1850.

Pettigrew, J. Bell—On the Mechanical Appliances by which Flight is attained in the Animal Kingdom (Trans. Linn. Soc.,) 1870.

Plateau, F.—Palpes des Insectes broyeurs (Bul. de la Soc. Zool. de France,) 1885.

—— Recherches exp. sur la vision chez les Arthropodes (Comptes Rendus de la Soc. Ent. de Belg.,) 1887, (Bull. de l'Acad. Roy. de Belgique,) 1888.

Planta, Dr. A. von—Die Brutdeckel der Bienen (Schweitz. Bienenzeitung and Bul. d'Apic. de la Suisse Romande,) 1884.

—— Coloration de la cire des Abeilles (Revue Internationale,) 1885.

—— Ueber die zugammensetzung einiger Nektar Arten (Brit. Bee Jour., Nectar and Honey,) 1886.

—— Ueber den Futtersaft der Bienen, 1888.

—— Nochmals uber den Futtersaft der Bienen (Schweitz. Bienenzeitung,) 1889.

Pollmann, Dr. A.—Die Biene und ihre Zucht, 1875.

Porter, C. J.—American Naturalist, XVII., p. 1238, 1883.

Quinby, M.—Mysteries of Bee-Keeping, 1885.

Ramdohr, T. C.—Kleine Abhandlungen aus der Anatom. und Physiol. der Insecten, 1811, 1813.

Ranvier—Lecons sur l'histologie du systeme nerveux, 1878.

Ratzeburg, Dr. J. T. C.—Untersuchung des Geschlechtszustandes bei den sogenannten Neutris der Bienen, 1833.

Reaumur, R. A. F.—Memoires pour servir a l'histoire des Insectes, 1734–1742. English Translation, 1744.

Reid, Dr.—The Honey-Bee, by E. Bevan, p. 388, 1838.

Rehberg, A.—Ueber die Entwicklung des Insectenflugels, 1886.

Rendu, V.—L'intelligence des Betes, 1864.

Rombouts, Dr. J. E.—Locomotion of Insects on smooth Surfaces (Amer. Mon. Mic. Jour.,) 1884.

Root, A. I.—A B C of Bee-Culture, 1890.

Root, L. C.—Quinby's Mysteries of Bee-Keeping, 1884.

Schiemenz, P.—Uber das Herkommen des Futtersaftes und die Speicheldrusen der Biene, nebst einem Anhange uber das Riechorgan, 1883.

Schindler, E.—Beitrage zur Kenntniss der Malpighi'schen Gefasse der Insekten, 1878.

Schirach, A. G.—Physikalische Untersuchung der bisher unbekannten aber nachher entdeckten Erzeugung d. Bienenmutter, 1767.

Schonfeld, Pastor—Bienenzeitung, 1854–1883.

—— Illustrierte Bienenzeitung, 1885–1890.

—— The Mouth of the Stomach in the Bee (British Bee Journ.,) 1883.

Schultze, M.—Untersuch. uber die zusammengesetzten Augen der Krebsen und Insekten, 1868.

Sedgwick–Minot—Recherches histologique sur les trachees de l'Hydrophilus piceus (Arch. de Physiol. Paris,) 1876.

Shuckard, W. E.—British Bees, 1866.

Siebold, Dr. C. T. E. von—On a True Parthenogenesis in Moths and Bees, 1857.

——— Bienenzeitung, 1872.

——— Ueber die Stimm und Gehororgane der Krebse und Insekten (Arch. fur Mic. Anat.,) 1860.

Simmermacher, G.—Untersuchungen uber Haftapparate an Tarsalgliedern von Insekten, 1884.

Smith, Dr. J. B.—Economic Entomology, 1896.

Straus–Durckheim, H.—L'Anatomie comparee des animaux articules, 1828.

Swammerdam, J.—Biblia Naturae, (in Dutch, German and English,) 1737–1752.

Tegetmeier, W. B.—On the Formation of Cells (Rep. Brit. Assoc., pp. 132, 133,) 1858.

——— On the Cells of the Honey-Bee (Trans. Ent. Soc. Lond., p. 34,) 1859.

Thorley, J.—Melissologia ; or the Female Monarchy, 1744–1765.

Tichomiroff—Development of the Silkworm. (In Russian) 1879.

Tinker, G. L.—Bee-Keeping for Profit, 1880.

Treviranus, G. R.—Vermischte Schriften, 1817, and Zeitsch. fur Physiol., 1829.

Treviranus, L. Ch.—Medizinische Zoologie, 1833.

Viallanes, H.—Recherches sur les terminaisons nerveuses motrices dans les muscles stries des Insectes, 1881.

Vogel, F. W.—Die Honigbiene und die Vermehrung der Bienenvolker, 1880.

Waterhouse, G. R.—On the Formation of the Cells of Bees and Wasps, 1864.

Weismann, A.—Zeitschrift f. Wissenschaft. Zool., 1863.

Westwood's Introduction to the Study of Insects, 1840.

Wolff, Dr. O. J. B.—Das Riechorgan der Biene (Nova acta der K.L. Arch. Deutsch. Akad. d. Naturf.,) 1875.

Wyman, Dr. J.—Notes on the Cells of the Bee, 1866.

Zoubareff, A.—Concerning an Organ of the Bee not yet described, (Brit. Bee Jour.,) 1883.

PART SECOND.

THE APIARY:

ITS CARE AND MANAGEMENT.

MOTTO :—" Keep all colonies strong."

INTRODUCTION TO PART II.

STARTING AN APIARY.

In apiculture, as in all other pursuits, it is all-important to make a good beginning. This demands preparation on the part of the apiarist, the procuring of bees, and location of the apiary.

PREPARATION.

Before starting in the business, the prospective bee-keeper should inform himself in the art.

READ A GOOD MANUAL.

To do this, he should procure some good manual, and thoroughly study, especially that portion which treats of the practical part of the business. If accustomed to read, think and study, he should carefully read the whole work, but otherwise he will avoid confusion by only studying the methods of practice, leaving the principles and science to strengthen, and be strengthened by, his experience. Unless a student, he would better not take a journal till he begins the actual work, as so much unclassified information, without any experience to correct, arrange and select, will but mystify. For the same reason he may well be content with reading a single work till experience, and a thorough study of this one, make him more able to discriminate ; and the same reasoning will preclude his taking more than one bee-journal until he has had at least a year's actual experience.

VISIT SOME APIARIST.

In this work of self-preparation, he will find great aid in visiting the nearest successful and intelligent apiarist. If successful, such a one will have a reputation ; if intelligent, he will take the journals, and will show by his conversation that he knows the methods and views of his brother apiarists, and, above all, he will not think *he knows it all*, and that his is the only way to success. If possible he should spend some weeks during the active season with such a bee-keeper, and should learn all he could of such a one, but always let judgment and common sense sit as umpire, that no plans or decisions may be made that judgment does not fully sustain.

TAKE A COLLEGE COURSE.

It will be *most wise* to take a course in some college, if age makes this practicable, where apiculture is thoroughly discussed. Here one will not only get the best training in his chosen business, as he will study, see and handle, and thus will have the very best aids to decide as to methods, system and apparatus, but will also receive that general culture which will greatly enhance life's pleasures and usefulness, and which ever proves the best capital in any vocation. At the Michigan Agricultural College there is a fully equipped apiary, and the opportunities for special study in bee-keeping and entomology are peculiarly good. Michigan is not exceptional.

DECIDE ON A PLAN.

After such a course as suggested above, it will be easy to decide as to location, hives, style of honey to produce, and general system of management. But here, as in all the arts, all our work should be preceded by a well-digested plan of operations. As with the farmer and the gardener, only he who works to a plan can hope for the best success. Of course, such plans will vary as we grow in wisdom and experience. A good maxim to govern all plans is, "Go slow." A good rule which will insure the above, "Pay as you go." Make the apiary pay for all improvements in advance. Demand that each year's credits exceed its debits; and that you may surely accomplish this keep an accurate account of all your receipts and expenses. This will be a great aid in arranging the plans for each successive year's operations.

Above all, avoid hobbies, and be slow to adopt sweeping changes. "Prove all things, hold fast that which is good."

HOW TO PROCURE FIRST COLONIES.

To procure colonies from which to form an apiary, as is in almost all kindred cases, it is always best to get them near at hand. We thus avoid the shock of transportation, can see the bees before we purchase, and in case there is any seeming mistake can easily gain a personal explanation and secure a speedy adjustment of any real wrong.

KIND OF BEES TO PURCHASE.

At the same price always take Italians or Carniolans, as they are certainly best for the beginner. If common black bees can be secured for three, or even for two dollars less per colony, by all means take them, as they can be Italianized at a profit for the difference in cost, and, in the operation, the young apiarist will gain valuable experience.

Our motto will demand that we purchase only strong colonies. If, as recommended, the purchaser sees the colonies before the bargain is closed, it will be easy to know that the colonies are strong. If the bees, as they come rushing out, remind you of Vesuvius at her best, or bring to mind the gush and rush at the nozzle of the fireman's hose, then buy. In the hives of such colonies all combs will be covered by the middle of May with bees, and in the honey season brood will be abundant. It is always wisest to begin in a small way. He will generally succeed best who commences with not more than four or five colonies.

IN WHAT KIND OF HIVES.

As plans are already made, of course it is settled as to the style of hive to be used. If bees can be procured in such hives they will be worth just as much more than though in any other hive, as it costs to make the hive and transfer the bees. This will certainly be as much as two or three dollars. *No apiarist will tolerate, unless for experiment, two styles of hives in his apiary.* Therefore, unless you find bees in such hives as you are to use, it will be best to buy them in box-hives if possible and transfer (see Chapter VII) to your own hives, as bees in box-hives can always be bought at reduced rates. In case the person from whom you purchase will take the hives back at a fair rate, after you have transferred the bees to your own hives, then purchase in any style of movable-comb hive, as it is easier to transfer from a movable-comb hive than from a box-hive. Some bee-keepers, who were willing to wait, have purchased a queen and bees by the pound, and thus secured colonies at very slight expense. A single pound of bees with a queen will develop into a good colony in a single year.

WHEN TO PURCHASE.

It is safe to purchase any time in the summer. In April or May (of course you purchase only strong colonies) if in the latitude of New York or Chicago—it will be earlier further south—you can afford to pay more, as you will secure the increase both of honey and bees. If you desire to purchase in autumn, that you may gain by the experience of wintering, either demand that the one of whom you purchase insure the safe wintering of the bees, or else that he reduce the selling price, at least one-third, from his rates the next April. Otherwise the novice would better wait and purchase in the spring. If you are to transfer at once, it is desirable that you buy in spring, as it is vexatious, especially for the novice, to transfer when the hives are crowded with brood and honey.

HOW MUCH TO PAY.

Of course the market, which will ever be governed by supply and demand, must guide you. But to aid you, I will append what at present would be a reasonable schedule of spring prices almost anywhere in the United States :

For box-hives, crowded with black bees—Italians would rarely be found in such hives—three dollars per colony is a fair price. For black bees in hives such as you desire to use, five dollars would be reasonable. For pure Italians in such hives, seven dollars is not too much.

If the person of whom you purchase will take the movable-comb hives after you transfer the bees, you can afford to pay three dollars for black bees, and five dollars for pure Italians. If you purchase in the fall, require 33⅓ percent discount on these rates. The above is, of course, only suggestive.

WHERE TO LOCATE.

If apiculture is an avocation, then your location will be fixed by your principal business or profession. And here I may state that, if we may judge from reports which come from nearly every section of the United States, from Maine to Texas, and from Florida to Oregon, you can hardly go amiss anywhere in our goodly land.

If you are to engage as a specialist, then you can select first with reference to society and climate, after which it will be well to secure a succession of natural honey-plants (Chapter XVII), by virtue of your locality. This suggestion is important, even in California, though it has far less weight than in other sections. If our location is along a river we shall find our honey harvest much prolonged, as the bloom on the upland will be early, while along the river flats it will be later. Who knows how much the many successful bee-keepers along the Mohawk Valley owe to their excellent location? The same holds true of the mouth of the canyons in California. The flowers of both mountain and valley will then contribute of their sweets. We also gain in the prolonged honey-flow, as the mountain bloom is much the later. It will also be well to look for reasonable prospects of a good home market, as good home markets are, and must ever be, the most desirable. It will be important, also, that your neighborhood is not over-stocked with bees. It is a well-established fact, that apiarists with few colonies receive relatively larger profits, especially in rather poor seasons, than those with large apiaries. While this may be owing in part to better care, much doubtless depends upon the fact that there is not an undue proportion of bees to the number of honey-plants, and consequent secretion of nectar. To have the undisputed monopoly of an area reaching at least two and one-half miles in every direction from your apiary, is unquestionably a great advantage.

If you desire to begin two kinds of business, so that your dangers from possible misfortune may be lessened, then a small farm—especially a fruit-farm—in some locality where fruit-raising is successfully practiced, will be very desirable. You thus add others of the luxuries of life to the products of

your business, and at the same time may create additional
pasturage for your bees by simply attending to your other
business. In this case, your location becomes a more complex
matter, and will demand still greater thought and attention.
Some of America's most successful apiarists are also noted as
successful pomologists. A dairy farm, especially where winter dairying is carried on, would combine well with bee-keeping. The alsike clover would please alike the cattle and the
bees. This is equally true in sections of California and
Arizona, etc., only alfalfa takes the place of alsike clover.

Bees are often taken "on shares." It is usual for one
party to furnish the bees, the other to perform all the labor.
The expenses are shared equally, as are the proceeds, both of
bees and honey. Where one has more colonies of bees than
will do well in one place—more than 100 East, more than 250
in California—then "out-apiaries" are often desirable. Such
men as Dr. Miller, Messrs. Manum, France, Dadant, Elwood,
Mendleson, and Hetherington, find these very profitable. Of
course, this is like running a railroad, and success will only
mate with brains, gumption and pluck. The out-apiaries
should be as convenient as bee-forage, roads and location will
permit. If possible, it is wise to locate on some farm, and
arrange so the farmer will have an interest that will insure
some oversight when the apiarist is away. A fruit-grower
may be wise enough to covet the presence of the bees, and so
give service to secure it.

Of course, convenient hives for moving, and a wagon
arranged with suitable rack, are very desirable. Great pains
must be taken that the bees are all secure. Horses stung may
mean great loss and harm. Mr. Manum makes assurance
doubly sure by covering his horses entirely with cotton blankets. One enterprising and energetic enough to found out-
apiaries will have the gumption to success, and fully meet
every emergency.

For position and arrangement of apiary see Chapter VI.

CHAPTER V.

HIVES AND SECTIONS.

An early choice among the innumerable hives is of course demanded ; and here let me state with emphasis, *that none of the standard hives are now covered by patents, so let no one buy rights.* It is in nearly all sections of our country, happily, unnecessary to decry patent hives. Our excellent bee-periodicals have driven from among us, for the most part, that excrescence—the patent-hive man. His wares were usually worthless, and his life too often a lie, as his representations were not infrequently false to the letter. As our bee-men so generally read the bee-papers, the patent-hive vendor will grow less and less, and will soon exist only in the past. It will be a blessed riddance.

Success by the skillful apiarist with almost any hive, is possible. Yet, without question, some hives are far superior to others, and for certain uses, and with certain persons, some hives are far preferable to others, though all may be meritorious. As a change in hives, after one is once engaged in apiculture, involves much time, labor and expense, this becomes an important question, and one worthy of earnest consideration by the prospective apiarist. I shall give it a first place, and a thorough consideration, in this discussion of practical apiculture.

BOX-HIVES.

I feel free to say that no person who reads, thinks and studies—and success in apiculture can be promised to no other —will ever be content to use the old box-hives. In fact, thought and intelligence, which imply an eagerness to investigate, are essential elements in the apiarist's character, and to such a one a box-hive would be valued just in proportion to the amount of kindling-wood it contained. I shall entirely ignore box-hives in the following discussions, for I believe no sensible, intelligent apiarists, such as read books, will tolerate them, and that, supposing they should, it would be an expen-

sive mistake which I have no right to encourage, in fact, am
bound to discourage, not only for the benefit of individuals,
but also for the art itself.

To be sure of success, the apiarist must be able to inspect
the whole interior of the hive at his pleasure, must be able to

FIG. 82.

The Munn Hive, after Munn.

exchange combs from one hive to another, and to regulate the
movements of the bees—by destroying queen-cells, by giving
or withholding drone-comb, by extracting the honey, by introducing queens, and by many other manipulations to be explained, which are only practicable with a movable-comb hive.

MOVABLE-COMB HIVES.

There are, at present, two types of the movable-comb hive
in use among us, each of which is unquestionably valuable, as
each has advocates among our most intelligent, successful, and
extensive apiarists. Each, too, has been superseded by the
other, to the satisfaction of the person making the change.

The kind most used consists of a box, in which hang the frames which hold the combs. The adjacent frames are so far separated that the combs, which just fill them, shall be the proper distance apart. In the other kind, the ends of the frames are wider than the comb, and when in position are close together, and of themselves form two sides of a box. When in use these frames are surrounded by a second box, without a bottom, which, with them, rests on a bottom-board. Each of these kinds is represented by various forms, sizes,

FIG. 83.

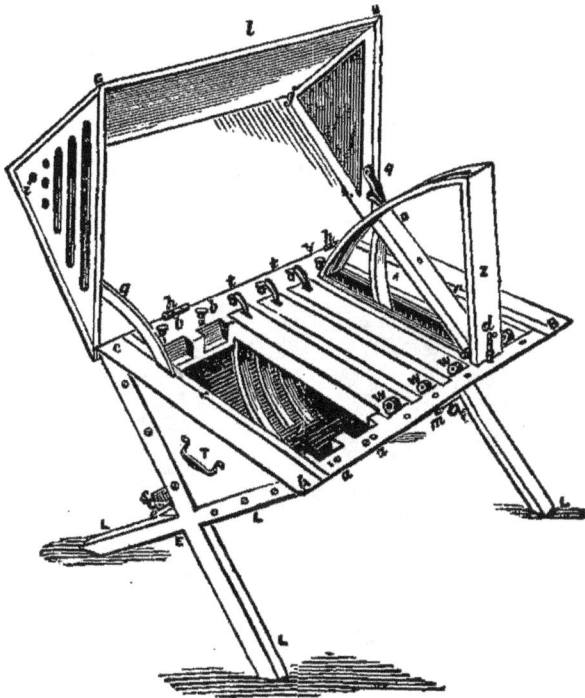

Munn's Improved Hive, after Munn.

etc., where the details are varied to suit the apiarist's notion. Yet, I believe that all hives in present use, worthy of recommendation, fall within one or the other of the above-named types.

EARLY FRAME HIVES.

In 1843, Mr. Augustus Munn, of England, invented a movable-comb hive (Fig. 82), which I need hardly say was not the Langstroth hive, nor a practical one. In 1851 this hive (Fig. 83) was improved (?). Well does Neighbour say in his valu-able hand-book, "This invention was of no avail to apiarists."

M. DeBeauvoys, of France, in 1847, and Schmidt, of Germany, in 1851, invented movable-comb hives. The frames were tight-fitting, and, of course, not practical. Dzierzon adopted the bar hive in 1838. In this hive each comb had to be cut loose as it was removed. It is strange that Mr. Cheshire speaks of Dzierzon's hive in connection with the Langstroth. It was a different type of hive entirely.

THE LANGSTROTH HIVE.

In 1851 our own Langstroth, without any knowledge of what foreign apiarian inventors had done, save what he could find in Huber, and edition 1838 of Bevan, invented the hive (Fig. 84) now in common use among the advanced apiarists of

FIG. 84.

Two-story Langstroth Hive.—From A. I. Root Co.

America. It is this hive, the greatest apiarian invention ever made, that has placed American apiculture in advance of that of all other countries. What practical bee-keeper of America could agree with H. Hamet, edition 1861, p. 166, who, in speaking of the DeBeauvoys' hive, says that the improved hives were without value except to the amateur, and inferior for practical purposes? Our apiarists not native to our shores, like the late Adam Grimm, Mr. C. F. Muth and Mr. Charles Dadant, always conceded that Mr. Langstroth was the inventor of this hive, and always proclaimed its usefulness. Well did the late Mr. S. Wagner, the honest, fearless, scholarly, truth-loving editor of the early volumes of the American Bee Journal, himself of German origin, say: "When Mr. Langstroth took up this subject, he well knew what Huber had done, and saw wherein he had failed—failing, possibly, only because he aimed at nothing more than constructing an observatory hive suitable for his purposes. Mr. Langstroth's object was other and *higher*. He aimed at making frames movable, interchangeable, and *practically* serviceable in bee-culture." And how true what follows: "*Nobody* before Mr. Langstroth ever succeeded in devising a mode of making and using a movable frame that was of any practical value in bee-culture." No man in the world, besides Mr. Langstroth, was so conversant with this whole subject as was Mr. Wagner. His extensive library and thorough knowledge made him a competent judge.

Mr. Langstroth, though he knew of no previous invention of frames contained in a case, when he made his invention, in 1851, does not profess to have been the first to have invented them. Every page of his book shows his transparent honesty, and his desire to give all due credit to other writers and inventors. He does claim, and very justly, to have invented the first practical frame hive, the one described in his patent, applied for in January, 1851, and in all three editions of his book.

For this great invention, as well as his able researches in apiculture, as given in his invaluable book, "The Honey-Bee," he has conferred a benefit upon our art which can not be overestimated, and for which we, as apiarists, can not be too grateful. It was his book—one of my old teachers, for which I have no word of chiding—that led me to some of the most delightful investigations of my life. It was his invention—the Langstroth hive—that enabled me to make those investigations. For one, I shall always revere the name of Langstroth, as a great leader in scientific apiculture, both in America and throughout the world. His name must ever stand beside those of Dzierzon and the elder Huber. Surely this hive, which left the hands of the great master in so perfect a form that even the details remain practically unchanged by many, I think most, of our first bee-keepers, should ever bear his name. Thus, though many use square frames like the Gallup, or deep frames, yet all are Langstroth hives.

CHARACTER OF THE HIVE.

The main feature of the hive should be simplicity, thereby excluding drawers and traps of all kinds. The hive should be made of good pine or whitewood lumber, thoroughly seasoned,

FIG. 85.

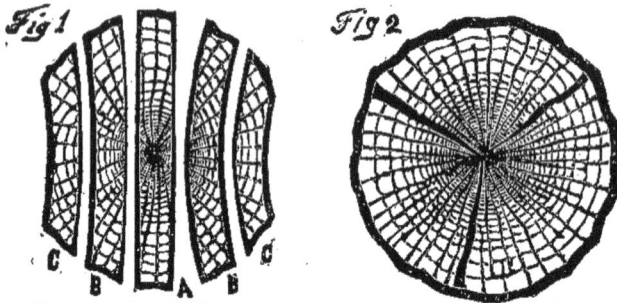

Principle of Warping.—From A. I. Root Co.

planed on both sides, and painted white on the outside. In making the hive nail the heart side of the board out, so as to prevent warping. To understand why see Fig. 85. Figure 84

FIG. 86.

One-story Langstroth Hive.—From A. I. Root Co.

represents a two-story Langstroth hive. As will be seen, this has a portico, and a bottom-board firmly nailed to the hive. Although Mr. Langstroth desired both these *features,* and many now are like-minded, many others omit both features.

This hive holds eight frames, which are as many as such bee-keepers as Messrs. Heddon, Taylor and Hutchinson desire.

Figure 86 represents the Simplicity one-story Langstroth hive as made by A. I. Root. This contains 10 frames, which, unfortunately, were slightly modified so that they are 17⅝ instead of 17⅜ inches long. Thus, this is not the Langstroth

FIG. 87.

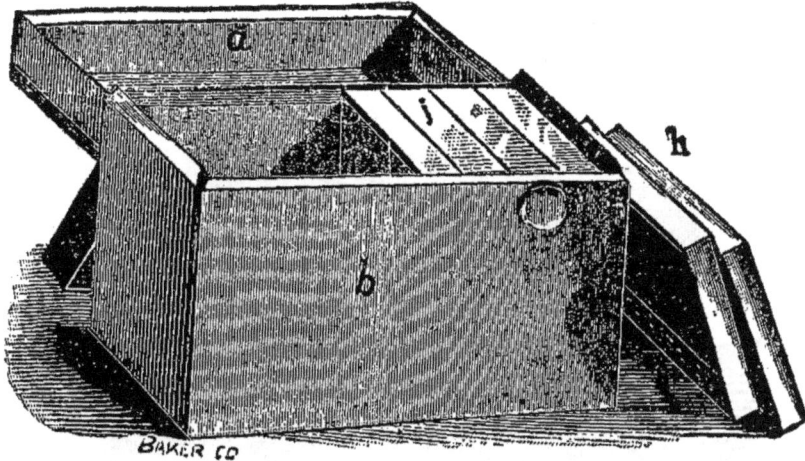

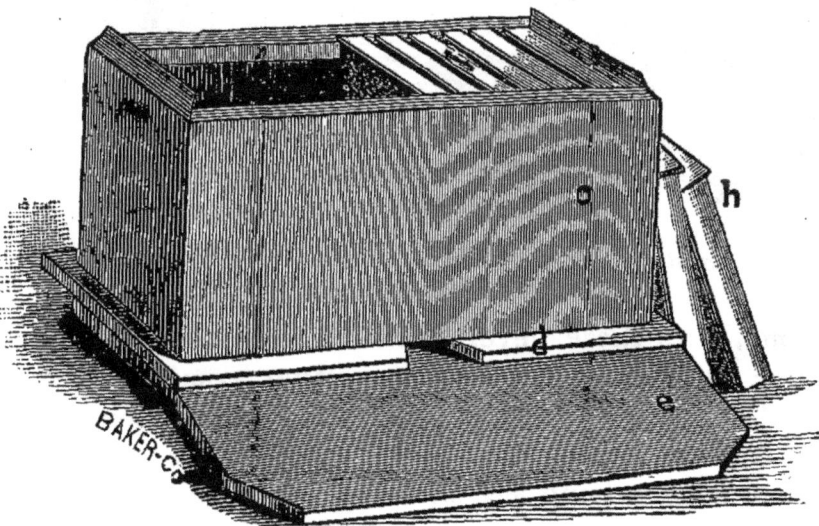

Two-story Langstroth Hive (Gallup Frame.)—Original.

a Cover hinged to hive.
c Brood-chamber.
e Alighting-board.
  Brood-frames.

b Upper story.
d Bottom-board.
i Wide section-frames.
h, h Frames outside hive.

frame, but the Simplicity-Langstroth. This style, one-story, is designed for securing comb honey, while the two-story (Fig. 84) is intended for use in obtaining extracted honey. Figure 87 represents a two-story Simplicity-Langstroth hive with Gallup frame; which is 11¼ inches square. This hive is pre- ferred by G. M. Doolittle. I have used it more than any other, and it has much to recommend it. The Simplicity feature invented by A. I. Root, I think, consists of a bevel union of hive with cover and bottom-board (Fig. 87). I think Mr. Root prefers this style no longer. Any Langstroth hive, with what-

FIG. 88.

Jones' Chaff-Hive, Frame, Frame for Sections, Division-Board and Perfor- ated-Zinc Division-Board.—From D. A. Jones.

ever frame, with these bevel connections is a Simplicity hive. This hive can be used to secure either comb or extracted honey. The bottom-board, d, and the alighting-board, e, may be separate from each other and from the hive; the opening may be made by cutting a V-shaped space in the bottom-board, while the cover, a, may or may not be hinged to the upper story. Mr. Root, in the original Simplicity, used the cover as a bottom-board, and formed an entrance by pushing the hive a little to one side. Many prefer to have the cover with a gable (Fig. 88), so made as to join the hive with a rabbet (Fig. 86 and 88), or to shut over the hive and rest on shoulders formed by nailing cleats about the hive near the top. These are heavy and costly. I much prefer a flat cover, and, if necessary to keep out water, we can follow Mr. Doolittle's plan and sheet with tin or zinc, though I think this unnecessary.

Figure 88 represents the Jones chaff hive. This takes a deep frame, and has double walls for chaff packing. These chaff hives are expensive, hard to handle and awkward to manage. After years of experience I discarded the chaff hives as no better in summer than the single-walled hives, and not so safe in winter as a good cellar. I have disposed of all of mine except three, which I keep for examples. Many, however, prefer such hives, and in some sections, and with some bee-keepers, they may be desirable.

WHAT STYLE TO ADOPT.

For many years I have used the Heddon-Langstroth, and like it so much that I recommend it above all others that I have tried. It is not only the simplest hive I have ever seen, but possesses many substantial advantages that are not possessed by any other hive so far as I know. It can be used with any size frame desired. I have it in use both with Langstroth and Gallup frames. I am free to express my preference for the Langstroth hive, with Langstroth frames. Its excellence warrants me in doing so, and the fact that it is by far the most used of any hive in the country, gives great advantage when one wishes to buy or sell bees. No beginner can make a mistake in adopting this hive. I will describe the hive for Langstroth frame, *but would advise any one to get a good hive as a pattern, if he is to adopt them, as much depends upon perfect exactness.*

The bottom-board and alighting-board (Fig. 87) may be separate if preferred, or not nailed to the hive. Mr. Heddon nails the bottom-board fast, and lets it project at one end, as seen in the figure (Fig. 89). A hive-stand is made by taking two boards (Fig. 89, *F*) six inches wide, and nearly as long as the bottom-board. Connect these at one end by a board 4½ inches wide, and as long as the hive is wide, nailed firmly at the bottom, and into the ends, and at the other end by a like board nailed the same way. We see (Fig. 94) this end-piece at the front of the hive nailed at the bottom so it rests on the ground. At the opposite end a like piece is nailed in the same way, so that all is even on the bottom. Figure 89 explains this better. The bottom of the hive (Fig. 89, *A*) is 13x19⅞ inches, outside measure, the sides made of six-eighths inch, bottom and cover of five-eighths, and ends of seven-eighths inch lumber. The height of this plain box is just 10 inches; that is, it is made of

FIG. 89

Heddon-Langstroth Hive.—From James Heddon.

F Bottom-board.   *A* Brood-chamber.
C Honey-board.   *D* Case with sections.
E Cover.

boards 10 inches wide. The side boards are 19⅞ inches long, so that they nail to the ends of the end-boards. If the corners are rabbeted, or, better, dovetailed (Fig. 90), they will be stronger, and less apt to separate with age and use. When used with the Gallup frame the ends of the hive project, and are nailed into the ends of the side-boards. The end-boards are rabbeted on top. This rabbet is cut three-eighths of an inch deeper than the thickness of the top-bar of the frame. With the Gallup frame (Fig. 96) we rabbet the side-boards. If the top-bar is three-eighths of an inch thick this rabbet should be six-eighths precisely. This is *very important*, as we *must have a three-eighths space exactly* between the top-bar and the top of the hive. If we make the hive ten and one-eighth (10⅛) inches high we give a space of half-inch between the bottom of the frame and bottom of hive. I like this wide space, and there is no objection to it. Near the top of the hive we will nail narrow cleats entirely around it; these strengthen the hive, and are convenient supports by which to lift the hive. Hand grooves (Fig. 90) can also be cut in end and side-boards for convenience in handling, if desired. Mr. Root favors these hand-holes always. They are easily cut, and are surely a convenience.

The entrance is cut in the end of the hive (Fig. 89), and the size is easily regulated by use of the Langstroth triangular

FIG. 90.

Dovetailed Hive.—From A. I. Root Co.

blocks (Fig. 89, *B, B*). Thus we may gauge the size to our liking. I would have the entrance the whole width of the hive, and seven-eighths of an inch high. This may aid to prevent the bees hanging out of the hive, and likewise may restrain the swarming impulse. The opening in the bottom-board (Fig. 87) is preferred by many. This is enlarged or restricted by simply pushing the hive forward or back, and, of

course, can only be used with loose bottom-boards. The fact
that most bee-keepers nail the bottom-board firmly and cut the
opening from the hive, argues that this on the whole is the
better style. For shipping and moving bees, which, with
"out-apiaries" and change of location to secure better pasturage, promises to be more and more the practice, the nailed

FIG. 91.

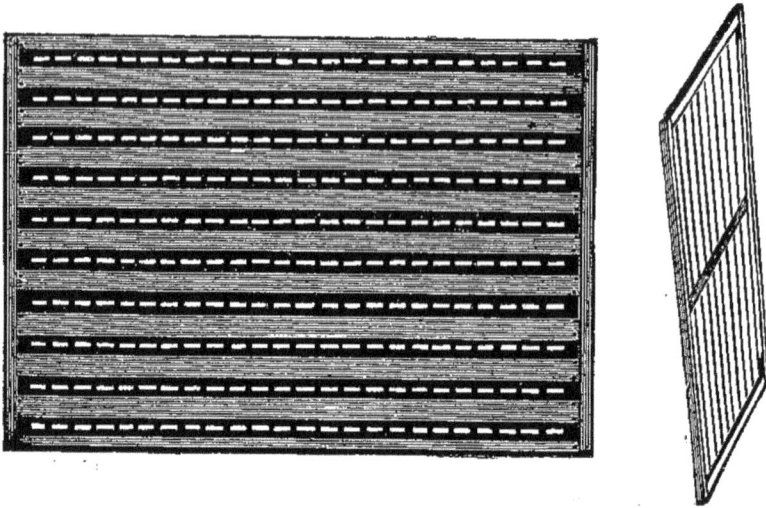

Queen-Excluding Honey-Board.—From D. A. Jones.

bottom-boards are very desirable; for quick cleaning of the
hives when spring opens, the movable bottoms are preferable.

There should never be but this one opening. Auger-holes
above, and openings opposite the entrance, are worse than
useless.

Except in very damp locations the hive should not rest
more than five or six inches from the ground. Tired and
heavily laden bees, especially on windy days, may fail to gain
the hive, if it is high up, as they return from the field.

For extracted honey, we use a second story precisely like
the body of the hive, except it is a half-inch less in depth;
that is, the sides are 9½ instead of 10 inches wide. Mr. Dadant
prefers half-story hives for the extracting frames, but he uses
the large Quinby frame (Fig. 95). If we wish we can *follow Dadant*, and use two or more of these upper stories, and tier up, in

which case we would not need to extract until the close of the harvest, when the honey would be ripened in the hive.

Upon the body of the hive rests the slatted honey-board (Fig. 91). It is seen in place (Figs. 89 and 93). This is also 13 by 19⅞ inches. The outer rim of this valuable invention and the slats are in one plane on the under surface, and the slats are three-eighths of an inch apart, leaving passages that width for the bees to pass through. On the upper surface the rim projects three-eighths of an inch above the slats, so that if a board be laid on the honey-board its lower surface will be three-eighths of an inch above the slats. When the honey-board is placed on the hive, the spaces between the slats must rest exactly over the center of the top-bars of the brood-frames below. In using hives with the Gallup or American frames the slats of course will run crosswise of the honey-board, and as before must break joints with the top-bars of the frames.

FIG. 92.

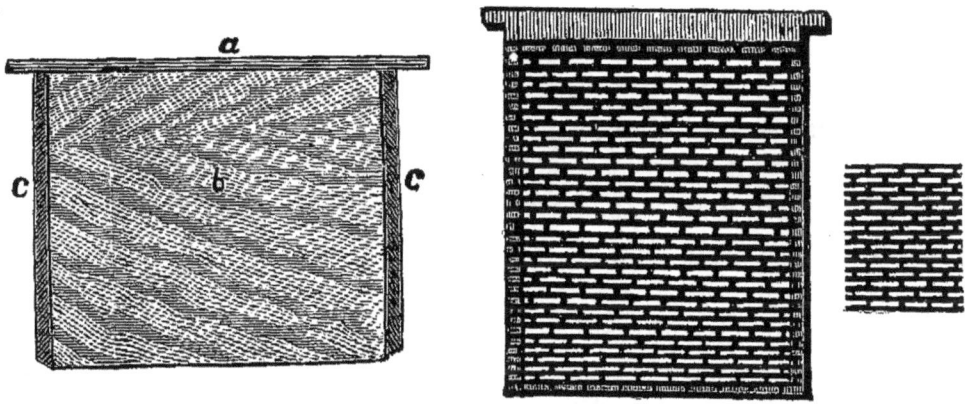

Plain Division-Board.   *Perforated-Zinc Division-Board.*
                        —From D. A. Jones.

The use of this prevents the bees from building brace-combs above the brood-frames, and keeps the sections very neat. No one after using this will do without it, I am sure. By tacking a piece of perforated-zinc (Fig. 92) on the under side of this honey-board it also becomes a queen-excluder. The grooves in the zinc must be *very exact*. They are .165 of an inch wide. It is cheaper, and so better, simply to place a narrow strip of the perforated-zinc between the slats of the honey-board (Fig.

91). By grooving the edges of the slats it is easy to insert the zinc strips when making the honey-board. The honey-board may be wholly of zinc with a wooden rim. The objection to this is the fact that the zinc is likely to sag and bend. Mr. Heddon suggests that a V-shaped piece of tin be soldered across the middle to strengthen the zinc and prevent sagging. The tin should be so placed as not to touch the frames below, but come between them. Mr. Heddon also suggests that the wooden rim be replaced by a narrow margin of the zinc itself, bent at right angles to the plane of the metal.

THE HEDDON SURPLUS-CASE.

As this admirable case is also a part of this hive, I will describe it right here, though it properly belongs to the subject of case for surplus honey. This case is just as long and broad as the hive, and three-eighths of an inch deeper than the height of the section to be used. (See Fig. 89, *D*.) Thus, on the hive described it will be 13 by 19⅞ inches, and if we use common 1-pound sections, which are 4¼ inches square, it will be 4⅝ inches deep. Partitions are fastened in by use of screws or nails just far enough apart to receive the sections; thus, in the 1-pound sections, 4¼ inches apart. These partitions are as wide as the crate or case is deep. Narrow strips of tin are nailed to the bottom of these partitions and to the bottom of the ends of the case, projecting enough to sustain the sections when they are placed in the case. It will be seen that when in place the sections reach to within three-eighths of an inch of the top of the case. This *must be just three-eighths of an inch*. It keeps the sections all clean, but will not if not JUST this bee-space.

THE COVER.

The cover of the hive (Fig. 89, *E*) is a plain board, a little wider and longer than the hive. The ends of this are fitted into a grooved cross-piece about twice as thick as the board, and firmly nailed. These cross-pieces prevent the top from warping and splitting. If preferred, the cover need be no longer or wider than the hive. In this case cross-pieces should be firmly nailed on the upper side to prevent warping or splitting. It will be seen that we have here no telescoping, and no beveling—simply one board rests upon another. At first I was much prejudiced against this simple arrangement. After giving it a thorough trial I wish nothing else. The only criticism I have for this hive after several years' experience is, that if the board cover is used in spring, the protection is insufficient. We break the propolis or glue in examining the bees, and then as the bees can not glue all close at this early season, the brood is apt to chill, and the bees to suffer, especially if the sides of the hives have shrunken, or the cover warped. By use of a quilt or warm woolen cloth just the size of the hive placed above, and a crate filled with dry sawdust above this, all is made snug and comfortable, and even this objection disappears. To adopt this style of hive is not expensive. We can use the same frames as before, and can make all new hives of this simple, plain pattern, and in time we will have only these hives.

To shade the hive nothing is so good as a shade-board made considerably wider than the hive, and nailed to two cleats five inches wide. Thus, when resting on the hive this shade-board will be five inches above the top of the hive. This has never blown off of my hives. Should it do so a brick could easily be fastened to the under side, out of sight, and thus make it entirely safe against winds.

Thus I have described the Heddon-Langstroth hive minutely, as with W. Z. Hutchinson, R. L. Taylor, and many others of our most able and intelligent apiarists, I find it, upon trial, as excellent as it is simple. Surely, when we can harness excellence and simplicity together we have a most desirable team. The simple union of parts by mere plain contact of the edges, or the cover simply lying on the hive, while it is just as acceptable to the bees, makes the hive far more simple of construction, and easy of manipulation. The honey-board and bee-spaces keep all so neat, that as one bee-keeper well says, their extra expense is very soon saved in the saving of time which their use insures. Any who may think of trying this hive better do as I did, try two or three at first, and see if in their judgment the "game is worth the candle."

All hives should be well painted with white paint. This color makes the heat less trying to the combs and bees. While

it may not be profitable to paint, yet when neatness and dura-
bility are both considered, surely painting pays well. For
paint I would use white lead, zinc and oil—about one-third as
much zinc as lead. Mr. Doolittle, whose opinion justly ranks
very high among American bee-keepers, thinks that white
paint makes shade unnecessary.

DIVISION-BOARD.

A close-fitting division-board (Fig. 92) is very important,
and no Langstroth hive is complete without it. Mr. Heddon,
in his excellent book, follows the English, and calls this a
dummy. It is especially useful in autumn, winter and spring
in contracting the hive, and thus economizing heat, and at
the harvest seasons in contracting the brood-chamber, so as to
secure the honey in the sections where it is desired. It is
made the same form as the frames, but is a little larger so
that it is close-fitting in the hive. It is easily made by nailing
a top-bar of the usual frame on top of a board that will just fit
in the hive, and reach to the top of the rabbet. If desired the
board may be beveled at the edges. When the division-board
is inserted in the hive it separates the brood-chamber into two
parts by a close partition. Many bee-keepers make them like
a close-fitting frame and cover with cloth, which is stuffed
with chaff. Others groove the edges and insert a strip of cloth
or rubber. The chaff board is for greater warmth, the rubber
to make the board fit closely, and yet give enough to make it
easy to withdraw the division-board when it swells from
dampness. Mr. Jones prefers that the division-board should
not reach quite to the bottom of the hive (Fig. 88). This enables the bees to pass under, and as heat rises there is very
little objection to this bee-space under the division-board.

We use the division-board to contract the chamber in winter,
to vary it so as to keep all combs covered with bees in spring, to
contract the brood-chamber when we wish to secure a full
force of bees in the sections, to convert our hives into nucleus
hives for queen-rearing, and in case we secure comb honey in
two-story hives, which, however, we do not practice now, to
contract the upper chamber when the season first opens.

CLOTH COVERS.

After the season is over, and the weather becomes cold, about the 20th of September, it is well to remove the honey-board, and to cover above the bees with a piece of heavy factory cloth, which thus forms the immediate cover for the bees in winter. The section-case full of dry, fine sawdust has now this cloth for its bottom, while the cover of the hive rests on the section-case.

It will be noted that I have made no mention in the above of metal rabbets, or, more correctly, metal supports. I have tried these for some years, and have usually recommended them, but for the past several years I have omitted them, and think I shall have no further use for them in my hives. If we wish them we have only to cut the rabbet a little deeper and tack inside the hive, just below the rabbet, a narrow strip of heavy tin, which shall project a little above the wooden rabbet, just enough to raise the top of the frame to within three-eighths of an inch of the top of the hive. The advantages of these are that they make a very narrow rest or support for the frames, and so the latter are more easily loosened, and in careless hands are less apt to kill bees when put into the hives. It is always easy, however, by means of a chisel to loosen frames, and if we are often manipulating our bees, as when extracting in summer, the frames are easily loosened without the metal supports. Some apiarists make hives without rabbets, making the frames to rest on the top of the hive. I have tried such hives thoroughly, and wish no more of them. Of course, with such hives the valuable honey-board and bee-spaces are impossible.

THE NEW HEDDON HIVE.

Mr. Heddon has patented and offered to the public a new hive which combines in principle the Langstroth and the Huber. I have tried this hive only for a short time, and so, guided by the rule I have always adopted, I do not recommend it. Yet the experienced bee-keeper can often judge correctly of what he has never tried, and I will add that I fully believe this hive and the method Mr. Heddon gives of manipulation in his valuable book, are well worth our attention. Mr. Heddon is so able that he rarely recommends what is not valuable. Several others have tried this hive, and speak in the highest terms of its value.   Among these are no less authorities than R. L. Taylor and W. Z. Hutchinson.   At the beginning of this chapter I caution all against patent hives.   This is necessary, as so many frauds have been committed under this guise; but if Mr. Heddon has given us something as valuable as it is unique and original, he well deserves a patent, which should be thoroughly respected, as should all worthy inventive effort. From my brief experience I fear the hive is too complicated for the average bee-keeper.   With a much longer experience (1900) I can not recommend it.   It works admirably if every-thing is perfectly exact; otherwise it is a vexation.   Absolute exactness is rare in our day and world.

I shall describe the hive only in brief, advising all who wish to investigate this newcomer, to procure Mr. Heddon's work, "Success in Bee-Culture," as this will be an excellent investment aside from the matter of the hive.

This hive (Fig. 93) has close-fitting frames fastened in a case by use of wooden thumb-screws.   The end-bars of the frames are wide like those of the Huber hives, and rest on tin supports.   The top and bottom bars of the hives are only as wide as the natural comb, seven-eighths of an inch.   The frames are only five and three-eighths (5⅜) inches deep, and this with the wide spaces between them makes it possible to do much without removing the frames.   There is a three-eighths inch space above the frames, and a honey-board as in the Heddon-Langstroth hive.

Thus, one or two shallow hives, can be used, and to con-tract the brood-chamber at any time we have only to remove one of them.   Figure 93 shows the hive, which, with two brood-chambers, gives about the capacity of a 10-frame Lang-stroth hive.   As all frames are securely held by the screws, any brood-chamber can be reversed, or any two can change places at the pleasure of the bee-keeper.   I have found the screws to swell and work with extreme difficulty.   I think Mr. Taylor excludes the screws, and wedges the frames instead. As the combs will all be firmly attached on all sides to the frames, there is no space for hiding, and the queen can gen-

erally be found without removing the frames. I have seen Mr.
Taylor find several queens with these hives in a few minutes
time.

FIG. 93.

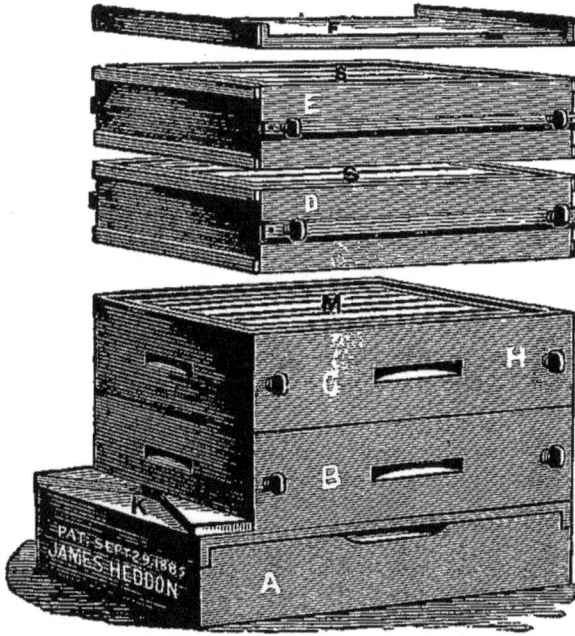

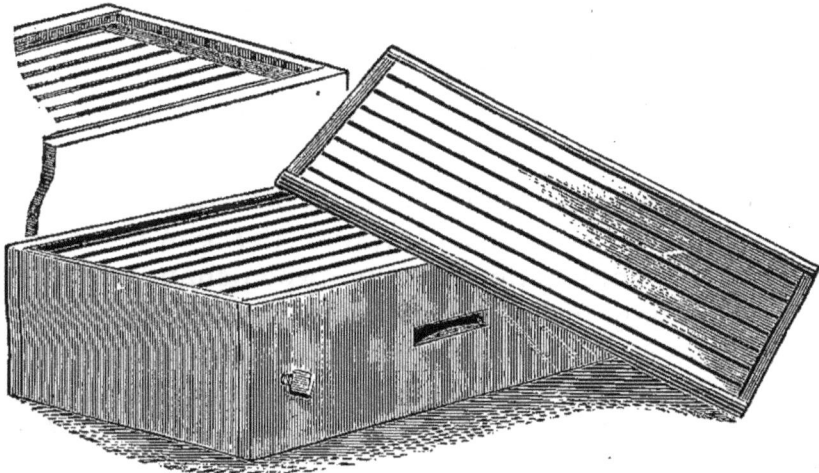

The New Heddon Hive.—From James Heddon.

A Stand.   D E Section-cases.   H Thumb-screw.
B C Two sections.   M Slatted honey-board.   F Cover.

The bottom-board (Fig, 94) has a raised rim. Thus the frames are one-half inch from the bottom. Of course, the bottom-board is loose. Mr. Heddon recommends single-story wide-frames with separators for the sections. These are also secured by the screws, and so any frame or the whole case can be reversed at will.

Of course, the old Heddon case without separators could be used, but could not be reversed. The points of excellence claimed for this hive, and I know from my experience that they are real, are easy contraction of brood-chamber, quick inversion of the brood-chamber or section-case, ease and quick-

FIG. 94.

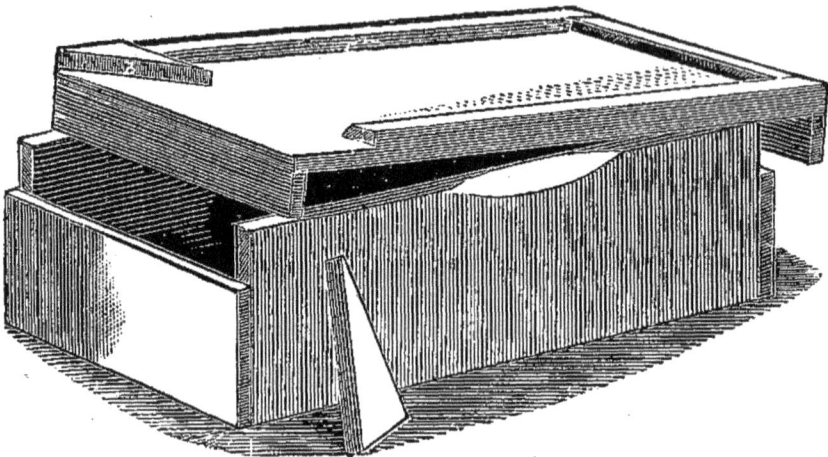

Heddon Bottom-Board.—From James Heddon.

ness of manipulation, and the interchangeableness of the brood-chambers forming the hive, and the power we have by quick and easy contraction of the brood-chamber to get all light-colored honey in the sections if we so desire.

Mr. J. M. Shuck has also patented a hive for which he claims the same advantages gained in the new Heddon hive. I have not worked with it enough to recommend it. I fear the hives are too complex for the general bee-keeper. The fact, too, that perfection of work and measurements despite our best care are very rare, urges against this hive, as it must be very accurate or it is a sore vexation. I advise all to go slow in adopting them, as we know the old, tried ones are excellent. I fear that in the hands of the general bee-keepers these new hives will not prove satisfactory.

THE FRAMES.

The form and size of frames, though not quite as various as the persons who use them, are still very different (Fig. 95). Some prefer large frames. I first tried the Quinby frame, and afterward the Langstroth (Fig. 95). The advantage claimed

FIG. 95.

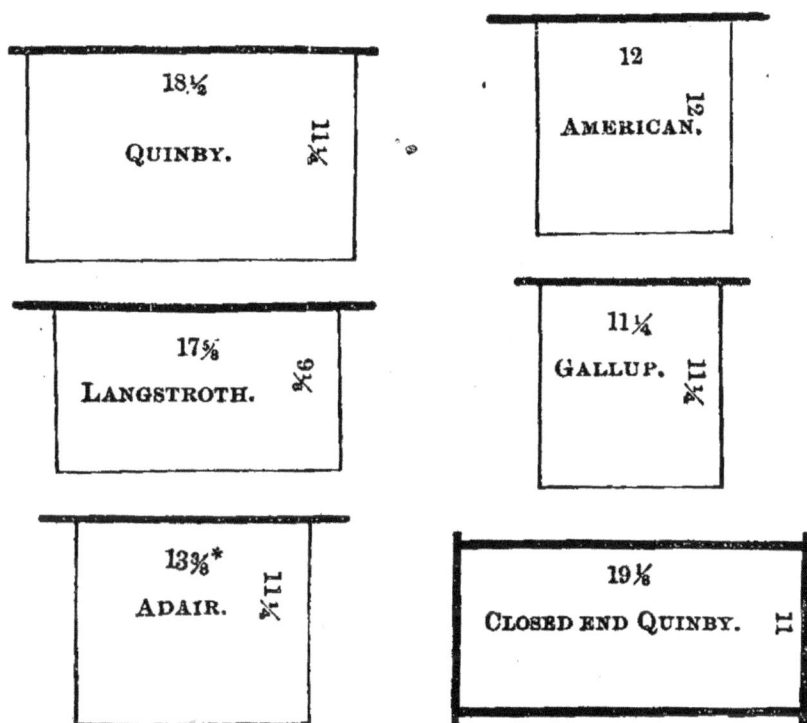

Brood-Frames.—From A. I. Root Co.

for large frames is that there are less to handle, and time is saved; yet may not smaller frames be handled so much more dextrously, especially if they are to be handled through all the long day, as to compensate, in part at least, for the number? The advantage of the shallow frame is, as claimed, that the bees will go into boxes more readily; yet they are not considered by some bee-keepers as safe for out-door wintering. This is the style recommended and used by Mr. Langstroth, which fact may account for its popularity in the United States.

Another frame in common use, is one about one foot square. I have long used one 11¼ inches square, and still think that this frame has much to commend it. It is light, easily handled, convenient for nucleus hives, and perhaps the best form for forming a compact winter cluster; and yet upon mature reflection I have decided to use in future, as already stated, the Langstroth frame, and advise all others to do so.

It is very desirable to have bees in hives such as others will wish in case we sell bees, as every bee-keeper is almost

FIG. 96.

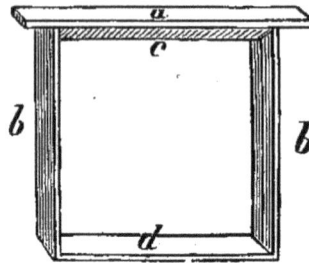

Gallup Frame.—Original.

a Top-bar.   c Comb-guide.
b, b Side-bars or uprights.   d Bottom-bar.

sure to do more or less each year. The Langstroth hive is used much more generally than any other, and that it is excellent is shown in the fact that most of our successful bee-keepers, from Canada to the Gulf, use it, and I am free to say that, taking the whole country through, it is doubtful if a better style or form exists than the regular Langstroth. The chief objection urged against its use, that it is not the best form to secure safe wintering, lacks force in view of the fact that many who have been most successful use this frame. Indeed, with thorough protection this frame is as good as any, and most bee-keepers are learning that in our Northern States protection is absolutely essential to success.

That we shall ever have a uniform frame used by all apiarists, though exceedingly desirable, is too much to be hoped. I do not think there is sufficient advantage in any form to warrant us in holding to it, if by yielding we could secure this uniformity. Nor do I think the form and size so material as to make it generally desirable for the apiarist to change all his hives, to secure a different style of frame.

To make a Langstroth frame I would use a top-bar (Fig. 96)—the figure illustrates a Gallup frame which is square, and will serve to make this explanation clearer, eighteen and seven-eighths (18⅞) inches long, seven-eighths (⅞) of an inch wide, and one-fourth (¼) of an inch thick. The end-bars (Fig. 96, *b*, *b*) should be eight and five-eighths (8⅝) inches long, and as wide and thick as the top-bar. The top-bar is fastened to the end-bars, as shown in the figure, by nailing through it into

FIG. 97.

Reversible Frame, Upper one hung in the Hive, Lower one partly reversed.
—From James Heddon.

the ends of the end-bars, so as to leave the top-bar projecting three-fourths (¾) of an inch. The bottom-bar is seventeen and three-eighths (17⅜) inches long, and as wide and thick as the other parts—though it may be only one-half as thick if preferred. It is also nailed to the ends of the end-bars, so that it is as long as the frame. The parts when made at the factory are often dovetailed so as to be more securely united.

For some years I have used the reversible frame (Fig. 97), which has valuable features which would warrant its use were it not for its complexity. With this frame there is no danger of the top-bar sagging, which is sure to enlarge the bee-space above and create mischief, and by inverting we secure the firm attachment of the comb to the frame along all its edges, and it helps to force our bees into the sections, simply by inverting the combs. This may not always succeed with the unskillful—some bee-keepers report failure—and it requires some time and attention. Figure 97 shows the character of the reversible frame as made by Mr. Heddon, and which I have found to work the best of any that I have used. As will be seen, the reversible part is a rectangle, pivoted in the center to the bottom of the short end-bars. These short end-bars at the top come within one-fourth (¼) inch of the side of the hive, and thin a little as they run down, so that the lower end is three-eighths (⅜) of an inch from the side of the hive. The bottom of the frame, indeed all below the short end-bar, is three-fourths (¾) of an inch from the side of the hive. This makes it easy to put in the frames without crushing the bees. It might be supposed that the bees would build combs between the lower end of the frame and the hive, but I have never seen a case of the kind, and I have used such frames now quite extensively for several years. These frames reverse very easily, and I do not know a single person who has thoroughly tried them, who does not value them highly. Here again let me suggest that in making changes, a few be tried first, and not all till we know we wish them.

As the use of comb foundation secures straight combs, with no drone-cells, it is very desirable. When this is fastened by merely pressing or sticking it to the top-bar, it is apt to sag and warp, hence it is becoming quite the custom to wire the frames (Fig. 97). This insures perfect safety if we wish to ship our bees, and secures against sagging or bulging of the foundation. If the foundation is put on with the Given press as the foundation is made, No. 36 wire is used; if pressed on by hand No. 30 wire is better.

The timber for frame should be thoroughly seasoned, and of the best pine or white wood. Care should be taken that the frame be made so as to hang vertically, when suspended on the rabbets of the hive. To secure this *very important* point—true frames that will always hang true—they should always be made around a guide.

A BLOCK FOR MAKING FRAMES.

This may be made as follows: Take a rectangular board (Fig. 98) eleven and one-eighth by thirteen and a quarter inches. On both ends of one face of this, nail hard-wood pieces (Fig. 98, *e, e*) one inch square and ten and three-fourths inches long, so that one end (Fig. 98, *g, g*) shall lack three-eighths inch of reaching the edge of the board. On the other face of the board, nail a strip (Fig. 98, *c*) four inches wide and eleven and three-eighths inches long, at right angles to it, and in such position that the ends shall just reach to the edges of

FIG. 98.

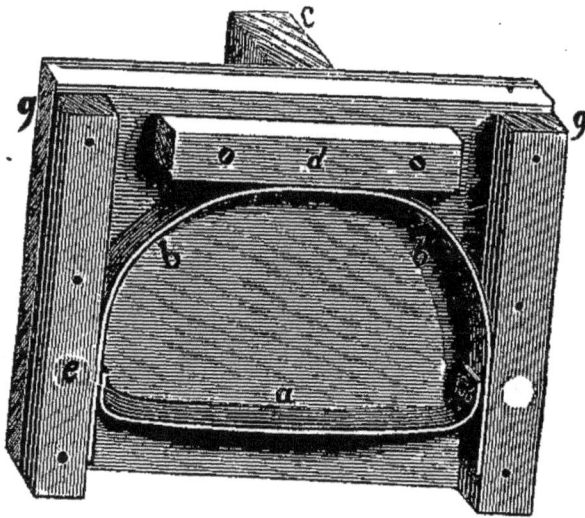

Block for making Gallup Frames.—Original.

the board. Midway between the one-inch-square pieces, screw on another hard-wood strip (Fig. 98, *d*) one inch square and four inches long, parallel with and three-fourths of an inch from the edge. To the bottom of this, screw a semi-oval piece of hoop-steel (Fig. 98, *b, b*), which shall bend around and press against the square strips. The ends of this should not reach quite to the bottom of the board. Near the ends of this spring fasten, by rivets, a leather strap an inch wide (Fig. 98, *a*), which shall be straight when thus riveted. These dimensions are for frames eleven and one-fourth inches square, outside measure, and must be varied for other sizes. Instead of the iron and strap, some use two pieces of wood with a central pivot. The upper ends of these levers are united by a strong elastic cord, so that the lower ends are constantly pressed against the side-pieces of the block. Recently we have used in such blocks, both for frame and section-making, a single hard-wood strip, a little shorter than the distance between the strips e and e. This is pivoted at the center to the center of the block. This is a very simple way to hold the side-pieces firmly against the strips e, e. We have only to turn this lever.

To use this block, we crowd the end-bars of our frames between the steel springs (Fig. 98, b, b) and the square strips (Fig. 98, e, e); then lay on our top-bar and nail, after which we invert the block and nail the bottom-bar, as we did the top-bar. Now press down on the strap (Fig. 98, a), which will loosen the frame, when it may be removed all complete and true. Such a gauge not only insures perfect frames, but demands that every piece shall be cut with great accuracy, and some such arrangement should always be used in making the frames.

The above description and Fig. 98 are for Gallup frames. For Langstroth frames the hard-wood strips would be eight and five-eighths (8⅝) inches long, and the distance between them would be sixteen and seven-eighths (16⅞) inches, that is, if the frames are made of pieces one-fourth of an inch thick. To make reversible frames we use two such guides. Wire nails are very excellent for making frames, and just the thing for the pivots in reversible frames.

When the frames are in the hive there should be at least a one-fourth or three-eighths inch space between the end of the frame and side of the hive. As before stated, the space below the frame may be one-half inch. A much wider space on the sides than that given above is likely to be filled with comb, and so prove vexatious. The wide space below gives no such trouble, and in winter it is desirable, as also in case the hive shrinks. It is very undesirable to have the frames reach to the bottom of the hive.

The distance between the frames may be one-half of an inch, or best one and three-eighths inches from center to center of the frames. This is better than one and one-half, as the brood is kept warmer, and worker-brood is more likely to be reared. A slight variation either way does no harm. Some men, of very precise habits, prefer nails or wire staples in the side and bottom of the frames. Mr. Cheshire calls these his suggestions, though Mr. Langstroth used them over twenty years ago, which, if I am correctly informed, was before Mr. Cheshire kept bees at all. These are to insure equal spacing of the frames. Mr. Jones prolongs the sides and bottom of the frame (Fig. 88) for the same purpose. These projections extend just a quarter of an inch, so as to maintain this unvarying distance. Some bee-keepers use frames with wide, close-fitting end-bars, or with top-bars wide and close-fitting

FIG. 99.

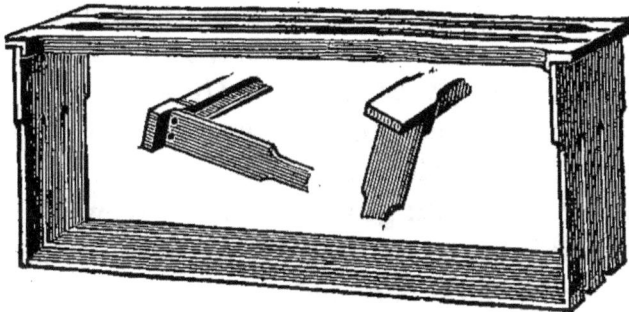

Hoffman Frames.—From A. I. Root Co.

at the ends. Mr. Root now favors the Hoffman frame (Fig. 99), as he calls it, which has the top-bar and upper ends of the end-bars wide and close-fitting. He claims more rapid handling, as the frames, he says, can be handled in groups. I have tried all these styles, and do not like them. It is easy for any bee-keeper to try them. "Prove all things ; hold fast that which is good," or that which pleases you.

COVER FOR FRAMES.

As before stated, a board covers the hive all through the honey season. This rests upon the upper story of the hive, or upon the upper section-case. From September to June, in the cold Northern climate, a piece of thick factory cloth should rest on the frames as before stated. This is just the size of the hive, and when properly adjusted no bee can pass above it. By cutting on three sides of an inch square, we form a flap in this cloth which may be turned back to permit the bees to enter the feeder, when feeding is desired. In fall, winter and spring, a section-case left on the hive and filled with fine sawdust or chaff is a most desirable substitute for a heavy, awkward chaff hive. Dr. Miller covers the year through with a cloth cover.

THE HUBER HIVE.

The other type of hives originated when Huber hinged several of his leaf or unicomb hives together so that the frames would open like the leaves of a book. In August, 1779, Huber wrote to Bonnet as follows: "I took several small fir-boxes, a foot square and fifteen lines wide, and joined them togther by hinges, so that they could be opened and shut like the leaves of a book. When using a hive of this description, we took care to fix a comb in each frame, and then introduced all the bees." (Edinburgh edition of Huber, p. 4.) Although Morlot and others attempted to improve the hive, it never gained favor with practical apiarists.

In 1866, Mr. T. F. Bingham, then of New York, improved upon the Huber hive, securing a patent on his triangular-frame hive. This, so far as I can judge, was the Huber hive made practical. Mr. Bingham now uses a modification of this hive (Fig. 101).

In 1868, Mr. M. S. Snow, then of New York, now of Minnesota, procured a patent on his hive, which was essentially the same as the hives now known as the Quinby and Bingham hives.

Soon after, the late Mr. Quinby brought forth his hive, which is essentially the same as the above, only differing in details. No patent was obtained by Mr. Quinby, whose great heart and boundless generosity endeared him to all acquaintances. Those who knew him best never tire of praising the unselfish acts and life of this noble man. If we except Mr. Langstroth, no other man, especially in the early days, did so much to promote the interest and growth of improved apicul-

ture in the United States. His hive, his book, his views of wintering, and foul brood, his introduction of the bellows-smoker—
a gift to apiarists—all speak his praise as a man and an apiarist.

The facts that the Bingham hive, as now made, is a great
favorite with those that have used it, that Mr. Quinby preferred this style or type of hive, that the Quinby form is used
by the Hetherington brothers—Capt. J. E., the prince of American apiarists, with his thousands of colonies, and O. J., whose
neatness, precision, and mechanical skill are enough to
awaken envy—are surely sufficient to excite curiosity and bespeak a description.

The Quinby hive (Fig. 100) as used by the Hetherington
brothers, consists of a series of rectangular frames (Fig. 100)

FIG. 100.

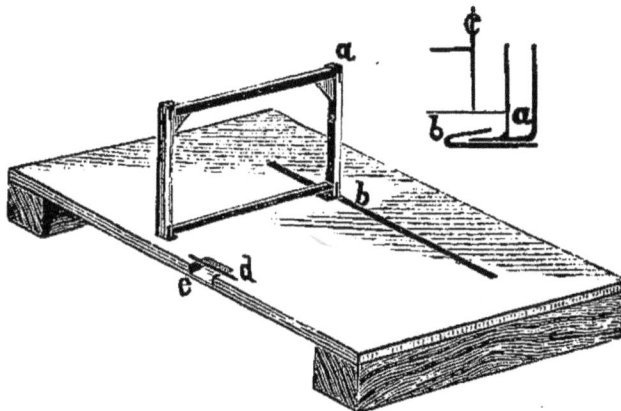

Frame, Bottom-Board and Frame-Support, of Quinby Hive.—Original.

twelve by seventeen inches, outside measure. The end-bars
of these frames are one and one-half inches wide, and half an
inch thick. The top and bottom one inch wide and half an
inch thick. The outer halves of the end-bar project one-fourth
of an inch beyond the top and bottom bars. This projection
is lined on the inside with sheet-iron, which is inserted in a
groove which runs one inch into each end of the end-pieces,
and is tacked by the same nails that fasten the end-bars to the
top and bottom bars. This iron at the end of the bar bends in
at right-angles (Fig. 100, a), and extends one-fourth of an inch
parallel with the top and bottom bars. Thus, when these

frames stand side by side, the ends are close, while half-inch openings extend between the top and bottom bars of adjacent frames. The bottom-bars, too, are one-fourth of an inch from the bottom-board. Tacked to the bottom-board, in line with the position of the back end-bars of the frames, is an inch strip of sheet-iron (Fig. 100, *b*, *b*) sixteen inches in length. One-third of this strip, from the front edge back, is bent over so it lies not quite in contact with the second third, while the posterior third receives the tacks which hold it to the bottom-board. Now, when in use, this iron flange receives the hooks on the corners of the frames, so that the frames are held firmly, and can be moved only back and sidewise. In looking at the bees we can separate the combs at once, at any place. The chamber can be enlarged or diminished simply by adding or withdrawing frames. As the hooks are on all four corners of the frames, the frames can be either end back, or either side up. This arrangement, which permits the inversion of the frames, is greatly praised by those who have tried it. It was claimed by the Hetheringtons years ago that by turning these frames bottom up the comb would be fastened above and below, and the bees, in their haste to carry the honey from the bottom of the frames, would rush at once into the sections. Boards with iron hooks close the side of the brood cavity, while a cloth covers the frames.

The entrance (Fig. 100, *e*) is cut in the bottom-board, as already explained, except that the lateral edges are kept parallel. A strip of sheet-iron (Fig. 100, *d*) is tacked across this, on which rest the ends of the front end-bars of the frames which stand above, and underneath which pass the bees as they come to and go from the hive. A box, without bottom and with movable top, covers all, leaving a space from four to six inches above and on all sides between it and the frames. This gives chance to pack with chaff in winter, and for side and top storing in sections in summer.

The Bingham hive (Fig. 101) is not only remarkably simple, but is as remarkable for its shallow depth, the frames being only five inches high. These have no bottom-bar. The end-bars are one and a half inches wide, and the top-bar square. The nails that hold the end-bars pass into the end of the top-bar, which is usually placed diagonally, so that an edge, not a face, is below; though some are made with a face below (Fig. 101, *f*), to be used when comb is transferred. The frames are held together by two wires, one at each end. Each wire (Fig. 101, *a*) is a little longer than twice the width of the hive when the maximum number of frames are used. The ends of each wire are united and placed about nails (Fig. 101, *b,b*) in the ends of the boards (Fig. 101 *c,c*) which form the sides of the brood-chamber. A small stick (Fig. 101, *a*) spreads

FIG. 101.

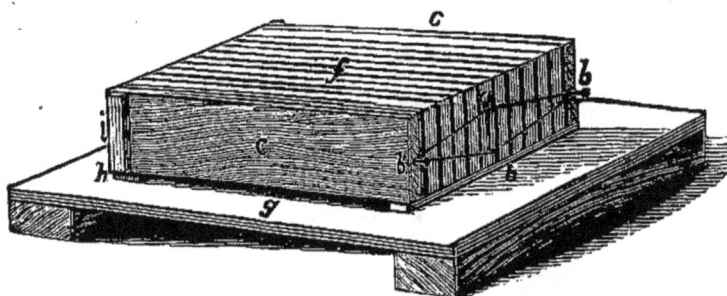

Frames and Bottom-Board of the Bingham Hive.—From A. I. Root Co.

these wires, and brings the frames close together. A box without bottom and with movable cover, is placed about the frames. This is large and high enough to permit of chaff packing in winter and spring. The bottom-board may be made like the one already described. Mr. Bingham does not bevel the bottom-board, but places lath under three sides of the brood-chamber, the lath being nailed to the bottom-board. He uses the Langstroth blocks to contract the entrance (Fig. 101, *g*).

The advantages of this hive are simplicity, great space above for surplus frames or boxes, capability of being placed one hive above another to any height desired, while the frames may be reversed, end for end, or bottom for top, or the whole brood-chamber turned upside down. Thus, by doubling, we may have a depth of ten inches for winter. It will be seen at once that this hive possesses all the advantages claimed for the new Heddon and Shuck hives, except the frames are not held so securely. Yet it is far more simple, which is greatly in its favor.

The objections which I have found in the use of such hives are the fact that so few use them, and danger of killing bees in rapid handling. They can be manipulated with rapidity if we care not how many bees we crush. It hurts me to kill a bee, and so I find the Langstroth style more quickly manipulated. Mr. Snow, too, who was the first to make the above style of hive, has discarded it in favor of the Langstroth. His objection to the above, is the fact that the various combs are not sure to be so built as to be interchangeable. Yet that such apiarists as those above named prefer these Huber hives, after long use of the other style, is certainly not without significance.

OBSERVATORY HIVE.

To study bees while they are at work, requires a hive so constructed that we can look in upon all the bees of the hive

FIG. 102.

Observatory Hive.—Original.

at pleasure. For this purpose I have used a small Langstroth hive (Fig. 102) containing one frame. Glass is used each side of the frame, and this is shaded by doors hung on hinges. We are able to look at the bees or make all dark inside at pleasure. To prevent the hive from becoming too crowded, we must every twenty-three or twenty-four days shake the bees from the frame, and replace the latter with another frame, which shall contain no brood. From such a hive, in my study window, I have received much pleasure and information.

APPARATUS FOR PROCURING COMB HONEY.

Although I feel sure that extracted honey will grow more and more in favor, yet it will never supersede the beautiful comb, which, from its exquisite flavor and attractive appearance, has always been, and always will be, admired and desired. So, no hive is complete without its arrangement of section frames and cases, all constructed with the view of securing this delectable comb honey in the form that will be most tempting to the eye and palate.

SURPLUS COMB HONEY IN SECTIONS.

Honey in several-pound boxes is no longer marketable, and is now almost wholly replaced by comb honey in sections. In fact, there is no apparatus for securing comb honey that promises so well as these sections. That they are just the thing to enable us to tickle the market is shown by their rapid growth in popular favor. Some years ago I predicted, at one of our State conventions, that they would soon replace boxes, and was laughed at. Nearly all who then laughed, now use these sections. They are cheap, and with their use we can get more honey, and in a form that will make it irresistible.

The wood should be white, the size small—two-pound sections are as large as the market will tolerate. One-pound sections are more salable, and in some markets even one-half pound sections are best of all. Of late, Mr. W. Harmer, of Manistee, Mich., is making and using successfully a two-ounce section. This is very neat and cheap. It is made of a shaving, and is glued. Such sections would be the thing to sell at fairs. The size of the sections has nothing to do with the amount of honey secured, and so the market and extra cost should guide the apiarist in this matter.

As early as 1877 I used veneer sections, which were essentially the same as the one-piece sections now so popular. After this I used nailed sections. At present only the very neatest sections can catch the market, and so we must buy our sections of those who can make them by machinery neater and cheaper than we possibly can by hand.

Dr. C. C. Miller, James Heddon, and many others, prefer sections made as are children's toy blocks—the sides fastened by a sort of mortise and tenon arrangement (Fig. 103). These are preferred, as they do not have the shoulder of the one-

FIG. 103.

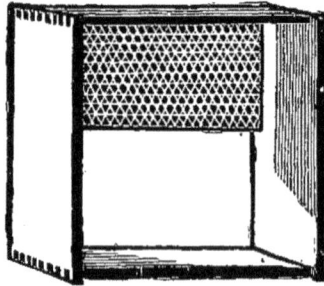

Dovetailed Section.—From A. I. Root Co.

piece section. They are objected to from the longer time required to put the pieces together, and their lack of rigidity when together, so that they are likely to get out of shape.

The Wheeler section—invented and patented by Mr. Geo. T. Wheeler, of Mexico, N. Y., in 1870—is remarkable for being

FIG. 104.

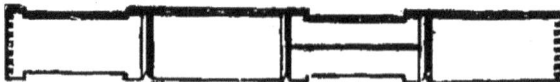

One-Pound Section.—From A. I. Root Co.

FIG. 105.

Prize Section.—From A. I. Root Co.

the first to be used with tin separators. Instead of making the bottoms narrower for a passage, Mr. Wheeler made an opening in the bottom.

Another style of section, termed the one-piece section (Fig. 104), is, as its name implies, made of a single piece of wood, with three cross cuts so that it can be easily bent into a square. The fourth angle unites by notches and projections, as before described. These one-piece sections are now, I think, the favorites among bee-keepers. I prefer these to the dovetailed. They are quickly and safely bent, if dampened slightly before bending, and are firm when in shape for use. Dr. Miller wets these quickly by pouring hot water at the to be corners while they are yet in the package. They must be even in the pack. If, as argued by Messrs. Dadant, Foster and Tinker, the sections open on all sides are superior, then we must perforce use these one-piece sections, rather than the dovetailed.

This last desirable feature is best secured in the plain section (Fig. 106), so-called in distinction from the bee-space

FIG. 106.

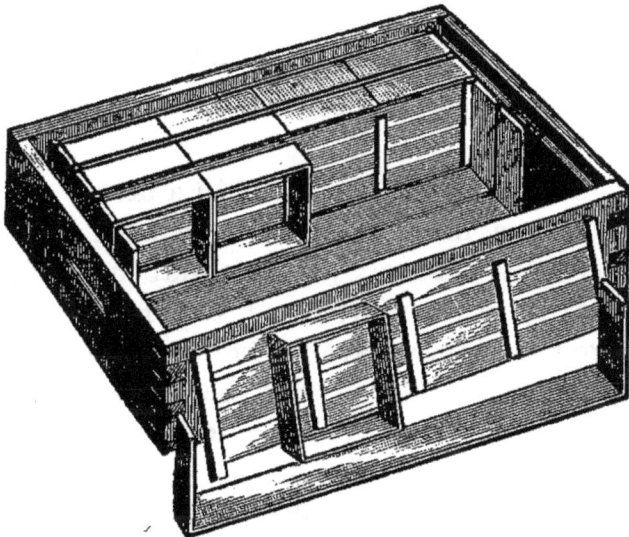

Plain Sections in Super, Showing Frame-Holders and Fence.
—From A. I. Root Co.

or bee-way sections just described. These are like the ends of the one-piece section all around (Fig. 106); that is, the bottom and top are not cut out to form bee-spaces. These plain sections give free communication, and thus are more readily filled, and as the honey projects to the very edge they look neater (Fig. 108). Of course, there is less wood than in the bee-space sections, and all edges are even. They are more easily and quickly scraped to remove propolis, etc. They are rapidly growing in favor. These are used with "fences," to be described, and in the ordinary supers (Fig. 106).

Heretofore there have been two prevailing sizes of sections in use in the United States—the prize section (Fig. 105), which is five and one-fourth by six and one-fourth inches, and the one-pound section (Fig. 104), which is four and one-fourth inches square. The latter is coming rapidly to the front, as

FIG. 107.

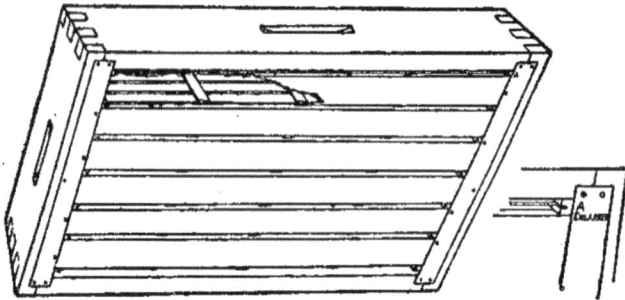

Plain Sections in Super, Showing Fence.—From A. I. Root Co.

honey in it sells more readily than if in a larger section. Even half-pound sections have taken the lead in the Boston and Chicago markets. It is barely possible that these small sections will rule generally in the markets of the future. They would often sell more readily, and are far better to ship, as the combs will seldom if ever break from the sections. If, in arranging our sections, we desire to have them oblong, we would better make them so that they will be longest up and down. Mr. D. A. Jones finds that if so made they are filled and capped much sooner (Fig. 108). Captain J. E. Hetherington prefers the oblong section, being one which is three and seven-eighths by five inches. Mr. Danzenbaker uses one which is four by five inches. He thinks honey in such sections (Fig. 108) sells for a higher price. In the depth of the section, which fixes the thickness of the comb, a change from the common style seems to be desirable. Heretofore they have been generally made two inches deep. With such sections we must use separators to secure perfect combs. Dr. Miller uses separators, and prefers a depth of one and five-sevenths, or two inches. By reducing the depth to from one and three-eighths to one and three-fourths inches, the expense of separators is found by some to be unnecessary. In feeding back to have sections completed, or where each section is removed as soon as capped, separators are indispensable. While I have never succeeded satisfactorily without separators—as the sections of comb would not be regular enough to ship well—yet I prefer the depth of my sections to be one and five-sevenths inches, or seven to the foot. These hold about three-fourths of a pound. I now believe that the best section for to-day is one four and one-quarter inches square and one and five-sevenths inches in

FIG. 108.

Oblong and Square Sections.—From A. I. Root Co.

depth. We secure nicer comb for the table, with the thinner combs, and more bees are able to work on a super or frame of sections, so that the foundation is more speedily drawn out. While a little more honey might be secured in two-pound sections, the market would, I think, make their use undesirable. Of course, any decided change in the form and size of our sections involves no small expense, as it requires that the supers or frames for holding the sections should also be changed. Often, however, by a little planning we can vary the form so as to reduce the size, without necessitating this expense.

HOW TO PLACE SECTIONS IN POSITION.

There are two methods, each of which is excellent, and has, as it well may, earnest advocates—one by use of frames, the other by supers.

SECTIONS IN FRAMES.

Frames for holding sections (Fig. 109) are made the same size as the frames in the brood-chamber. The depth of the

FIG. 109.

Gallup Section-Frame.—Original.

frame, however, is the same as the depth of the sections. The bottom-bar is three-eighths of an inch narrower than the remainder of the frame, so that when two frames are side by side, there is three-eighths of an inch space between the bottom-bars, though the top and side pieces are close together. In case sections are used that are open on all sides, then the ends of the section-frames must also be narrow. I should fear such an arrangement would be objectionable from the amount of propolis that would be used by the bees to make all secure.

The sections are of such a size (Fig. 110) that four, six or nine, etc., will just fill one of the large frames. Nailed to one side of each large frame are two tin, or thin wooden, strips (Fig. 110, *t, t*) in case separators are to be used, as long as the frame, and as wide into one inch as are the sections. These are tacked half an inch from the top and the bottom of the

FIG. 110.

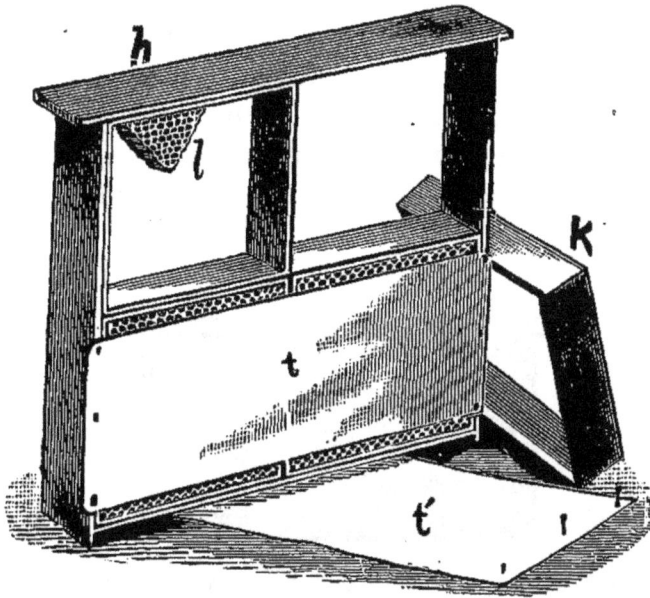

Gallup Frame with Sections.—Original.

large frames, and so are opposite the sections, thus permitting the bees to pass readily from one tier of sections to another, as do the narrower top and bottom bars of the sections, from those below to those above. Captain Hetherington tells me that Mr. Quinby used these many years ago. It is more trouble to make these frames if we have the tins set in so as just to come flush with the edge of the end-bars of the frames, but then the frames would hang close together, and would not be so stuck together with propolis. These may be hung in the second story of a two-story hive, and just enough to fill the same—my hives will take nine—or they can be put below, beside the brood-combs. Mr. Doolittle, in case he hangs these below, inserts a perforated division-board, so that the queen will not enter the sections and lay eggs.

The perforated-zinc division-board (Fig. 92) would serve
admirably for this purpose. A honey-board (Fig. 91) of the
same material keeps sections, either in supers or frames, that
are above the hive, neat, and also keeps the queen from entering them. The workers enter just as freely.

In long hives, the "New Idea"—which, though I would
not use, nor advise any one else to use, I have found quite satisfactory, after several years' trial, especially for extracted
honey—I have used these frames of sections, and with good
success. The Italians enter them at once, and fill them even
more quickly than other bees fill the sections in the upper
story. In fact, one great advantage of these sections in the
frames is the obvious and ample passageways, inviting the

FIG. 111.

Langstroth Frame with One-Pound Sections.—From A. I. Root Co.

bees to enter them. But in our desire to make ample and inviting openings, caution is required that we do not overdo the
matter, and invite the queen to injurious intrusion. So we
have Charybdis and Scylla, and must, by study, learn so to
steer between as to avoid both dangers.

Mr. Jones finds that by using the division-board made of
perforated-zinc (Fig 92), the queen is kept from the sections,
and they can be safely placed in one end of the body of the
hive.

Figure 111 shows a Langstroth frame full of one-pound
sections. As already stated, Mr. Heddon recommends the use
of one-story wide-frames, with separators, and so made as to
admit of inversion (Fig. 93). At first I used these deep frames
exclusively. The great objection to them is the daubing with
propolis, and difficulty of removing the sections from the wide

frames. This has led me to replace the wide frames by the more convenient and desirable section-case or crate.

CRATES OR RACKS.

These (Fig. 112) are to be used in lieu of large frames, to hold sections, and are very convenient, as we can use one tier

FIG. 112.

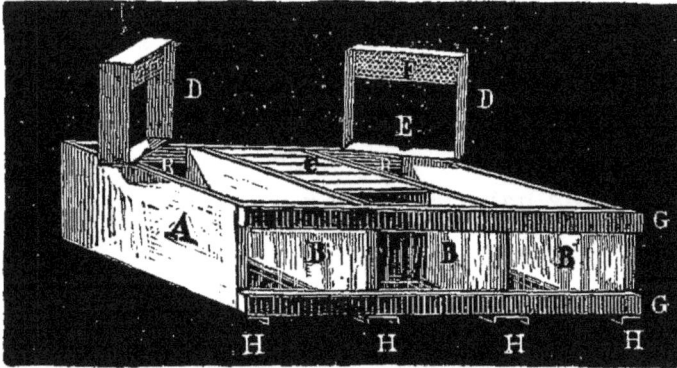

Crate for Sections.—Original.

at first, and as the harvest advances tier up, or "storify," as our British friends would say, until we may use three, or even

FIG. 113.

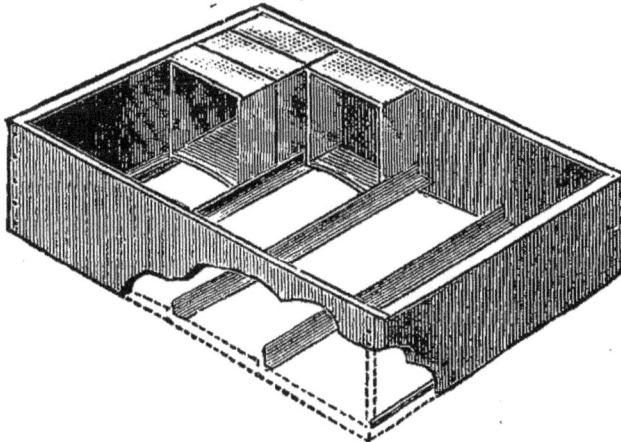

⊥ Super.—From A. I. Root Co.

four, tiers of sections on a single hive. I think this far the best arrangement for securing comb honey.

Southard and Ranney, of Kalamazoo, have long used a very neat rack, as seen in Fig. 112.

It will be seen that the Heddon case (Fig. 93), already described (page 225) as a part of the Heddon-Langstroth hive, is only a modification of the Southard crate. This crate does not permit the use of separators.

The case or super preferred and used by Dr. C. C. Miller (Fig. 113) is one with ⊥ shaped tin supports, on which rest the sections. This is just like the Heddon case, except the partitions are omitted. Projecting tin strips are tacked on the bottom of the sides as well as ends. These strips on the ends help hold the end rows of sections, while those on the sides hold the ⊥ shaped tins, which in turn support the sections. As the vertical part of the ⊥ supports the separator, it should not be more than one-half inch high. As most of us use—must use—separators, this is probably one of the best section-honey

FIG. 114.

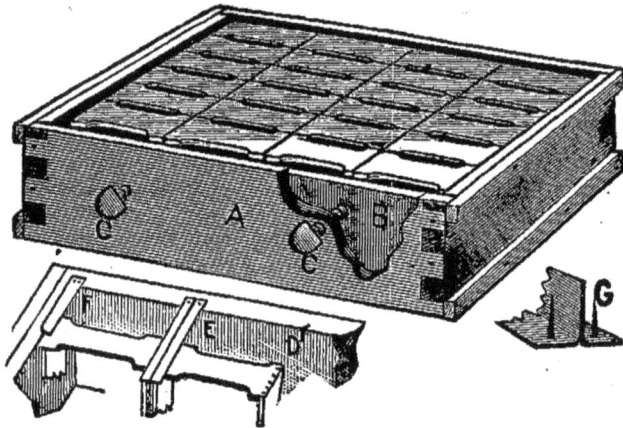

Hilton **T** *Super.—From A. I. Root Co.*

cases for us, and so one of the best arrangements for securing comb honey. Mr. Hilton (Fig. 114), of Michigan, does not like the movable ⊥ supports, and so he omits the projecting tin pieces, and tacks the ⊥ tins at the ends to the bottom of the side of the case.

Mr. Heddon has a case (Fig. 93) which permits inversion, through the use of wide frames and thumb-screws. Still another method to support sections (Fig. 115) has many advocates. The case is like the one used with the ⊥ tins, but has projecting tin supports tacked to the ends only. On these rest plain frames with no top-bar (Fig. 115), which in turn support the sections. If bee-space sections are used, then the bottom-bar of these frame-supports must have bee-ways or spaces cut

FIG. 115.

Dovetailed Super with Frames and Section-holders.—From A. I. Root Co.

D Wooden Separator.
E Sections with Foundation Starters.

in them. These are also used to hold the plain sections (Fig. 106), in which case, as the fence (Fig. 116) always used with these sections furnishes a bee-way, the frames, like the sections, are entirely plain. Of course, separators can be used with these supports, in case we use the bee-space sections.

FIG. 116.

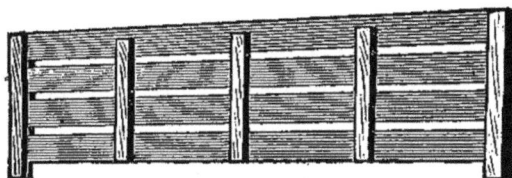

Fence for Plain Sections.—From A. I. Root Co.

FENCES.

The fence is simply a slatted separator made by nailing three boards (Fig. 116) three-sixteenths of an inch apart to end posts, which project three-eighths of an inch below the lowest board; cross-pieces of the same thickness as the corner posts, three-sixteenths of an inch, are like the corner posts nailed on each side connecting the boards of the fence. They do not reach below the lowest board. Thus, these fences permit very free communication (Figs. 106, 107). The whole distance at the bottom of the sections has a wide bee-way which also reaches part way up the ends. Of course, the cross-pieces are exactly opposite the ends of the sections which they separate. As these separators have spaces, they give ample connection between sections, and favor rapid comb-building and honey-storing. Fences are also placed outside the last row of sections. They secure added warmth by the double wall of bees, and so better filled sections. No wonder that these plain sections and fences are rapidly coming into use. Their use, of course, necessitates the use of cases with frames having no top-bars to hold the sections and fences (Fig. 106).

If we discard separators the old Heddon case is excellent; if we must use separators then the case with $\bot$ shaped tin supports is perhaps the best in the market. The plain sections are so admirable that they will be largely used; then the frame supports must be used. In any case a follower (Fig. 115, D) should be used to crowd the sections with separators close together. This may be pushed by use of a thumb-screw (Fig. 114), wedge, or steel spring.

Mr. Adam Grimm once wrote that boxes above the hive should not be closely covered. As already stated, Mr. Heddon puts no close cover over his sections. Mr. Hasty is pleased with simply a cloth, cheap muslin, above his sections, and a board cover to protect from rains. Such ventilation of the sections is scientific as well as practical.

All apiarists who desire to work for comb honey that will sell, will certainly use the sections, and adjust them by use of either frames or cases. Each method has its friends, though I think cases or supers are justly taking the lead.

SEPARATORS.

These may be of wood or tin. While the tin were first used, and do work well, the wood seem to be growing in favor, and seem likely wholly to replace the tin. The wood are poorer conductors of heat, and also give a foothold for the bees, both of which are desirable qualities.

FOOT-POWER SAW.

Every apiarist, who keeps only a few bees, will find, if he makes his own hives. a foot-power saw very valuable. I have used, with great satisfaction, the admirably combined foot-power saw of W. F. & John Barnes Co. It permits rapid work,

FIG. 117.

Horse-Power.—From A. I. Root Co.

insures uniformity, and enables the apiarist to give a finish to his work that would rival that of the cabinet-maker.

Those who procure such a machine should learn to file and set the saw, and should *never* run the machine when not in perfect order.

When just beginning the business it will generally be wise to secure a fully equipped hive of some bee-keeper or dealer in supplies. If there is a hive factory near at hand, it may pay to buy all hives ready made ; otherwise high freights may make this unprofitable, If a person wishes to manufacture hives by the score, either for himself or others, even the foot-power saw will soon become too slow and wearying. In this case some use wind-power, which is too uncertain to give full satisfaction; others use horse-power, and still others procure a small steam-engine.

Mr. M. H. Hunt, a very thoughtful apiarist, uses a very convenient horse-power (Fig. 117). The large wheel is fifteen feet in diameter, the horse is inside the rim, and the band consists

FIG. 118.

Saw-Table.—From A. I. Root Co.

of a chain, that it may not slip. To get the horse in position, the wheel is simply lowered.

I have used a tread-power which pleases me much. It is safe, can be used under shelter, and if one has colts or young horses it serves well to quiet them. As gasoline engines are now so cheap, and convenient; and as crude oil for steam engines is so cheap, such engines will generally be preferred when one's business is at all extensive. In case we use other than foot or hand power, our saw-table must be firm and heavy. The one illustrated here (Fig. 118) is recommended by Mr. A. I. Root.

CHAPTER VI.

POSITION AND ARRANGEMENT OF APIARY.

As it is desirable to have our apiary grounds so fixed as to give the best results, and as this costs some money and more labor, it should be done once for all. As plan and execution in this direction must needs precede even the purchase of bees, this subject deserves an early consideration. Hence, we will proceed to consider position, arrangement of grounds, and preparation for each individual colony.

POSITION.

Of course, it is of the first importance that the apiary be near at hand. In city or village this is imperative. In the country, or at suburban homes, we have more choice, but close proximity to the house is of much importance. In a city it may be necessary to follow Mr. Muth's example, and locate on the house-tops, where, despite the inconvenience, we may achieve success. The lay of the ground is not important, though, if a hill, it should not be very steep. It may slope in any direction, but better any way than toward the north. Of course, each hive should stand perfectly level.

ARRANGEMENT OF GROUNDS.

Unless sandy, these should be well drained. If a grove offers inviting shade, accept it, but trim high to avoid damp. Such a grove could soon be formed of basswood and tulip trees, which, as we shall see, are very desirable, as their bloom offers plenteous and most delicious honey. Even Virgil urged shade of palm and olive, also that we screen the bees from winds. Wind-screens are very desirable, especially on the windward side. Such a screen may be formed of a tall board fence, which, if it surrounds the grounds, will also serve to protect against thieves. Yet these are gloomy and forbidding, and will be eschewed by the apiarist who has an eye to esthetics. Evergreen screens, either of Norway spruce, Austrian or other pine, or arbor vitæ, each or all are not only very effective, but are quickly grown, inexpensive, and add greatly to the beauty of the grounds. In California eucalyptus is very desirable shade. The species grow vigorously, stand drouth, and if wisely selected afford much honey. Such a fence or hedge is also very desirable if the bees are near a street or highway. It not only shuts the bees away, as it were, but it so directs their flight upward that they will not trouble passers-by. If the apiary is large, a small, neat, inexpensive house in the center of the apiary grounds is indispensable. This will serve in winter as a shop for making hives, frames, etc., and as a store-house for honey, while in summer it will be used for extracting, transferring, storing, bottling, etc. In building this, it will be well to construct a frost-proof, *thoroughly drained*, dark and well-ventilated cellar. (See Chapters XVIII and XIX.)

PREPARATION FOR EACH COLONY.

Virgil was right in recommending shade for each colony. Bees are forced to cluster outside the hive, if the bees are subjected to the full force of the sun's rays. By the intense heat the temperature inside becomes like that of an oven, and the wonder is that they do not desert entirely. I have known hives, thus unprotected, to be covered with bees, idling outside, when, by simply shading the hives, all would go merrily to work. The combs, too, and foundation especially, are liable, in unshaded hives, to melt and fall down, which is very damaging to the bees, and very vexatious to the apiarist. The remedy for all this is always to have the hives so situated that they will be entirely shaded all through the heat of the day. This might be done, as in the olden time, by constructing a shed or house, but these are expensive and very inconvenient, and, therefore, to be discarded.

If the aiarist has a convenient grove this may be trimmed high, so as not to be damp, and will fulfill every requirement. So arrange the hives that while they are shaded through all the heat of the day, they will receive the sun's rays early and late, and thus the bees will work more hours. I always face my hives to the east. Such a grove is also very agreeable to the apiarist who often must work all the day in the hottest

FIG. 119.

Nucleus and Simplicity Hive Shaded by Grape-vine.—From A. I. Root Co.

sunshine. If no grove is at command, the hives may be placed on the north of a Concord grape-vine (Fig. 119), or other vigorous variety, as the apiarist may prefer. This should be trained to a trellis, which may be made by setting two posts, either of cedar or oak. Let these extend four or five feet above the ground, and be three or four feet apart. Two or three supporting arms of narrow boards can be nailed at right angles to a single post on which to train the vines, or we may connect them at intervals of eighteen inches with three galvanized wires, the last one being at the top of the posts. Thus we can have shade and grapes, and can see for ourselves that bees do not injure grapes. These should be at least six feet apart. A. I. Root's idea of having the vine of each succeeding row divide the spaces of the previous row, in quincunx order (Fig. 120), is very good ; though I should prefer the rows in this case to be four instead of three feet apart. I have tried grape-vines and evergreens to shade hives, and do not like them. They are too much in the way. Unless I can have a grove trimmed high up I much prefer a simple shade-board as already suggested. This is simply a wide board nailed to the edge of two cross-boards, which are about four inches wide. I make these eighteen inches wide by two feet long. I have some even larger. If one cross-board is a little narrower it gives a slant that insures a rapid removal of the water in a rain. I have never known these shade-boards to blow off. Should they do so a second board parallel to the shade-board could be nailed to the cross-boards. A brick placed on this would make all secure. This shade-board is inexpensive, always out of the way, and ready for service.

Many apiarists economize by using fruit-trees for shade, which, from their spreading tops, serve well, though often from their low branches they are not pleasant to work under. Mr. Doolittle thinks if hives are painted white shade is unnecessary. Mr. A. I. Root's idea of having sawdust under and about the hives has much to recommend it. The objection to sawdust is the danger from fire. I have used sawdust, cement, asphalt, etc. I think on the whole a fine grass lawn kept closely and smoothly mown is as convenient as any plan, and it certainly has taste and beauty to recommend it. If closely mown, one will rarely lose a queen. While ashes or sawdust make a queen walking upon them more conspicuous, I much prefer the beautiful grass plat.

FIG. 120.　*Grape-Vine Apiary.—From A. I. Root Co.*

CHAPTER VII.

TO TRANSFER BEES.

As the prospective bee-keeper may have purchased his bees in box-hives, barrels, or hollow logs, and so, of course, will desire to transfer them immediately into movable-frame hives, or, as already suggested, may wish to transfer from one movable-frame to another, I will now proceed to describe the process.

Among the many valuable methods which Mr. Heddon has given to the bee-keeping public, not the least valuable is that of transferring. This method should be used only at or just before the swarming season—the best time to transfer. After blowing a little smoke into the hive, sufficient to alarm the bees, we set it a little aside, and put in its place the new hive full of wired foundation. We now turn the old hive, whatever it may be, bottom side up, and place a box over it. If the bees are sufficiently smoked, it will make no difference even if the box is not close-fitting to the old hive. Yet the beginner will feel safer to have it so; and in this case no stinging can take place. We then with a stick or hammer rap on the hive for from ten to twenty minutes. The bees will fill with honey and go with the queen into the upper box and cluster. If towards the last we carefully set the box off once or twice, and vigorously shake the hive, and then replace the box, we will hasten the emigration of the bees, and make it more complete. I got this last suggestion from Mr. Baldridge. A few young bees will still remain in the old hive, but these will do no harm.

We next take the box, which contains the queen and nearly all the bees, and shake the bees all out in front of the hive already placed on the old stand. The bees will at once take possession, draw out, or better, build out, the foundation in a surprisingly short time, and will give us a set of combs which will surpass in beauty those procured in any other way. Should the bees be unable to gather any honey for some days, which at this season is not likely to occur, of course we must feed them.

We set the old hive aside for twenty-one days, when the young bees will all come from the cells. Should the weather be cold, it might be well to put this in a warm room, so the brood will not chill. At the time of swarming this will rarely be necessary. We now drum out these bees as before, kill the queen, which has been reared, and unite the bees with the others, or form a separate colony as before, as the number of bees determines. We can now split out the corners of the old hive, split the gum, or separate the staves of the barrel, so as not to break the comb. This should be carefully cut loose, and the honey extracted by use of the wire comb-holder (Fig. 150), and the comb melted into wax for foundation. The only loss in this method is the time which the bees require to build out the foundation, and this is far more than made up in the superior combs which are secured. I think the time expended in melting up the combs, etc., is more than made up by the time saved in transferring.

THE OLD METHOD.

If one has no foundation, or desires to give the bees the comb and honey at once, even at the cost of less shapely combs, he then should drum the bees out as before, on a warm day when they are busy at work, and put the box containing the bees on the old stand, leaving the edge raised so that the bees which are out may enter, and so all the bees can get air. This method is difficult, except in early spring, and is best done about noon, when the bees are busy on the fruit-bloom. It is not safe to transfer on a hot day, when the bees are idle, as the risks from robbing are too great. If other bees do not trouble, as they usually will not if busily gathering, we can proceed in the open air. If they do, we must go into some room. I have frequently transferred the comb in my kitchen, and often in a barn.

Now knock the old hive apart, as already described, cut the combs from the sides, and get the combs out of the old hive with just as little breakage as possible. Mr. Baldridge, if transferring in spring, saws the combs and cross-sticks loose from the sides, turns the hive into the natural position, then strikes against the top of the hive with a hammer till the fastenings are broken loose, when he lifts the hive, and the combs are all free and in convenient shape for rapid work.

We now need a barrel, set on end, on which we place a board fifteen to twenty inches square, covered with several thicknesses of cloth. Some apiarists think the cloth useless, but it serves, I think, to prevent injury to comb, brood or honey. We now place a comb on this cloth, and set a frame on the comb, and cut out a piece of the comb the size of the inside of the frame, taking pains to save all the worker-brood. Now crowd the frame over the comb, so that the latter will be in the same position that it was when in the old hive; that is, so the honey will be above—the position is not very important —then fasten the comb in the frame, by winding about all one or two small wires, or pieces of wrapping-twine. To raise the frame and comb before fastening, raise the board beneath till

FIG. 121.

FIG. 122.

Transferring-Clasp.—
From American Bee Journal.

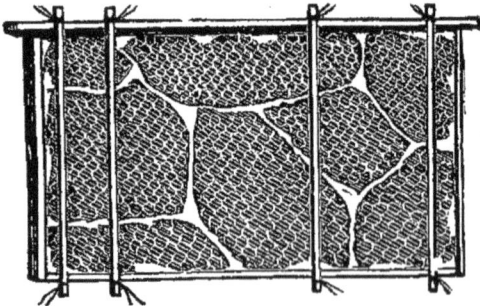

Transferred Comb.—From American Bee Journal.

the frame is vertical. Set this frame in the new hive, and proceed with the others in the same way till we have all the worker-comb—that with small cells—fastened in. To secure the pieces, which we shall find abundant at the end, take thin pieces of wood, one-half inch wide, and a trifle longer than the frame is deep, place these in pairs either side the comb, extending up and down, and enough to hold the pieces secure till the bees shall fasten them (Fig. 121), and secure the strips by winding with small wire, just below the frame (*Fig. 122*), or by use of small rubber bands, or else tack them to the frame with small tacks. Some bee-keepers use U-shaped pieces of wire or tin to hold the comb in the frame.

Captain Hetherington has invented and practices a very neat method of fastening comb into frames. In constructing his frames, he bores small holes through the top, side and bottom bars of his frames, about two inches apart; these holes are just large enough to permit the passage of the long spines of the hawthorn. Now, in transferring comb, he has but to stick these thorns through into the comb to hold it securely. He can also use all the pieces, and still make a neat and secure frame of comb. He finds this arrangement convenient, too, in strengthening insecure combs. In answer to my inquiry, this gentleman said it paid well to bore such holes in all his frames, which are eleven by sixteen inches, inside measure. I discarded such frames because of the liability of the comb to fall out.

Having fastened all the nice worker-comb into the frames —of course, all other comb will be melted into wax—we place all the frames containing brood together in the center of our new hive, especially if the colony is weak, or the weather cool, and confine the space by use of the division-board, adding the other frames as the bees may need them. We now place the new hive on the stand, opening the entrance wide, so that the bees can enter anywhere along the alighting-board. We then shake all the bees from the box, and any young bees that may have clustered on any part of the old hive, or on the floor or ground, where we transferred the comb, immediately in front of the hive. They will enter at once and soon be at work, all the busier for having passed "from the old house into the new." In two or three days remove the wires, or strings or sticks, when we shall find the combs all fastened and smoothed off, and the bees as busily engaged as though their present home had always been the seat of their labors.

In practicing this method, many proceed at once to transfer without drumming out the bees. In this case the bees should be well smoked, should be driven, by the use of the smoker, away from the side of the old hive where the combs are being cut loose, and may be brushed direct from the old combs into the new hive. This method will only be preferred by the experienced. The beginner will find it more easy and pleasant first to drum out all the bees before he commences to cut out the combs.

Of course, in transferring from one frame to another, the matter is much simplified. In this case, after thoroughly smoking the bees, we have but to lift the frames and shake or brush the bees into the new hive. For a brush, a chicken or turkey wing, a large wing or tail feather from a turkey, goose or peacock, or a twig of pine or bunch of asparagus twigs serves admirably. Cheap and excellent brushes (Fig. 154) are now for sale by all supply-dealers. Now cut out the comb in the best form to accommodate the new frames, and fasten as already suggested. After the combs are all transferred, shake all remaining bees in front of the new hive, which has already been placed on the stand previously occupied by the old hive.

Sometimes bees from trees in the forest are transferred to hives and the apiary.

HUNTING BEE-TREES.

Except for recreation, this is seldom profitable. It is slow and uncertain work. The tree, when found, is not our own, and though the owner may consent to our cutting it, he may dislike to do so. The bees, when found, are difficult to get alive ; it is even more difficult to get the honey in good condition, and, when secured, the honey and bees are often almost worthless.

The principle upon which bees are " lined " is this : That after filling with honey, a bee always takes a direct course— " a bee-line "—to its hive. To hunt the bee-trees we need a bottle of sweetened water, a little honey-comb, unless the bees are gathering freely from forest flowers, and a small bottom-less box with a sliding glass cover, and a small shelf attached to the middle of one side on the inside of the box. A shallow tray, or piece of honey-comb, is to be fastened to this shelf. If the bees are not found on flowers, we can attract them by burning a piece of honey-comb. If on a flower, set the box over them after turning a little of the sweetened water in the comb or tray on the shelf. It is easy to get them to sipping this sweet. Then slide the glass, and, when they fly, watch closely and see the direction they take. By following this line we come to the bee-tree, or more likely to some neighbor's apiary. By getting two lines, if the bees are from the same tree, the tree will be where the lines meet. We should be careful not to be led to neighboring apiaries, and should look very closely when the bees fly, to be sure of the line. Experience makes a person quite skillful. It need hardly be said that in warm days in winter, when there is snow on the ground, we may often find bee-trees by noting dead bees on the snow, as also the spotting of the snow, as the bees void their feces. When a tree is found, we must use all possible ingenuity to get the combs whole if we wish to transfer the bees. We may cut in and remove the comb; may cut out the section of tree containing the bees and lower this by use of a rope; or we may fell the tree. In this last case we may make the destruction less complete if we fall the tree on other smaller trees to lessen the jar.

CHAPTER VIII.

FEEDING AND FEEDERS.

As already stated, it is only when the worker-bees are storing that the queen deposits to the full extent of her capability, and that brood-rearing is at its height. In fact, when storing ceases, general indolence characterizes the hive. This is peculiarly true of the German and Italian races of bees. Hence, if we would achieve the best success, we must keep the workers active, even before gathering commences, as also in the interims of honey-secretion by the flowers ; and to do this we must feed sparingly before the advent of bloom in the spring, and whenever the workers are forced to idleness during any part of the season, by the absence of honey-producing flowers. For a number of years I have tried experiments in this direction by feeding a portion of my colonies early in the season, and in the intervals of honey-gathering, and always with marked results in favor of the practice. Of course it is not well to feed unless we expect a honey harvest the same season. Thus, I would not feed after clover or basswood bloom unless I expected a fall harvest. The fact that honey seasons are uncertain, makes the policy of feeding merely to stimulate questionable.

Mr. D. A. Jones has truly said that if feeeding in the autumn be deferred too long, till the queen ceases laying, it often takes much time to get her to resume, and not infrequently we fail entirely.

Every apiarist, whether novice or veteran, will often receive ample reward by practicing stimulative *feeding early* in the season ; then his hive at the dawn of the white clover era will be redundant with bees, well filled with brood, and in just the trim to receive a bountiful harvest of this most delicious nectar.

Feeding is often necessary to secure sufficient stores for

winter—for no apiarist, worthy of the name, will suffer his faithful, willing subjects to starve, when so little care and expense will prevent it. This is peculiarly true in Southern California, where severe drouths often prevent any harvest, and these may occur on two successive years.

If we only wish to stimulate, the amount fed need not be great. A half pound a day, or even less, will be all that is necessary to encourage the bees to active preparation for the good time coming. For information in regard to supplying stores for winter, see Chapter XVIII.

Bees, when very active, especially in very warm weather, like most higher animals, need water. This very likely is to permit evaporation in respiration, and the necessary cooling of the body. At such times bees repair to pool, stream or watering-trough. As with other animals, the addition of salt makes the water more appetizing, and doubtless more valuable. Unless water is near, it always ought to be furnished to bees. Any vessel containing chips or small pieces of boards to secure against drowning will serve for giving water. In case bees trouble about watering-troughs, a little carbolic acid or kerosene-oil on the edge of the trough will often send them away.

WHAT TO FEED.

For this purpose I would feed granulated sugar, reduced to the consistency of honey, or else extracted honey kept over from the previous year. If we use two-thirds syrup and one-third good honey we save all danger of crystallization or granulation. We add the honey when the syrup is hot, and stir. The price of the honey will decide which is the more profitable. The careful experiments of R. L. Taylor show that nearly three times as much honey as syrup will be consumed. This argues strongly for the syrup. Dark, inferior honey often serves well for stimulative feeding, and as it is not salable, may well be used in this way. To make the syrup, I use one quart of water to two of sugar, and heat till the sugar is dissolved. Mr. R. L. Taylor first boils the water, hen stirs in the sugar till all boils, when he says it will not granulate even with no acid added. This also removes all danger of burning the syrup, which must never be done. By stirring till all the sugar is dissolved we may make the syrup without any heat. We use equal parts of sugar and water, and may easily stir by using the honey extractor. We put in the water and add the sugar as we turn the machine. A little tartaric acid—an even teaspoonful to fifteen pounds of syrup—or even a little extracted honey, will also prevent crystallization. If fed warm in early spring it is all the better.

Many advise feedin the poorer grades of sugar in spring. My own experience makes me question the policy of ever using such feed for bees. The feeding of glucose or grape sugar is even worse policy. It is bad food for the bees, and its use is dangerous to the bee-keeper's reputation, and injurious to our brother bee-keepers. Glucose is so coupled with fraud and adulteration that he who would "avoid the appearance of evil" must let it severely alone.

In all feeding, unless extracted honey is what we are using, we can not exercise too great care that such feed is not carried to the surplus boxes. Only let our customers once taste sugar in their comb honey, and not only is our own reputation gone, but the whole fraternity is injured. In case we wish to have our combs in the sections filled or capped, we must feed extracted honey, which may often be done with great advantage. I have often fed extracted honey back to the bees, after the honey-flow ceased, when it would be quickly stored in the sections. More frequently, however, I have utterly failed of success.

HOW TO FEED.

The requisites of a good feeder are : Cheapness, a form to admit quick feeding, to permit no loss of heat, and so arranged that we can feed at all seasons without in any way disturbing the bees. The feeder (Fig. 123), which I have used with good satisfaction, is a modified division-board, the top-bar of which (Fig. 123, b) is two inches wide. From the upper central portion, beneath the top-bar, a rectangular piece the size of an oyster-can is replaced with an oyster-can (Fig. 123, g), after the top of the latter has been removed. A vertical piece of wood (Fig. 123, d) is fitted into the can so as to separate a space about one inch square, on one side, from the balance of the chamber. This piece does not reach quite to the bottom of the can, there being a one-eighth inch space beneath. In the top-bar there is an opening (Fig. 123, e) just above the smaller space below. In the larger space is a wooden float (Fig. 123, f) full of holes. On one side opposite the larger chamber of the can, a half-inch piece of the top (Fig. 123, e) is cut off, so that the bees can pass between the can and top-bar on to the float, where they can sip the feed. The feed is turned into the hole in the top-bar (Fig. 123, e), and without touching a bee, passes down under the vertical strip (Fig. 123, d) and raises the float (Fig. 123, f). The can may be tacked to

FIG. 123.

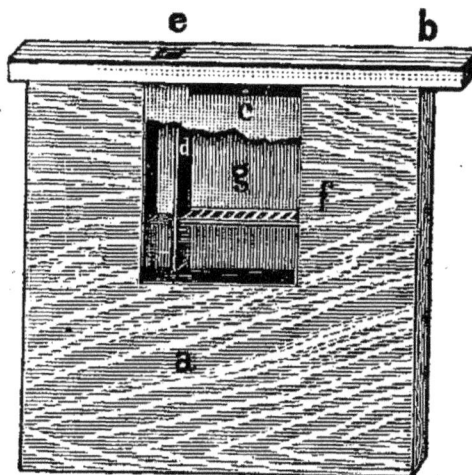

Division-Board Feeder.—Original.

Lower part of the face of the can removed, to show float, etc.
—Original.

the board at the ends near the top. Two or three tacks through the can into the vertical piece (Fig. 123, d) will hold the latter firmly in place; or the top-bar may press on the vertical piece so that it can not move. Crowding a narrow piece of woolen cloth between the can and board, and nailing a similar strip around the beveled edge of the division-board, makes all snug. The objection to this feeder is that it can not be placed just above the cluster of bees. On very cold days in spring the bees can not reach their food in any other position. The feeder is placed at the end of the brood-chamber, and the

top-bar covered by the quilt. To feed, we have only to fold
the quilt over, when with a tea-pot we pour the feed into the
hole in the top-bar. If a honey-board is used, there must be a
hole in this just above the hole in the division-board feeder.
In either case no bees can escape, the heat is confined, and our
division-board feeder is but little more expensive than a
division-board alone.

Some apiarists prefer a quart can set on a block (Fig.
124), or it may be used with a finely perforated cover. This is

FIG. 124.

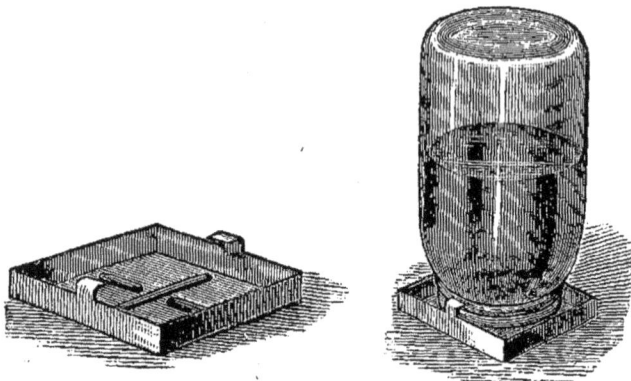

Fruit-Jar Feeder.—From A. I. Root Co.

filled with liquid, the cover put on, and the whole quickly
inverted and set above a hole in the cover just above the bees.
Owing to the pressure of the air, the liquid will not descend so
rapidly that the bees can not sip it up. The objections to this
feeder are, that it is awkward, raises the cushions so as to permit the escape of heat, and must be removed to receive the
feed. Mr. A. I. Root recommends the little butter-trays sold
at the groceries, for feeding. These cost only one-third of a
cent. "Need no float, and work admirably." I have tried
these, and think they have only their cheapness to recommend them. They raise the cover, can not be filled without
disturbing the bees, leak, and daub the bees. Even paper
sacks of good quality, with small holes in them, have been
used. They are laid on the frames, and cost very little. As
feeders last for a lifetime, I prefer to pay more and get good
ones.

The Simplicity feeder (Fig. 125), invented by A. I. Root, is shown on its side in the illustration. This is used at the entrance, and so is not good for cold weather. As the feed is

FIG. 125.

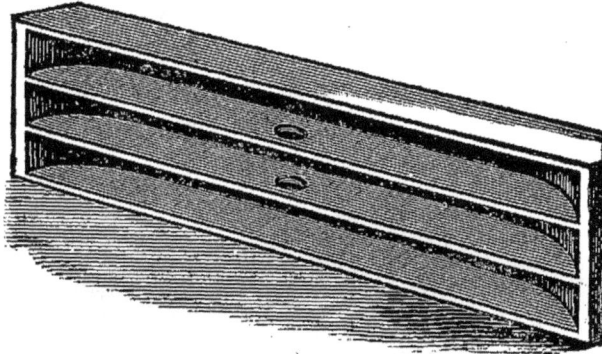

Simplicity Bee-Feeder.—From A. I. Root Co.

exposed it can only be used at night, when the bees are not flying. It is never, I think, desirable to feed outside the hive.

The Shuck feeder (Fig. 126) is a modification of the Simplicity, and a great improvement. This is used at the entrance of the hive, or by nailing two together, so that the sides marked *D* will face each other, we can use it above the bees. We then would place the opening *D* above a hole in the cloth

FIG. 126.

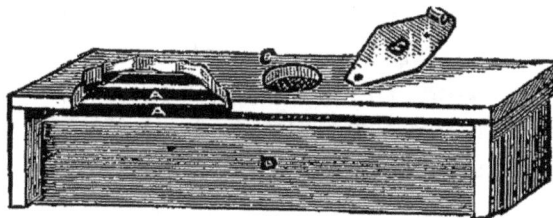

Shuck's Boss Bee-Feeder.—From American Bee Journal.

cover, or honey-board, turn the feed in at *C*, and the bees would come up at *D*, pass under the cover, and down into the saw-cuts (Fig. 126, *A, A*), when they would sip the feed, and then crawl up on the partitions. This feeder works admirably. but it is patented, costs too much, and is improved in the

SMITH FEEDER.

This feeder (Fig. 127) is larger than the Shuck—I make them eight by twelve inches—and is covered all over with wire gauze (Fig. 127, *a*), which is raised by the wooden rim so that the bees can pass readily over the partitions (Fig. 127). The central saw-cuts (Fig. 127) do not reach the end of the feeder, so there is a platform left (Fig. 127, *b*) through which a hole (Fig. 127, *c*) is made. This rests above a hole in the cloth

FIG. 127.

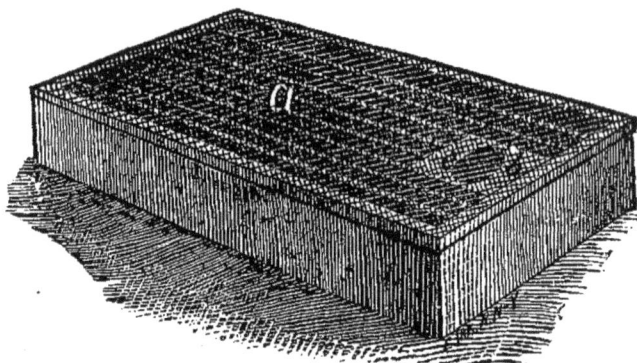

Smith Bee-Feeder.—Original.

below, and is the door through which the bees reach the feed. When in position just above the bees it may be covered by a shingle or piece of pasteboard, to prevent daubing the cloth or cushion, and all by the chaff cushion. To feed, we have only to raise the cushion and the pasteboard, and turn the food through the gauze. No bees can get out, there is no disturbance, no danger from the robbers, and we can feed at any time, and can feed very rapidly if desired. I like this feeder the best of any I have ever tried. I make them out of two-inch plank.

The Heddon feeder (Fig. 128) is much the same in principle as the Smith, and has all the advantages. It is the size of a section-crate, and so holds many pounds. The figure makes it plain. The spaces in this are not saw-cuts, but are formed by thin boards nailed in a box vertically, and a space on one or both sides (Fig. 128) does not connect with the food reservoir, but serves as a passage-way for the bees from hive to feeder. In the center is a passage (Fig. 128, *c*) which connects with the food reservoir, but is not accessible to the bees. In this the food is poured when feeding, which makes it unnecessary to have the wire gauze above, or to smear the top when feeding, as in case of the Smith feeder, yet this feeder does not retain the heat in spring. The center of the cover slides back, so the whole cover need not be removed when feeding is done. The vertical partitions, except the one next to the space (Fig. 128) where the food is added, do not run quite to the board which covers the feeder, and so the bees can pass into

FIG. 128.

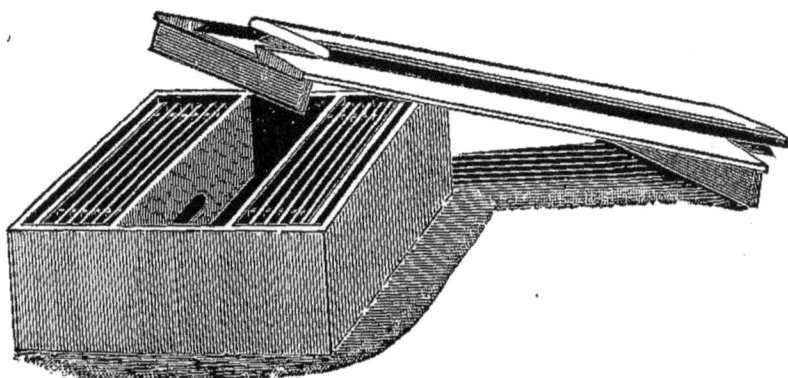

Heddon Bee-Feeder.—From James Heddon.

all the spaces except where we pour in the food. No partition except the one next to the space where the bees pass to and from the hive runs quite to the bottom, so the food will pass readily from one space to the other, and will always be equally high in all.

Mr. D. A. Jones and many others having tight bottom-boards to their hives use no feeder, but turn the feed right into the hive. Dr. C. C. Miller, like L. C. Root, prefers to feed by filling frames of empty comb with the syrup or honey. The empty combs are laid flat, in a deep box or tub, under a colander or finely perforated pan. The syrup, as it falls, fills the cells of comb. After the comb is filled on both sides, we have only to hang it in the hive. I have found that by use of a fine spray-nozzle and force-pump we can fill frames very fast.

The best time to feed is just at nightfall. In this case the

feed will be carried away before the next day, and the danger
to weak colonies from robbing is avoided.

In feeding during the cold days of April, all should be
close above the bees to economize heat. In all feeding, care is
requisite that we may not spill the feed about the apiary, as
this may, and very generally will, induce robbing.

If, through neglect, the bees are found to be destitute of
stores in mid-winter, it is not best to feed liquid food, but solid
food, like the Viallon candy or the Good mixture of honey and
sugar, which will be described under the head of shipping
queens. Cakes of either of these should be placed on the
frames above the cluster of bees. Mr. Root has had excellent
success in feeding cakes of hard candy made as follows:
Granulated sugar is put in a pan and a very little water
added. This is heated by placing on a stove, but *never* in
direct contact with the fire. In the latter case it may be
burned, as shown by the taste, odor, or from the fact that it
kills the bees. If the pan is placed on the stove, the contents
will never be burned. It must be boiled until if dropped on a
saucer in cold water, or if the finger is wet in cold water, then
dipped in the hot sugar, and again in water, the hard sugar is
brittle. It must be boiled until the hardened product is brittle,
or else it will be too soft and will drip. It can now be stirred
until it begins to thicken and then molded in dishes, or in the
regular comb frames. In this last case we lay the frame close
on a board covered with thin paper, and turn the thickening
sugar into it. By adding one-fourth rye-meal we have a good
substitute for pollen, which may be used in case of a scarcity
of the latter. Of course, frames of this hard candy may be
hung right in the hive. In a cellar or on warm days outside
frames of honey may be given to the bees.

CHAPTER IX.

QUEEN-REARING.

Suppose the queen is laying two thousand eggs a day, and that the full number of bees is forty thousand, or even more—though as the bees are liable to so many accidents, and as the queen does not always lay to her full capacity, it is quite probable that this is about an average number—it will be seen that each day that a colony is without a queen there is a loss equal to about one-twentieth of the working force of the colony, and this a compound loss, as the aggregate loss of any day is its special loss augmented by the several losses of the previous days. Now, as queens are liable to die or to become impotent, and as the work of increasing colonies demands the absence of queens, unless the apiarist has extra ones at his command, it is imperative, would we secure the best results, ever to have at hand extra queens. Queen-rearing for the market is often very remunerative, and often may well engage the apiarist's exclusive attention. So the young apiarist must learn early

HOW TO REAR QUEENS.

As queens may be needed early in the spring, preparations looking to the rearing of queens must commence early. As soon as the bees are able to fly regularly, we must see that they have a supply of bee-bread. If there is not a supply from the past season, and the locality of the bee-keeper does not furnish an early supply, then place unbolted flour (that of rye or oats is best) in shallow troughs near the hives. It may be well to give the whole apiary the benefit of such feeding before the flowers yield pollen. If the bees are not attracted to this we need not add honey, etc., to induce them to take it. This is a sure sign that it is not needed. I found that in Central Michigan bees can usually gather pollen by the first week of April, which, I think, is as early as they should be allowed to fly, and, in fact, as early as they will fly with sufficient regularity to make it pay to feed the meal. I much question, after some years of experiment, if it is ever necessary at this place to give the bees a substitute for pollen. In case of long storms, the bee-bread may be exhausted. I have never known such a case, when the hard candy frames with rye meal described at the close of the last chapter may be hung in the hive.

The best colony in the apiary—or if there are several colonies of equal merit, one of these—should be stimulated to the utmost, by daily feeding with warm syrup, and by increase of brood taken from other colonies. As this colony becomes strong, a comb containing drone-cells should be placed in the center of the brood-nest. Very soon drone-eggs will be laid. I have often had drones flying early in May. As soon as the drones commence to appear, remove the queen and all eggs and uncapped brood from some good, strong colony, and replace it with eggs or brood just hatched from the colony containing the queen from which it is desired to breed. By having placed one or two bright, new, empty combs in the midst of the brood-nest of this colony four days beforehand, we shall have in these combs just such eggs and newly hatching brood as we desire, with no brood that is too old.

If we have more than one colony whose excellence warrants their use to breed from, then these eggs should be taken from some other than the one which has produced our drones. This will prevent the close in-breeding which would necessarily occur if both queens and drones were reared in the same colony; and which, though regarded as deleterious in the breeding of all animals, should be practiced in case one single queen is of decided superiority to all others of the apiary. The queen and the brood that have been removed may be used in making a new colony, in a manner soon to be described under "Dividing or Increasing the Number of Colonies." This queenless colony will immediately commence forming queen-cells (Fig. 93). Sometimes these are formed to the number of fifteen or twenty, and in case of the Syrian and Cyprian races fifty or sixty, and they are started in a full, vigorous colony; in fact, under the most favorable conditions. Cutting off edges of the comb, or cutting holes in the same where there are eggs or larvæ just hatched, will almost always insure the starting of queen-cells in such places. It will be noticed that our queens are started from eggs, or from larvæ but just hatched, as we have given the bees no other, and so they are fed the royal pabulum from the first. Thus we have met every possible requisite to secure the most superior queens. As we removed all the brood the nurse-bees will have plenty of time, and be sure to care well for these young queens. By removal of the queen we also secure a large number of cells, while if we waited for the bees to start the cells preparatory to natural swarming, in which case we secure the two desirable conditions named above, we shall probably fail to secure so many cells, and may have to wait longer than we can afford.

Even the apiarist who keeps black bees and desires no others, or who has only pure Italians, will still find that it pays to practice this selection, for, as with the poultry fancier, or the breeder of our larger domestic animals, the apiarist is ever observing some individuals of marked superiority, and he who carefully selects such queens to breed from, will be the one whose profits will make him rejoice, and whose apiary will be worthy of all commendation. It occurs to me that in this matter of careful selection and improvement of our bees by breeding, rests our greatest opportunity to advance the art of bee-keeping. As will be patent to all, by the above process we exercise a care in breeding which is not surpassed by the best breeders of horses and cattle, and which no wise apiarist will ever neglect. Nor do I believe that Vogel can be correct in thinking that drones give invariably one set of character and the queens the others. This is contrary to all experience in breeding larger animals.

It is often urged, and I think with truth, that we shall secure better queens if we wait for the queen-cells to be started naturally by the bees, under the swarming impulse; and by early feeding and adding brood from other colonies we can hasten this period; yet, if we feed to stimulate, whenever the bees are not storing, and keep the colony redundant in bees of all ages by adding plenty of capped brood from other colonies, we shall find that our queens are little, if any, inferior, even if their production is hastened by removal of a queen from the hive. If these directions are closely followed, there will be little brood for the bees to feed, and the queen-cells will not suffer neglect. Mr. Quinby not only advised this course, but he recommended starting queen-cells in nuclei; but he emphasized the importance of giving but very little brood, so nearly all the strength of the nurse-bees would be expended on the queen-cells.

After we have removed all the queen-cells, in a manner soon to be described, we can again supply eggs, or newly-hatched larvæ—always from those queens which close observation has shown to be the most vigorous and prolific in the apiary—and thus keep the same queenless colony or colonies engaged in starting queen-cells till we have all we desire. Yet we must not fail to keep this colony strong by the addition of *capped* brood, which we may take from any colony as most convenient. It is well also to feed a little each day in case the bees are not gathering. We must be cautious that our cells are started from only such brood as we take from the choicest queen. I have good reason to believe that queen-cells should not be started after the first of September, as I have observed that late queens are not only less prolific, but shorter lived. In nature, late queens are rarely produced, and if it is true that they are inferior, it might be explained in the fact that their ovaries remain so long inactive. As queens that are so long unmated are utterly worthless, so, too, freshly mated queens long inactive may become enfeebled. However, some of our queen-breeders think late queens just as good. Possibly they may be, if reared with the proper cautions.

In eight or ten days the cells are capped, and the apiarist is ready to form his nuclei. For the rearing of a small number of queens, the above is very satisfactory. If, however, we are rearing queens for the market, in which we must have numerous cells at our command, and to avoid cutting comb and to secure better spacing better methods have been devised. Mr. Henry Alley cut narrow single-celled strips of worker-comb with newly-hatched larvæ, fastened these to the top-bar of his frame, or to bars inserted parallel to the top-bar, and by inserting the brimstone end of a match and turning it destroys each alternate larva. These put in a colony dequeened, but with many young bees and much hatching brood, gave him good cells rightly spaced. Others have used drone comb cut in the same way, and in each alternate cell have inserted a little royal jelly from a queen-cell about ready to be capped, and then added a worker-larva. This accomplishes the same purpose, and mutilates no worker-comb.

Mr. Doolittle, who has given much time to research in this line, first used the partially built queen-cells always to be found in every hive. These could be fixed to comb or cross-bars at pleasure, and by placing in each a particle of royal jelly and a newly-hatched larva, he secured good queen-cells. If these were in a queenless colony with abundant young bees, the best of queens were reared. Mr. Doolittle found, what I am sure is true, that the best queens, bred naturally, were those reared before the natural swarm issued, or were always started as queens very early, if not from the egg itself, were reared with plenty of nurse or young bees in the hive, and in times, usually, of rapid gathering of honey. Mr. Doolittle found that he could not always get his queen-cups or incipient queen-cells when needed, and soon invented the valuable

FIG. 129.

Form for making Cups.—From A. I. Root Co.

method of dipping and producing artificial cups at pleasure. He describes the whole method of discovery in his valuable and very interesting book. The mould, or dipping-stick (Fig. 129), is like a rake-tooth with one end fashioned so as just to fit into a good, normal queen-cell. This is immersed first in water, then for nine-sixteenths of an inch into melted wax which is kept melted by use of a lamp. It is inserted seven or eight times alternately in the water and in the wax, but for a less and less distance each time in the latter. This makes the cup heavy and thick at the bottom and thin at the top. A twirling motion, when held at various angles, makes the walls

of the cup uniform.   At least a little pressure loosens the cell
from the stick, when it is dipped once more and stuck to the
strip (Fig. 130), which will hold it in the frame.   Usually there
are twelve or fourteen to one strip.   This can be fastened
close below the comb in a partly filled frame.   A little royal
jelly from a queen-cell just ready to be capped is now inserted
in each cup, and a larva less than one day old, always with

FIG. 130.

Doolittle Cell-Cups.—From George W. York & Co.

food about it, is transferred to this in precisely the same position
it had in the worker-cell.   An ear-spoon or quill toothpick,
cut and bent into a spoon-like form, or hard-wood stick of
similar shape, is excellent to transfer the jelly and larvæ.
One queen-cell will furnish enough jelly for from eight to
twelve or fourteen cells.   Of course, the larvæ will be taken
from the best queen in the apiary.   To get these cells cared
for, the frame is put in an upper story of a strong colony with
a queen-excluding honey-board (Fig. 91) between two frames
full of brood in all stages.   They can be built out and finished
below by using a perforated-zinc division-board (Fig. 88, 92),
which will surely keep the queen away.   It should be placed
between the same kind of frames as when put above.   In ten
or twelve days we have probably twelve very fine capped
queen-cells which can be easily removed.

Mr. W. H. Pridgen, of North Carolina, has improved Mr.
Doolittle's scheme by a wholesale method of forming the cups.
He fastens twelve or more of the dipping-sticks to a strip of

wood and dips all of them at once. He even suggests that these may be mounted on the circumference of a wheel which carries them alternately through the water and wax and auto- matically raises so as to preserve the right depth in the melted wax each time. They may be inserted in close-filling holes in a narrow board so as to be quite easily moved up and down. These are dipped till the cups are satisfactory, then all dipped once more at the end, touched to a narrow board (Fig. 131) to which they will adhere. Then by wetting the tips and

FIG. 131.

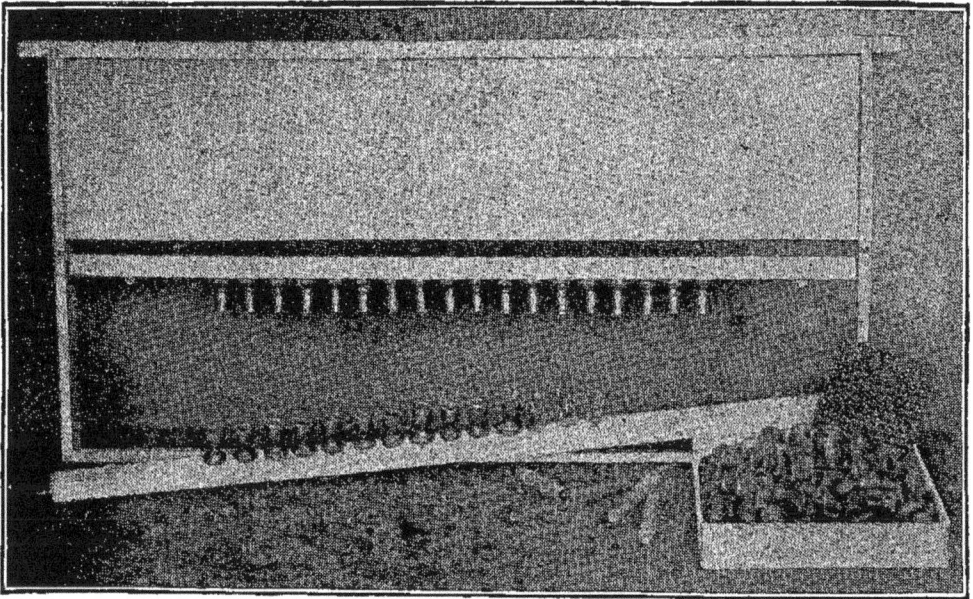

Pridgen Cell-Cups.—From George W. York & Co.

board, the dipping-sticks are easily removed one at a time (Fig. 131). Each dipping-stick is five-eighths of an inch in diameter. It commences to taper five-sixteenths of an inch from the end, tapers strongly one-eighth of an inch, then grad- ually to the end. The strips with cells adhering are one-half inch square, and are fastened in frames by a single wire nail at each end passing through the side of the frame and into the end of the square piece. Comb may be close above them. As already explained, each worker brood-cell is lined with a sec- ond cell consisting of many cocoons. By cutting off the walls of old dark comb to within an eighth of an inch of the base by use of a sharp, warmed knife, these inner cells, which Mr. Pridgen and others call cocoons, may be easily loosened by bending the comb. These were first used by the Atchleys. He loosens them in this way, when they contain larvæ about a day old, from his best queen. By pushing into these a transferring stick, concave at the end (Fig. 132), he can raise the inner cell-larva, food and all, and insert them into a cup. This is a quick way to people the cells with larvæ. Mr. Pridgen often bores small five-sixteenth inch holes nearly through the stick to receive the cups, waxes the stick, and then presses the newly-formed cups into these. In this case he pushes them in with a stick much like the dipping-sticks, only longer and a trifle smaller. In these may be placed a little jelly and the larvæ as already described. Mr. Pridgen places these for a

FIG. 132.

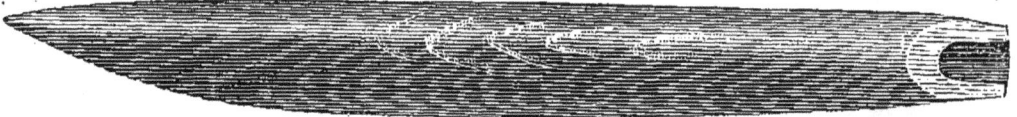

Pridgen Transferring-Stick (full length and size.)
—From George W. York & Co.

few hours in a hive which was filled with brood twelve days before, and placed with a queen-excluder on another colony. When he wishes to give the cups and larvæ, he removes the upper hive, shakes the bees that they may soon find that they are queenless, shuts them in over a broadly ventilated bottom-board, and in a few hours gives them the cups. They accept the care of these at once. He has had thirty-six received and fed in this way. He soon removes these to an upper story over a colony, with the queen-excluder, of course, between them. In from ten to twelve days he has a fine lot of cells for the nuclei. Mr. Pridgen puts a comb partly-filled with water in the hive that is shut up. As we have seen, this would be a time when water would be very essential. The bees are confined and worried. While some queen-breeders still use the Alley method, most now use the Doolittle, and most will soon adopt the Pridgen improvement, as many have done already.

NUCLEI.

A nucleus is simply a miniature colony of bees—a hive and colony on a small scale—for the purpose of rearing and keeping queens. We want the queens, but can afford to each nucleus only a few bees. The nucleus hive, if we use frames not more then one foot square, need be nothing more than an ordinary hive, with chamber confined by a division-board to the capacity of three frames. If our frames are large, then it may be thought best to construct special nucleus hives. These are small hives, which need not be more than six inches each way, that is, in length, breadth and thickness, and made to contain from four to six frames of corresponding size. These frames are filled with comb. I have for many years used the first-named style of nucleus hive, and have found it advantageous to have a few long hives made, each to contain five chambers; while each chamber is entirely separate from the one next to it, is five inches wide, and is covered by a separate, close-fitting board, and the whole by a common cover. The entrances to the two end chambers are at the ends near the same side of the hive. The middle chamber has its entrance at the middle of the side near which are the end entrances, while the other two chambers open on the opposite side, as far apart as possible. The outside might be painted different colors to correspond with the divisions, if thought necessary, especially on the side with two openings. Yet I have never taken this precaution, nor have I been troubled much by losing queens. They have almost invariably entered their own apartments when returning from their wedding-tour. It seems from observation that the queen is more influenced by position than by color of hive in returning to it from mating. Who that has watched his bees after moving a hive a little one side of its previous position—even if only a few inches—can doubt but that the same is true of the worker-bees. These hives I use to keep queens in during the summer. Except the apiarist engage in queen-rearing extensively as a business, I doubt the propriety of building such special nucleus hives. The usual hives are good property to have in the apiary, will soon be needed, and may be economically used for all nuclei. In

spring I make use of my hives which are prepared for prospective summer use, for my nuclei.

Mr. E. M. Hayhurst, one of our best queen-breeders, uses
the full-size Langstroth frame, in full-sized hives, for queen-
rearing, while Mr. Root uses the same frames in small special
hives which hold three frames. These (Fig. 119) he fastens
high up on his grape-vine trellises, just back of his other hives,
which can be used for seats as he works with the nuclei.

We now go to different hives of the apiary, and take out
three frames for each nucleus, at least one of which has brood,
and so on, till there are as many nuclei prepared as we have
queen-cells to dispose of. The bees should be left adhering to
the frames of comb, only *we must be certain that the queen is
not among them,* as this would take the queen from where she
is most needed, and would lead to the sure destruction of one
queen-cell. To be sure of this, we never take such frames till
we have seen the queen, that we may be *sure* she is left behind.
It is well to close the nucleus for at least twenty-four hours,
so that enough bees will surely remain to cover the combs,
and so prevent the brood from becoming chilled. Another
good way to form nuclei, is to remove the queen from a full
colony, and as soon as she is missed use all the frames and
bees for nuclei. We form them as already described. In this
way we are not troubled to find but one queen. If any desire
the nuclei with smaller frames, these frames must of course
be filled with comb, and then we can shake bees immediately
into the nuclei, till they have sufficient to preserve a proper
temperature. Such special articles about the apiary are costly
and inconvenient. I believe that I should use hives even with
the largest frames for nuclei. L. C. Root, who uses the large
Quinby frame, uses the same for his nuclei. In this case we
should need to give more bees. Twenty-four hours after we
have formed this nucleus, we are ready to insert the queen-cell.
We may do it sooner, even at once, but always at the risk of
having the cell destroyed. To insert the queen-cell—for we
are now to give one to each nucleus, so we can never form
more nuclei than we have capped queen-cells—the old way was
to cut it out, using a sharp thin-bladed knife, commencing to
cut on either side the base of the cell, at least one-half inch

distant, for *we must not in the least compress the cell*, then cutting up and out for two inches, then across opposite the cell. This leaves the cell attached to a wedge-shaped piece of comb (Fig. 133), whose apex is next to the cell. If we get our cells by the Doolittle or other improved methods, we can easily cut down and pry each cell off. A similar cut in the middle frame of the nucleus, which, in case of the regular frames, is the one containing brood, will furnish an opening to receive the wedge containing the cell. The comb should also be cut away

FIG. 133.

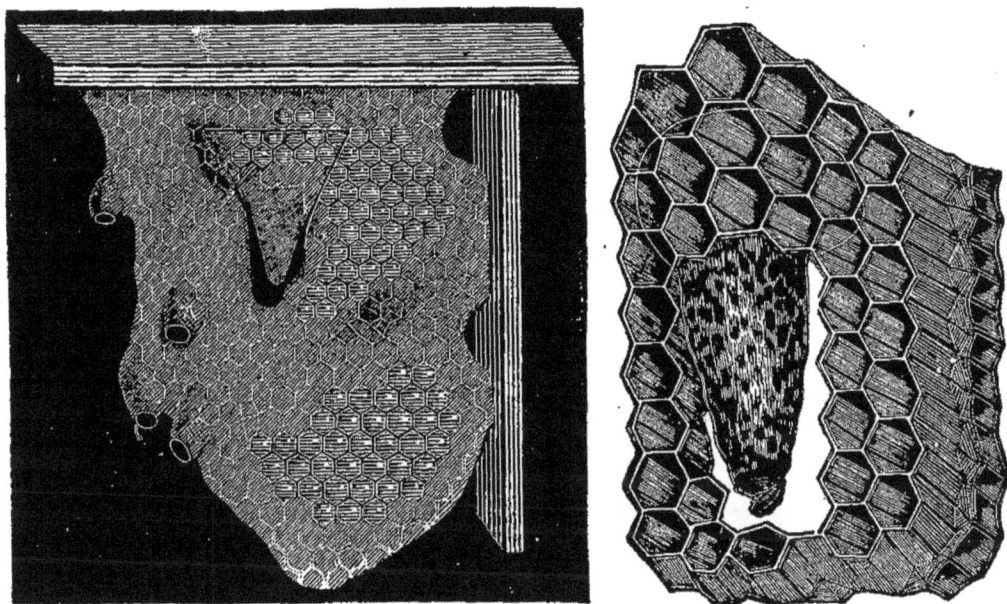

Grafted Queen-Cell.—From A. I. Root Co.   *Queen-Cell with Hinged Cap. From A. I. Root Co.*

beneath (Fig. 133), so that the cell can not be compressed. Mr. Root advises a circular cut (Fig. 133). Of late I have just placed the cell between two frames, and succeed just as well. If two or more fine cells are so close together that separation is impossible, then all may be inserted in a nucleus. By close watching afterward we may save all the queens. If we have used bright new comb as advised above, we can see the queen move in the cell if she is ready to come out, by holding it between us and the sun, and may uncap such cells, and let the queen run in at the entrance of any queenless hive or nucleus at once. In selecting combs for queen-cells, we should reject any that have drone-comb. Bees sometimes start queen-cells over drone-larvæ. Such cells are smoother than the others, and of course are worthless.

After all the nuclei have received their cells and bees, they have only to be set in a shady place and watched to see that

FIG. 134.

Entrance-Guard.

sufficient bees remain. Should too many leave, give them more by removing the cover and shaking a frame loaded with bees over the nucleus; keep the opening nearly closed, and cover the bees so as to preserve the heat. The main caution

FIG. 135.

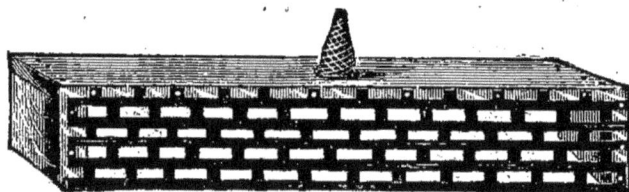

Drone-Trap.—From A. I. Root Co.

in this *is to be sure not to get any old queen in a nucleus.* In two or three days the queens will appear, and in a week longer will have become fecundated, and that, too, in case of the first queens, by selected drones, for as yet there are no others in the apiary. I can not over-estimate the advantage of always having extra queens. To secure mating from selected drones, later, we must cut all drone-comb from inferior colonies, so that they shall rear no drones. If drone-larvæ are in uncapped cells, they may be killed by sprinkling the comb with cold water. By giving the jet of water some force, as may be easily done by use of a fountain pump, they may be washed out, or we may throw them out with the extractor, and then use the comb for starters in our sections.

It is very important that those who rear queens to sell shall have no near neighbors who keep bees, and shall keep only very superior bees, that undesirable mating may be prevented. If one has neighbors who keep bees, he can see that they keep only the best, or possibly he can rear his queens before others have drones flying. He can also get his neighbor to use the Alley drone-trap (Fig. 135). If drones are flying from undesirable colonies, they can be kept from leaving the hive by use of the entrance-guards (Fig. 134), or may be captured or destroyed by use of the Alley drone-trap (Fig. 135). These are made of the perforated-zinc, and while they permit the passage of the workers, they restrain the queen and drones.

FIG. 136.

Queen-Cage.—From A. I. Root Co.

The spaces in these are .165. In England they make them .180 of an inch, but small queens may pass through these larger spaces. By shaking all the bees in front of the hive, we can, by use of these, soon weed out all the drones. With these in front of hive, we can keep the queen from leaving with a swarm. Occasionally, however, a queen will pass through unless the smaller spaced zinc is used. By keeping empty frames and empty cells in the nuclei, the bees may be kept active; yet with so few bees one can not expect very much from the nuclei. After cutting all the queen-cells from our old hive, we can again insert eggs, as above suggested, and obtain another lot of cells, or, if we have a sufficient number, we can leave a single queen-cell, and this colony will soon be the happy possessor of a queen, and just as flourishing as if the even tenor of its ways had not been disturbed. If it is preferred, the bees of this colony may be used in forming the nuclei, in which case there is no danger of getting a queen in any nucleus thus formed, or of having the queen-cells destroyed. We can thus start seven or eight nuclei very quickly. Mr. Doolittle forms nuclei by disturbing the bees—jarring the hive—till they fill with honey, then shakes them into a hive or box and sets them in a dark room or cellar for twenty-four hours. Then they will always, he says, accept a queen-cell or a virgin queen of any age at once. A full colony may be usually re-queened in the same way.

QUEEN LAMP-NURSERY.

This aid to bee-keeping was first used by F. R. Shaw, of Chatham, Ohio. The double wall enclosing water was the invention of A. I. Root. It is substantially a tin hive, with two walls enclosing a water-tight space an inch wide, which, when in use, is filled with water through a hole at the top. Each nursery may hold from six to eight frames. Some prefer to have special frames for this nursery, each of which contains several close chambers. The queen-cells are cut out and put in these chambers.

By use of a common kerosene lamp placed under this nursery, the temperature must be kept from 80 degrees F. to 100 degrees F. By placing the frames with capped queen-cells in this, the queens develop as well as if in a hive or nucleus. If the young queens, just from the cell, are introduced into a queenless colony or nucleus, as first shown by Mr. Langstroth, they are usually well received. Unless one is rearing a great many queens, this lamp-nursery is not desirable, as we still have to use the nucleus to get the young queens fecundated, have to watch carefully to get the young queens as soon as they appear, must guard it carefully as moths are apt to get in, and, finally, unless great pains are taken, this method will give us inferior queens. Mr. W. Z. Hutchinson, one of our best queen-breeders, thinks very highly of the lamp-nursery.

Some bee-keepers use a cage (Fig. 136) with projecting pins which are pushed into the comb, so that they hold the cage. A cell is put into each of these, and then they may be put into any hive. Of course the bees can not destroy the cell, as they can not get at it. Dr. Jewell Davis' queen-nursery

FIG. 137.

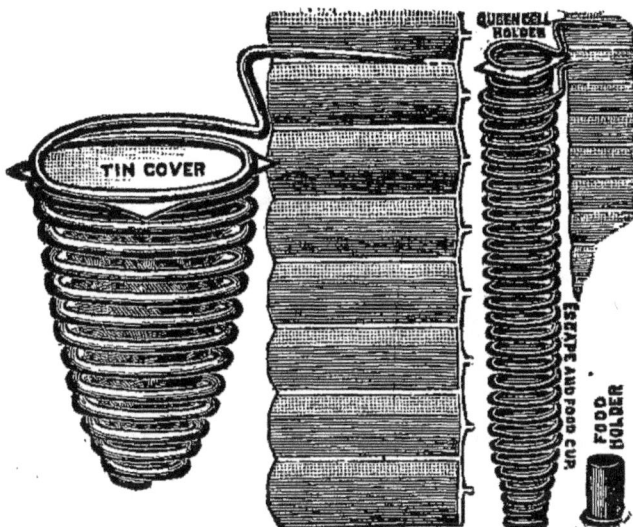

West Cell-Protector.—From A. I. Root Co.

consists of a frame filled with such cages, which can be hung in any hive. I have tried both, and prefer this to the lamp-nursery. The West cell-protector (Fig. 137) is excellent. The cell can not be destroyed, and as the protector is open at the end the queen comes forth into the nucleus, and is almost sure to be well received. This is an excellent way to insert queen-cells. Mr. Root recommends putting a little honey at the end of the cell, so the queen will get this at once. Mr. Doolittle, to introduce virgins, puts them in a cage with candy, and covers the opening with paper, as well as candy, so as to delay her egress. Rarely they fail to eat through this, when they must be liberated.

SHALL WE CLIP THE QUEEN'S WING ?

In the above operation, as in many other manipulations of the hive, we shall often gain sight of the queen, and can, if we desire, clip her wing, *if she has met the drone* ; *but never before*, that in no case she shall lead the colony away to parts unknown. This is an old practice, for Virgil speaks of retaining the bees by tearing off the wings of "the king." This does not injure the queen, as some have claimed. General Adair once stated that such treatment injured the queen, as it cut off some of the air-tubes, which view was approved by so excellent a naturalist as Dr. Packard. Yet I am sure that this is all a mistake. The air-tube and blood-vessel, as we have seen, go to the wings to carry nourishment to these members. With the wing goes the necessity of nourishment and the need of the tubes. As well say that the amputation of the human leg or arm would enfeeble the constitution, as it would cut off the supply of blood.

Many of our best apiarists have practiced this clipping of the queen's wing for years. Yet these queens show no diminution of vigor; we should suppose they would be even more vigorous, as useless organs are always nourished at the expense of the organism, and, if entirely useless, are seldom long continued by nature. The ants set us an example in this matter, as they bite the wings off their queens, after mating has transpired. They mean that the queen shall remain at home, *nolen volens*, and why shall not we require the same of the queen-bee ? Were it not for the necessity of swarming in nature, we should doubtless have been anticipated in this matter by Nature herself.

Some of our first apiarists think that queens with wings clipped are not as acceptable to the other bees. I have now had experience for thirty years in this practice, and have yet to see the first indication that the above is true. Still, *if the* queen essays to go with the swarm, and if the apiarist is not at hand, she will sometimes be lost, never regaining the hive ; but in this case the bees will be saved, as *they* will return without fail. Many of our farmers are now keeping bees with marked success and large profits, who could not continue at all except for this practice. Mr. George Grimm kept about eighty colonies of bees, and said he worked only ten days in the year. But he clipped the queens' wings, and his wife did the hiving.

Some apiarists clip one primary wing the first year, the secondary the second year, the other primary the third, and, if age of the queen permits, the remaining wing the fourth year. Yet, such data, with other matters of interest and importance, better be kept on a slate or card, and firmly attached to the hive, or else kept in a record opposite the number of the hive. The time required to find the queen is sufficient argument against the "queen-wing record." This is not an argument against the once clipping of the queen's wings, for, in the nucleus hives, queens are readily found, and even in full colonies this is not very difficult, especially if we keep Italians or any other races of yellow bees. It will be best, even though we have to look up black queens in full colonies. The loss of one good colony, or the vexatious trouble of separating two or three swarms which had clustered together, and finding each queen, or the hiving of a colony perched high up on some towering tree, would soon vanquish this argument of time.

To clip the queen's wing, which we must never do until she commences to lay eggs, take hold of her wings with the right thumb and index finger—never grasp her body, *especially her abdomen*, as this will be very apt to injure her—raise her off the comb, then turn from the bees, place her gently on the left hand, and press on her feet with the left thumb sufficiently to hold her. Now with the right hand, by use of a small, delicate pair of scissors, cut off about one-half of one of the front or primary wings. This method prevents any movement of legs or wings, and is easy and quick. I think Mr. Root advises grasping the queen by the thorax. I prefer the method given here.

Some bee-keepers—inexperienced they must be—complain that queens thus handled often receive a foreign scent, and are destroyed by the worker-bees. I have clipped hundreds and never lost one.

LAYING WORKERS.

We have already described laying workers. As these can

only produce unimpreginated eggs, they are, of course, value-
less, and unless superseded by a queen will soon cause the
destruction of the colony. As their presence often prevents
the acceptance of cells or a queen, by the common workers,
they are a serious pest. The absence of worker-brood, and
the abundant and careless deposition of eggs—some cells being
skipped, while others have received several eggs—are pretty
sure indications of their presence. The condition that favors

FIG. 138.

Hive-Scraper.—Original.

these pests, is continued absence of a queen or means to produce one ; thus they are very likely to appear in nuclei. They
seem more common with the Cyprian and Syrian bees.

   To rid a colony of these, unite it with some colony with a
good queen, after which the colony may be divided if very
strong. Simply exchanging places of a colony with a laying
worker, and a good, strong colony will often cause the destruction of the wrong-doer. In this case, brood should be given to
the colony which had the laying worker, that they may rear a
queen ; or better, a queen-cell or queen should be given them.
Caging a queen in a hive, with a laying worker, for thirty-six
hours, will almost always cause the bees to accept her. We
may also use the Doolittle candy cage with the opening covered
with paper. Her escape is so tardy that she will be safe.
Shaking the bees off the frames two rods from the hive, will
often rid them of the counterfeit queen, after which they will
receive a queen-cell or a queen. But prevention is best of all.
We should never have a colony or nucleus without either a
queen or means to rear one. It is well to keep young brood in
our nuclei at all times. Queens reared from brood four days
from the egg are often drone-layers, and never desirable.

In all manipulation with the bees we need something to loosen the frames. Many use a chisel or small iron claw. I have found an iron scraper (Fig. 138), which I had made by a blacksmith, very convenient. It serves to loosen the frames, draw tacks, and scrape off propolis. It would be easy to add the hammer.

QUEEN-REGISTER OR APIARY REGISTER.

With more than a half dozen colonies it is not easy to know just the condition of each colony. Something to mark the date of each examination, and the condition of the colony

FIG. 139.

QUEEN REGISTER.
EGGS.

No._____

MISSING. BROOD.

NOT APPROVED. o CELL.

APPROVED. HACTHED.

MARCH.

OCT. APRIL.

LAYING.

SEPT. o MAY.

DIRECTIONS.—Tack the card on a conspicuous part of the Hive or Nucleus; then, with a pair of plyers, force a common pin into the center of each circle, after it is bent in such a manner that the head will press securely on any figure or word. These Cards mailed free, at 6c. per doz. or 40c. per hundred.
Use tinned or galvanized tacks; they will stand rain, &c.

AUG. JUNE

JULY.

A. I. ROOT, MEDINA, O.

at that time is very desirable. Mr. Root furnishes the Queen-Register (Fig. 139). With this it is very easy to mark the date of examination of each hive, and the condition of the colony at the time. Mr. Hutchinson prefers this. Mr. Newman furnished an Apiary Register which served admirably for the same purpose. Each hive is numbered. Dr. Miller tacks a small square piece of tin bearing the number in black paint to

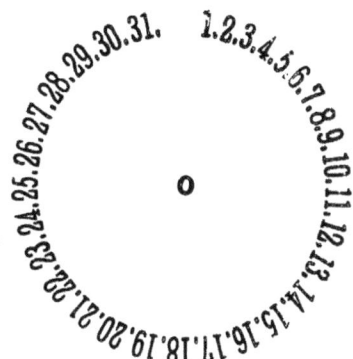

each hive. A corresponding number in the Register gives us all desired facts. We have only to note down at the time the condition of each colony and date of examination in the Register. Mr. Root prefers a slate whose position on the cover of the hive shows the condition of colony, and dates can be written on it.

CHAPTER X.

INCREASE OF COLONIES.

No subject will be of more interest to the beginner than that of increasing colonies. He has one or two, he desires as many more, or, if very aspiring, as many hundred, and if a Jones, a Hetherington, or a Harbison, as many thousand. This is a subject, too, that may well engage the thought and study of men of no inconsiderable experience. I believe that many veterans are not practicing the best methods in obtaining an increase of colonies.

Before proceeding to name the ways, or to detail the methods, let me state and enforce that it is always safest, and generally wisest, especially for the beginner, to be content with doubling, or certainly with tripling, his number of colonies each season. Especially let all remember the motto: "Keep all colonies strong."

There are two ways to increase: The natural, known as swarming, already described under natural history of the bee; and the artificial, improperly styled artificial swarming. This is also called, and more properly, "dividing."

SWARMING.

In case of the specialist, or in case some one can be near by to watch the bees, swarming is without doubt the best way to increase. Therefore, the apiarist should be always ready with both means and knowledge for immediate action. Of course, necessary hives were all secured the previous winter, *and will never be wanting.* Neglect to provide hives before the swarming season is convincing proof that the wrong pursuit has been chosen.

If, as I have advised, the queen has her wing clipped, the matter becomes very simple, in fact, so much simplified that were there no other argument, this would be sufficient to recommend the practice of clipping the queen's wing. Now, if several swarms cluster together, we have not to separate them; they will usually separate of themselves and return to their old homes. To migrate without the queen means death, and life is sweet even to bees. and is not willingly to be given up except for home and kindred. Even if they all enter one hive, the queens are not with them, and it is very easy to divide them as desired. Neither has the apiarist to climb trees, to secure his bees from bushy trunks, from off the lattice-work or pickets of his fence, from the very top of a tall, slender, fragile fruit-tree, or other most inconvenient places. Nor will he even be tempted to pay his money for patent non-swarming hivers or patent swarm-catchers. He knows his bees will return to their old quarters, so he is not perturbed by the fear of loss or plans to capture the unapproachable. It requires no effort " to possess his soul in patience." If he wishes to increase, he steps out, takes the queen by the remaining wings, as she emerges from the hive, soon after the bees commence. their hilarious leave-taking, puts her in a cage, opens the hive, destroys, or, if he wishes to use them, cuts out the queen-cells as already described, gives more room —either by adding a super of sections or taking out some of the frames of brood, as they may well be spared—places the cage enclosing the queen under the quilt, and leaves the bees to return at their pleasure. At nightfall the queen is liberated, the hive may be removed to another place, and very likely the swarming-fever is subdued for the season.

If it is desired to unite the swarm with a nucleus, exchange the places of the old hive with the caged queen, as soon as the swarm is out, and the nucleus hive, to which, of course, the swarm will now come. The queen-cells should be removed at once from the old hive, and the queen liberated. The nucleus colony, now strongly enforced, should have empty frames, but always with starters, added, making five in all; and a super of sections with thin foundation added at once. The five frames Langstroth size, more if smaller—are put on one side and the rest of the space filled by division-boards. Here the nucleus is at once transformed into a large, strong colony.

If it is desired to hive the swarm separately—and usually

this gives the best results, even if we do not care for increase —we remove the old hive to one side, and turn it entirely around, so that the entrance that was east is now west. We now place a new hive with five or six empty frames, which have narrow starters, right where the old hive previously stood, in which the caged queen is put. We fill the extra space in this hive with division-boards, and set on it the super of sections previously placed on the old hive; or in case this colony that just swarmed had not previously received a super of sections, we place a super with a queen-excluding honey-board on the hive where the new swarm is now to enter.

As this colony has no comb in the brood-chamber, only foundation starters, and has sections with comb or thin foundation, the bees will commence to work vigorously in the sections, especially as the brood-chamber is so restricted. This idea originated with Messrs. Doolittle and Hutchinson, and is fully explained in "Advanced Bee-Culture," Mr. Hutchinson's excellent book, which should be in the hands of every comb-honey producer.

The hive from which the swarm issued—now close beside the hive with the new colony—should be turned a little each day so that by the eighth day the entrance will be as before to the east, or close to that of the other hive. On the eighth day this hive is carried to some distant part of the bee-yard. Of course all the bees that are gathering—and by this time they are numerous—will go to the other hive, which will so weaken the still queenless colony that they will not care to send out another or second swarm, and so will destroy all remaining queen-cells and queens after the first queen comes forth. This is a quick, easy way to prevent after, or second, swarms. It originated with Mr. James Heddon, and I find that, with rare exception, it works well. I believe where one is with his bees, this last-described plan is the most profitable that the bee-keeper can adopt. Sometimes the mere introducing of a new queen into the old hive will prevent any further swarming. The queen at once destroys the queen-cells.

Some extensive apiarists, who desire to prevent increase of colonies, when a colony swarms, cage the old queen, destroy all queen-cells, and exchange this hive—after taking out three or four frames of brood to strengthen nuclei, replacing these with frames with starters of foundation—with one that recently swarmed, which was previously treated the same way. Thus a colony that recently sent out a swarm, but retained their queen, has probably, from the decrease of bees, loss of brood, and removal of queen-cells, lost the swarming-fever, and if we give them plenty of room and ventilation, they will accept the bees from a new swarm, and spend their future energies in storing honey. If the swarming-fever is not broken up, we shall only have to repeat the operation again in a few days.

Still another modification, in case no increase of bees but rather comb honey is desired, is recommended by such apiarists as Doolittle, Davis, and others. The queen is caged seven days, the queen-cells in the hive are then destroyed, the queen liberated, and everything is arranged for immense yields of comb honey. In this case the queen is idle, but the bees seem to have lost not one jot of their energy. I tried this plan many years ago with great success, and recommended it to Mrs. L. B. Baker, who prized it highly. Dr. C. C. Miller, instead of caging the queen, places her with a nucleus on top of the old hive, thus keeping her at work, by exchange of frames. After seven days he destroys the queen-cells in the old hive and unites the nucleus with it. Here the queen is kept at work, the swarming impulse subdued, and a mighty colony made ready for business. This plan slightly modified has the sanction of such admirable apiarists as Messrs. Elwood and Hetherington.

Two objections are sometimes raised right here. Suppose several swarms issue at once, one of which is a second swarm, which of course has a virgin queen, then all will go off together, and our loss is grevous indeed. I answer that second swarms are unprofitable, and should never be permitted. We should be so vigilant that this fate would never befall us. If we will not give this close attention without such stimulus, then it were well to have this threatening danger hanging over us. Again, suppose we are not right at hand when the swarm issues, the queen wanders away and is lost. Yes, but if unclipped the whole colony would go, now it is only the queen. Usually the queen gets back. If not, a little looking will generally find her not far away within a ball of friendly workers. At nightfall smoke these bees, and by watching we learn the colony which swarmed, as the bees about the queen will repair at once to it. Mr. Doolittle suggests that we may always find what colony swarmed when a swarm is out. If we take a portion of the bees from the cluster into a pail and swing them around lively, then throw them out, they will at once, he says, fly to the old home. When a swarm first issues, young bees, too young to fly, crawling about the hive, will often reveal the colony that swarmed.

HIVING SWARMS.

But in clipping wings, some queens may be omitted, or, from taste, or other motive, some bee-keepers may not desire to "deform her royal highness." Then the apiarist must possess the means to save the would-be rovers. The means are : good hives in readiness ; some kind of a brush—a turkey-wing will do ; a basket with open top, which should be at least eighteen inches in diameter, and so made that it may be attached to the end of the pole ; and two poles, one very long and the other of medium length.

Now, let us attend to the method : As soon as the cluster commences to form, place the hive in position where we wish the colony to remain, leaving the entrance wide open. As soon as the bees are fully clustered, we must manage as best we can to empty the whole cluster in front of the hive. As the bees are full of honey they are not likely to sting, but will sometimes. I have known bees, when clustered in a swarm, to be very cross. This, however, is not usual. Should the bees be on a twig that could be sacrificed, this might be easily cut off with either a knife or saw (Fig. 140), and so carefully as hardly to disturb the bees, then carry (Fig. 76) and shake the bees in front of the hive, when with joyful hum they will at once proceed to enter. If the twig must not be cut, shake them all into the basket, and empty before the hive. Should they be on a tree trunk, or a fence, then brush them with the wing into the basket, and proceed as before. If they are high up on a tree, take the pole and basket, and perhaps a ladder will also be necessary. Many devices like a bag on a hoop, a

FIG. 140.

Whitman's
Fountain Pump.
—From
A. I. Root Co.

suspended wire-basket, with a tripod to sustain it, etc., are often recommended. These are not much seen in the apiaries of our best bee-keepers. Always let ingenuity have its perfect work, not forgetting that the object to be gained is to get just as many of the bees as is possible on the alighting-board in front of the hive. Carelessness as to the quantity might involve the loss of the queen, which would be serious. The bees *will not* remain unless the queen enters the hive. Should a cluster form where it is impossible to brush or shake them off, they can be driven into a basket, or hive, by holding it above them and blowing smoke among them. All washes for the hive are more than useless. It is better that it be clean and pure. With such, if they are shaded, bees will generally be satisfied. But assurance will be made doubly sure by giving them a frame of brood, in all stages of growth, from an old hive. This may be inserted before the work of hiving is commenced. Mr. Doolittle thinks this does little or no good, and tends to induce the building of drone-comb. Mr. Betsinger says they are even more apt to go off ; but I think he will not be sustained by the experience of other apiarists. He certainly is not by mine. I never knew but one colony to leave uncapped brood ; I have often known them to swarm out of an empty hive once or twice, and to be returned, after brood had been placed in the hive, when they accepted the changed conditions, and went at once to work. We should expect this, in view of the attachment of bees for their nest of brood, as also from analogy. How eager the ant to convey her larvæ and pupæ—the so-called eggs—to a place of safety, when the nest has been invaded and danger threatens. Bees doubtless have the same desire to protect their young, and as they can not carry them away to a new home, they remain to care for them in one that may not be quite to their taste. Of course if swarming is permitted either with or without clipped queens, the bees must be closely watched at the swarming season. Dr. Miller secures a bright, active girl or boy to watch. He says the watcher can sit in the shade and go and look once in every four or five minutes. For 100 colonies it takes the whole time of one person, as the noise made by so many flying bees makes actual inspection of all hives necessary. This watching is necessary from 8 a.m. till 2 p.m.; or in very best weather from 6 a.m. to 4 p.m., or even later. Of course there is relief on rainy days.

Farmers can keep many colonies and attend to their farm work as usual. They have only to have a boy or girl to catch and cage the queens—the "gude wife" may do this—and inform at noon or night what colonies have swarmed. When a colony swarms, the impulse seems to be general, and often a half dozen colonies will be on the wing in a trice. These will very often—generally, in truth—cluster together. In this case, to find the queens is well nigh impossible, and we can only divide up the bees into suitable colonies, and as soon as we find any starting queen-cells, give them a queen. Of course we may lose every queen but one. In view of this trouble, and the expense and doubtful practicality of the various swarm-catchers in vogue, I would say, *Clip the queen's wing.*

If no more colonies are desired, the swarm may be given to a colony which has previously swarmed, after removing from the latter all queen-cells, and adding to the room by putting on the sections, and removing some frames of brood to strengthen nuclei. These frames may be replaced with empty combs, foundation, or frames with starters. We may even return the bees to their old home, by taking the same precautionary measures, with a good hope that storing and not swarming will engage their attention in future; and if we change their position, or better, exchange their position with that of a nucleus, we shall be still more likely to succeed in overcoming the desire to swarm. A swarm may be given to a colony that was hived as a swarm a day or two before with great safety, by shaking all the bees of both in front of the hive. Some seasons, usually when honey is being gathered each day for long intervals, but not in large quantities, the desire and determination of some colonies to swarm is implacable. Room, ventilation, changed position of hive, each and all will fail. Then we can do no better than to gratify the propensity by giving the swarm a new home, and make an effort.

TO PREVENT SECOND SWARMS.

The Heddon method of preventing second swarms has already been explained. This method is valuable because it requires no looking up of queen-cells, and thus saves time.

As already stated, the wise apiarist will always have on hand extra queens. Now, if he does not desire to form nuclei (as already explained), and thus use these queen-cells, he will at once give the old colony a fertile queen. At the same time this practice secures only carefully reared queens from his best colonies. As the queen usually destroys all queen-cells, farther swarming is prevented. The method of introduction will be given hereafter, though in such cases there is very little danger incurred by giving them a queen at once. If desired, the queen-cells can be used in forming nuclei, in manner before described. If extra queens are wanting, we have only to look carefully—*very carefully*, as it is easy to miss a small, worthless cell in some cranny or corner of the comb—through the old hive and remove all but one of the queen-cells. A little care will certainly make sure work, as after swarming the old hive is so thinned of bees that only carelessness will overlook queen-cells in such a quest. Mr. Doolittle waits till the eighth day, or till he hears the piping of the young queen ; then cuts out all queen-cells, when, of course, he certainly inhibits second swarms. When this practice fails, as it very rarely does, it is because two cells were left.

TO PREVENT SWARMING.

As yet we can only partly avert it. Mr. Quinby offered a large reward for a perfect non-swarming hive, and never had to make the payment. Mr. Hazen attempted it, and partially succeeded, by granting much space to the bees, so that they should not be impelled to vacate for lack of room. The Quinby hive, already described, by the large capacity of the brood-chamber, and ample opportunity for top and side storing, looks to the same end. Mr. Simmins, of England, thinks he can prevent swarming by keeping unoccupied cells between the brood-nest and entrance to the hive. Mr. Muth says if we always have empty cells in the brood-nest swarming will seldom occur. Yet he says "seldom." We may safely say that a perfect non-swarming hive or system is not yet before the bee-keeping public. The best aids toward non-swarming are shade, ventilation, and roomy hives. But as we shall see in the sequel, much room in the brood-chamber, unless we work for extracted honey—by which means we may greatly repress the swarming fever—prevents our obtaining honey in a desirable style. If we add sections, unless the connection is quite free—in which case the queen is apt to enter them and greatly vex us—we must crowd some to send the bees into the sections. Such crowding is almost sure to lead to swarming. I have, by uncapping the combs of honey in the brood-chamber, as suggested to me years ago by Mr. M. M. Baldridge—causing the honey to run down from the combs—sent the bees crowding to the sections, and thus deferred or prevented swarming. Those who have frames that can be turned upside down, or invertible hives, may often secure the same results by simple inversion. By placing the sections in the brood-chamber till the bees commence to work on them, and then removing them above, or by carrying brood up beside the sections, the bees are generally induced to commence working in sections. Some sections with combs in them often aids much. This requires too much manipulation, and so is not practical with the general bee-keeper.

It is possible that by extracting freely when storing is very rapid, and then by freely feeding the extracted honey in the interims of honey-secretion, we might prevent swarming, secure very rapid breeding, and still get our honey in sections. My experiments in this direction have not been as successful as I had hoped, and I can not recommend the practice, though some apiarists claim to have succeeded. Even if this could be made to work it involves too much labor to make it advisable.

The keeping of colonies queenless, in order to secure honey without increase, is practiced and advised by some even of our distinguished apiarists. As already stated, I have done this with excellent results. Dr. C. C. Miller's method, already described, accomplishes the same object, and keeps all the queens at work all the time.

ARTIFICIAL INCREASE.

While, as already remarked, there is no better way than to allow swarming as just described, when one's circumstances make it possible to do so, yet it is true that some of our best bee-keepers prefer to divide. In some cases the bee-keeper can be with his bees only at certain times—often early in the morning, late in the afternoon, or perchance at the noontide hour; then, of course, artificial dividing becomes necessary. It is practiced to secure any desired increase of colonies, also to prevent loss from swarming when no one is by. This requires more time than swarming, as detailed above, and may not—probably often doos not—secure quite as good results. Yet I am very sure, from a long experience, that, with sufficient care, artificial colonies may be formed that will fully equal natural swarms in the profits they bring to their owners. I am sure I could get ten colenies from one in a season, and if I had combs and should feed I think I could nearly double these figures.

HOW TO DIVIDE.

Mr. Cheshire argues against natural swarming and in favor of dividing, as the former tends, through selection, to develop the swarming habit. I do not accept his reasoning, as, unless we permit swarming, we can not tell what colonies to breed from, as we have no way to know their tendencies. Often, too, swarming only indicates great prolificness. By the process already described, we have secured a goodly number of fine queens, which will be in readiness at the needed time. Now, as soon as the white clover harvest is well commenced, early in June, we may commence operations. If we have but one colony to divide, it is well to wait till they become pretty populous, but not until they swarm. Take one of our waiting hives, which now holds a nucleus with laying queen, and place the same close alongside the colony we wish to divide. This must be done on a warm day, when the bees are active, and better be done while the bees are busy, in the middle of the day. Remove the division-board of the new hive, and then remove five combs well loaded with brood, and. of course, containing some honey, from the old colony, bees and all, to the new hive. Also take the remaining frames and shake the bees into the new hive; *only be sure that the queen still remains in the old hive.* Fill both the hives with empty frames—if the frames are filled with empty comb it will be still better; if not, it will always pay to give at least starters of comb or foundation—and return the new hive to its former position. The old bees will return to the old colony, while the young ones will remain peaceably with the new queen. The old colony will now possess at least seven frames of brood, honey, etc., the old queen, and plenty of bees, so that they will work on as though naught had transpired, though perhaps moved to a little harder effort, by the added space and five empty combs or frames with starters or full sheets of foundation. These last may all be placed at one end, or placed between the others, though not so as greatly to divide brood. The new colony will have eight frames of brood, comb, etc., three from the nucleus and five from the old colony, a young laying queen, plenty of bees (those of the previous nucleus and the young bees from the old colony), and will work with a surprising vigor, often even eclipsing the old colony.

If the apiarist has several colonies, it is better to make the new colony from several old colonies, as follows: Take one frame of brood-comb from each of six old colonies, or two from each of three. and carry them, bees and all, and place with the nucleus. *Be sure that no queen is removed.* Fill all the hives with empty combs, or frames with starters of foundation, as before. In this way we increase without in the least disturbing any of the colonies, and may add a colony every day or two, or perhaps several, depending on the size of our apiary, and can thus almost always, so experience says, prevent swarming.

By taking only brood that is all capped, we can safely add one or two frames to each nucleus every week, without adding any bees, as there would be no danger of loss by chilling the brood. In this way, as we remove no bees, we have to spend no time in looking for the queen, and may build up our nuclei into full colonies, and keep back the swarming impulse with great facility.

These are unquestionably the best methods to divide, and so I will not complicate the subject by detailing others. The only objection that can be urged against them—and even this does not apply to the last—is that we must seek out the queen in each hive, or at least be sure that we do not remove her, though this is by no means so tedious if we have Italians or other races of yellow bees, as, of course, we all will. I might give other methods which would render unnecessary this caution, but, to my mind, they are inferior, and not to be recommended. If we proceed as above described, the bees will seldom prepare to swarm at all, and, if they do, they will be discovered in the act, by such frequent examinations, and the work may be cut short by at once dividing such colonies, as first explained, and destroying their queen-cells, or, if desired, using them for forming new nuclei.

CAPTURING ABSCONDING SWARMS.

Sometimes swarms break cluster and take wing for their prospective home before the bee-keeper has hived them. Throwing dirt among them will sometimes cause them to alight again. Throwing water among them in the form of a fine spray (Fig. 140) will almost always do this. For such purpose some hand pump is very desirable. Whitman's fountain pump is one of the most convenient. It costs about $7.00. Another important use for such a pump in the apiary is this: If a swarm, when clustered, is sprinkled occasionally, it will remain clustered indefinitely. This permits us to retain a swarm in case it is more convenient to hive it later. While most customs have a reasonable basis, the common one of horns and bells and beating of pans to stop a swarm is a notable exception. It does not do the least good.

CHAPTER XI.

ITALIANS AND ITALIANIZING.

The history and description of Italian bees have already been considered, so it only remains to discuss the subject in a practical light.

The superiority of the Italians seems no longer a mooted question. I now know of no one among the able apiarists in our country who takes the ground that a thorough balancing of qualities will make as favorable a showing for the German as for the Italian bees, though I think that the late Baron of Berlepsch held to this view.

I think I am capable of acting as judge on this subject. I have never sold a dozen queens in my life, and so have not been unconsciously influenced by self-interest. In fact, I have never had, if I except six years, any direct interest in bees at all, and all my work and experiments had only the promotion and spread of truth as the ultimatum. Again, I have kept both blacks and Italians side by side, and carefully observed and noted results during eight years of my experience. I have carefully collected data as to increase of brood, rapidity of storing, early and late habits in the day and season, kinds of flowers visited, amiability, etc., and I am more than persuaded that the general verdict, that they are superior to the German race, is entirely correct. The Italians are *far* superior to the German bees in many respects, and, though I am acquainted with all the works on apiculture printed in our language, and have an extensive acquaintance with the leading apiarists of our country from Maine to California, yet I know hardly a man that has opportunity to form a correct judgment, does not give strong preference to the Italians. The black bees are in some respects superior to the Italians, and if a bee-keeper's methods cause him to give these points undue importance, in forming his judgments, then his conclusions may be wrong. Faulty management, too, may lead to wrong conclusions.

The Italians certainly possess the following points of superiority :

First. They possess longer tongues, and so can gather from flowers which are useless to the black bee. This point has already been sufficiently considered. How much value hangs upon this structural peculiarity I am unable to state. I have frequently seen Italians working on red clover. I never saw a black bee thus employed. It is easy to see that this might be, at certain times and certain seasons, a very material aid. How much of the superior storing qualities of the Italians is due to this lengthened ligula, I am unable to say. Mr. J. H. Martin has a very ingenious tongue measurer by which the length of the tongues of bees in the several hives can be quickly and accurately compared. I have made a very simple and convenient instrument to accomplish the same end ; two rectangular pieces, one of glass and the other of wire gauze, are so set in a frame that the glass inclines to the gauze. At one end they touch ; at the other they are separated three-fourths of an inch. Honey is spread on the glass and all set in the hive. The bees can only sip the honey through the gauze. The bees that clean the farthest from the end where it touches the gauze have the longest tongues. This gives only relative lengths, while Mr. Martin's register tells the absolute length.

Second. They are more active, and, with the same opportunities, will collect a good deal more honey. This is a matter of observation, which I have tested over and over again. Yet I will give the figures of another : Mr. Doolittle secured from two colonies 309 pounds and 301 pounds, respectively, of *comb honey*, during the one season. These surprising figures, the best he could give, were from his best Italian colonies. Similar testimony comes from Klein and Dzierzon over the sea, and from hosts of our own apiarists.

Third. They work earlier and later. This is not only true of the day, but of the season. On cool days in spring I have seen the dandelions swarming with Italians, while not a black bee was to be seen. On May 7, 1877, I walked less than half a mile, and counted sixty-eight bees gathering from dandelions, yet only two were black bees. This might be considered an undesirable feature. Yet, from careful observation covering thirty years, I think that Italian bees are quite as apt to winter well and pass the spring months without harm as are black bees.

Fourth. They are far better to protect their hives against robber-bees. Robbers that attempt to plunder Italians of their hard-earned stores soon find that they have "dared to beard the lion in his den." This is so patent that even the advocates of black bees are ready to concede it.

Fifth. They are proof against the ravages of the bee-moth's larvæ. This is also universally conceded. This is no very great advantage, as no respectable bee-keeper would dread moths, even with the black bees.

Sixth. The queens are decidedly more prolific. This is probably in part due to the greater and more constant activity of the workers. This is observable at all seasons, but more especially when building up in the spring. No one who will take the pains to note the increase of brood will long remain in doubt on this point.

Seventh. They are less apt to breed in winter, when it is desirable to have the bees very quiet.

Eighth. The queen is more readily found, which is a great advantage. In the various manipulations of the apiary, it is frequently desirable to find the queen. In full colonies I would rather find three Italian queens than one black one. Where time is money this becomes a matter of much importance.

Ninth. The bees are more disposed to adhere to the comb while being handled, which some might regard a doubtful compliment, though I consider it a desirable quality.

Tenth. They are, in my judgment, less liable to rob other bees. They will find honey when the blacks gather none, and the time for robbing is when there is no gathering. This may explain the above peculiarity.

Eleventh. In my estimation, a sufficient ground for preference, did it stand alone, is that the Italian bees are *far more amiable.* Years ago I got rid of my black bees because they were so cross. A few years later I got two or three colonies, that my students might see the difference, but to my regret; for, as we removed the honey in the autumn, they seemed perfectly furious, like demons, seeking whom they might devour, and this, too, despite the smoker, while the far more numerous Italians were safely handled, even without smoke. The experiment at least satisfied a large class of students as to superiority. Mr. Quinby speaks in his book of their being cross, and Capt. Hetherington tells me that if not much handled they are more cross than the blacks. From my own experience, I can not understand this. Hybrids, between blacks and Italians, are ofttimes even more cross than are the pure black bees, but otherwise are nearly as desirable as the pure Italians.

I have kept these two races side by side for years ; I have studied them most carefully, and I feel sure that none of the foregoing eleven points of excellence are too strongly stated.

The black bees will go into sections more readily than Italians, yet the skillful apiarist will find it easy to overcome this objection in the manner already described.

There is no question but that the German bees produce nicer, whiter comb honey than do the Italians. This superiority is due in part to thicker cappings, and to a wider air-space between honey and capping. This, however, is too nice a point to count very greatly in their favor. The comb honey produced by Italians does not have to go begging in the markets.

The advantages of the Italians, which have been considered thus fully, are more than sufficient to warrant the exclusion of the German bees from the apiary. I say truly, no one needs to be urged to a course that adds to the ease, profit and agreeableness of his vocation. Darwin showed, years ago, that introduced or newly imported plants or animals usually possessed more vigor than those " to the manor born." Hence the wisdom of frequent and repeated importations from Italy. This is the more desirable unless the queen-breeder works carefully and scientifically to improve his stock. No doubt we have American queen-breeders whose bees are superior to any they might secure from Italy. Yet even these may find it desirable to bring occasional fresh stock from the Ligurian hills.

THE NEW RACES OF BEES.

All of the valuable characteristics of the Italian bees are exaggerated in the Syrian bees, except that of amiability. This feature—irritability—would not be an objection to an experienced bee-keeper. I believe, after several years' experience with the Syrians, that they would soon become as pleasant to manage and handle as are the Italians. They are not so readily subdued with smoke as are the Italians, and require careful handling. They are astonishingly prolific, and keep up the brood-rearing whether there are nectar-secreting flowers or not. As the queen fills a comb before leaving it, the brood is kept very compact. They start a large number of queen-cells, and so for queen-rearing they are super-excellent. The comb honey of these bees is said to be quite inferior, because of thin caps—a point I have failed to observe. I think the honey about equal to that of the Italians in appearance. The Cyprian bees are in no way superior to the Syrians, so far as I can learn, though I have had no experience with them, and they are considerably more irritable. The Carniolans (Figs. 11, 12, 13) are much praised by European bee-keepers. They are certainly very amiable, and so excellent for the beginner. From seeing Mr. Frank Benton handle his Carniolans the summer of 1899, in Washington City, I am persuaded that their amiability has not been exaggerated. The bees were not gathering, it was about sunset, and yet Mr. Benton handled them with no smoke or protection, and that very roughly, yet we received no stings. Mr. Benton, whose experience surely makes him a competent judge, values the Carniolan bees as of superior excellence.

WHAT BEES SHALL WE KEEP ?

The beginner certainly better keep Italians or Carniolans. The Italians are so excellent that the Syrians, good as they are, did not take root among us.

HOW TO ITALIANIZE.

From what has been already explained regarding the natural history of bees, it will be seen that all we have to do to change our bees is to change our queens. Hence, to Italianize a colony, we have only to procure and introduce an Italian queen. The same, of course, is true of Cyprianizing or Syrianizing. If we change the queen we soon change the bees.

HOW TO INTRODUCE A QUEEN.

In dividing colonies, where we give our queen to a colony composed wholly of young bees, it is safe and easy to introduce a queen in the manner explained in the section on artificial increase of colonies. To introduce a queen to a colony composed of old bees requires more care. First, we should seek out the old queen and destroy her, then cage our Italian queen in a wire cage (Fig. 141), which may be made by wind-

FIG. 141.

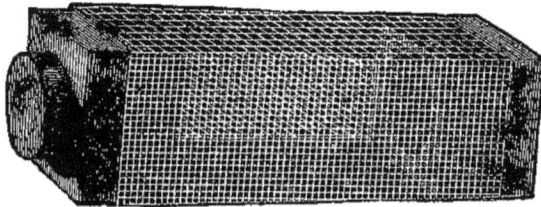

Queen-Cage.—From A. I. Root Co.

ing a strip of wire-cloth, three and one-half inches wide, and containing fifteen to twenty meshes to the inch, about the finger. Let it lap each way one-half inch, then cut it off. Ravel out the half-inch on each side, and weave in the ends of the wires, forming a tube the size of the finger. We now have only to put the queen in the tube and pinch the ends together, and the queen is caged. The cage containing the queen should be inserted between two adjacent combs containing honey, each of which will touch it. The queen can thus sip honey as she needs it. If we fear the queen may not be able to sip the honey through the meshes of the wire, we may dip a piece of clean sponge in honey and insert it in the upper end of the cage before we compress this end. This will furnish the queen with the needed food. In forty-eight hours we again open the hive, after a thorough smoking, and also the cage, which is easily done by pressing the upper end at right angles to the direction of the pressure when we closed it. In doing this do not remove the cage. Now keep watch, and if, as the bees enter the cage, or as the queen emerges, the bees attack her, secure her immediately and re-cage her for another forty-eight hours. I have introduced many queens in this manner, and have very rarely been unsuccessful. At such times if the queen is not well received by the bees, then she is "balled," as it is termed. By the expression, "balling the queen," we mean that the worker-bees press about her in a compact cluster, so as to form a real live ball as large as a good-sized peach. Here the queen is held till she dies; or at least I have

FIG. 142.

Queen-Cage.—From A. I. Root Co.

repeatedly had queens balled and the next day would find them in front of the hive dead. By smoking the ball or throwing it into water the queen may be speedily liberated. Mr. Dadant stops the cage with a plug of wood (Fig. 141), and when he goes to liberate the queen replaces the wooden stopple with one of comb, and leaves the bees to liberate the queen by eating out the comb. Mr. Betsinger uses a larger cage, open at one end, which is pressed against the comb till the mouth of the cage reaches the middle of it. If I understand him, the queen is thus held by comb and cage till the bees liberate her. It is a better way to form a nucleus, all of young bees, and let the bees liberate the queen from a cage with the opening stopped with candy.

If, upon liberating the queen, we find that the bees "ball" her, that is, gather so closely about her as to form a compact cluster, we must at once smoke the bees off and re-cage the queen, else they will hold her a prisoner till she is dead.

The Peet cage (Figs. 136 and 142), which is not only an introducing but a shipping cage, is a most valuable invention. The back of the cage is tin (Fig. 142), and may be drawn out, which leaves the back of the cage entirely open. The pieces in front (Fig. 136) are to be tacked on in shipping. They prevent the accompanying bees from stinging any one who may handle the cage, and also secure ventilation. The tin points, which turn easily, are turned at right angles to the cage, as shown in the figure. The cage is pressed close up to a smooth piece of comb containing both brood and honey, where it is held by the tin points, and then the tin back is withdrawn. The bees will soon eat under the comb and thus liberate the queen and almost always accept her. I have had such admirable success with this cage that I heartily recommend it. The food in the cage will keep the queen, even though the bees do not feed her through the wire, and there is no honey in the comb. The Benton cage (Fig. 144) is a modification of the Peet cage, and as it is now almost universally used for shipping and introducing, it must be an improvement. Here candy holds the queen a prisoner, and she is safe from starvation until the bees liberate her by eating the candy, which ought, and usually does, make them sweet and amiable.

Judge Andrews, of Texas, states a valuable point in this connection, which, though I have not tried, I am glad to give. The reputation of Judge Andrews and the value of the suggestion alike warrant it. He says the queens will be accepted just as quickly when caged in a hive with a colony of bees, even though the old queen is still at large in the hive. Such caged queens, says the Judge, after two or three days, are just as satisfactory to the worker-bees as though " to the manor born," and even more safe when liberated—of course the old queen is first removed—as the bees start no queen-cells, if the old queen has remained in the hive until this time, and the presence of queen-cells agitates the newly liberated queen, which is pretty sure to cause her destruction. Here, then, we may cage and keep our queens after they have been fecundated in the nuclei, and at any time can take one of these, or the old queen, at pleasure, to use elsewhere, though, if the latter, we must liberate one of the caged queens, which, says the Judge, "will always be welcomed by the bees." Mr. Doolittle, as already stated, causes the bees to fill themselves with honey, then shakes them into a box, which is set for a day in a cool, dark room, when the new queen can be given them at once, even though she be a virgin. It is also stated that if we remove a queen at noonday, and after dark smoke the colony, after keeping the queen fasting for half an hour, we may safely introduce her at once. I have tried neither of these methods. I think this is the method of Mr. Simmins, of England.

When bees are not strong, especially if robber-bees are abundant, it is more difficult to succeed, and at such times the utmost caution will occasionally fail of success if the bees are not all young. Sometimes a queen may be safely introduced into a queenless colony by simply shaking the bees all down in front of the hive, and as they pass in, letting the queen run in with them. If the queen to be introduced is in a nucleus, we can almost always introduce her safely by taking the frame containing the queen, bees and all, and setting it in the middle of the hive containing the queenless colony; though it is well to smoke the colony well.

A young queen, just emerging from a cell, can always be safely given at once to a colony, after destroying the old queen.

A queen-cell is usually received with favor, especially if the colony has been queenless for twenty-four hours. If we use a cell we must be careful to destroy all other queen-cells that may be formed; and if the one we supply is destroyed, wait twenty-four hours and introduce another. If we wait seven or eight days, and then destroy all their queen-cells, the bees are sure to accept a cell. If we use the West cell-protector (Fig. 137) then there is of course no danger.

If we are to introduce an imported queen, or one of very great value, we might make a new colony, all of young bees. We simply place two or three combs of fully matured brood in a hive, and the queen on them. By nightfall there will be a goodly cluster of young bees. Unless the day and night are warm the hive must be set in a warm room. The entrance should be closed, in any case. This keeps the queen from leaving, and robber-bees from doing harm. As the number of bees warrant it, more brood may be added, and by adding capped brood alone we may very soon have a full-sized colony.

By having a colony thus Italianized in the fall, we may commence the next spring, and, as described in the section explaining the rearing of queens, we may control our rearing of drones, queens, and all, and ere another autumn have only

FIG. 143.

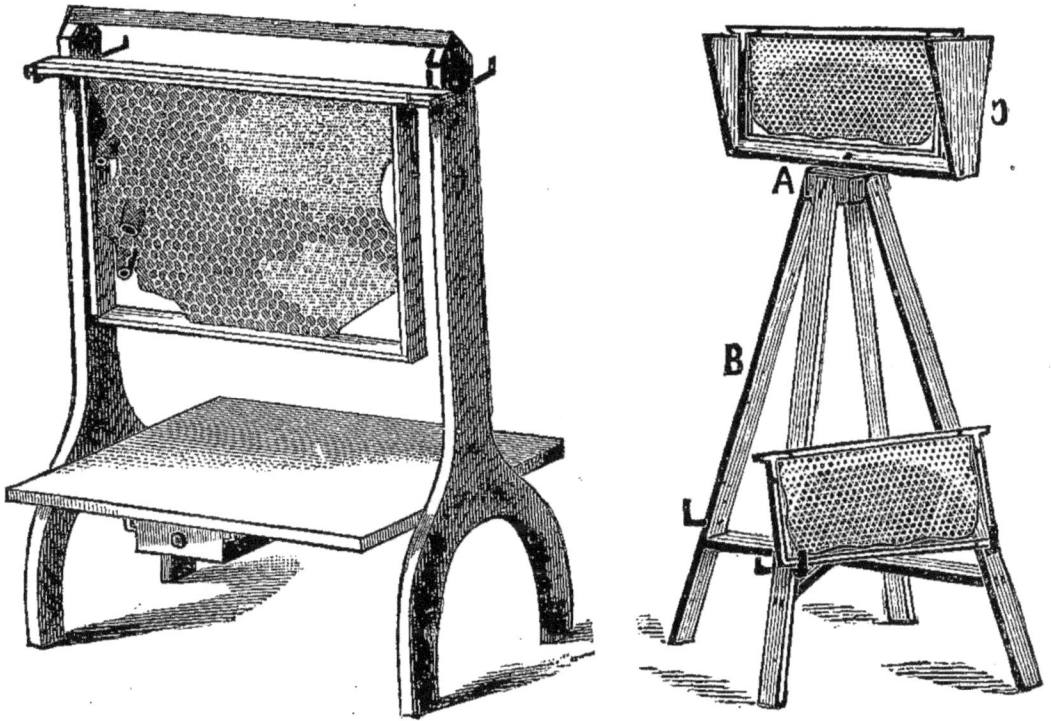

Valentine's Comb-Stand.
—From A. I. Root Co.

Young's Easel.

the beautiful, pure, amiable, and active Italians. I have done this several times, and with the most perfect satisfaction. I think by making this change in blood, we add certainly two dollars to the value of each colony, and I know of no other way to make money so easily and pleasantly. Newly introduced queens will often commence laying at once; almost always within two or three days; although if introduced in late autumn, when the bees have ceased activity for the winter, they may not lay until spring.

VALENTINE'S COMB-STAND.

In the work of finding queens, and in other manipulations, it is often desirable to take out frames. If these are set down beside the hive they are liable to injury. J. M. Valentine has given us a "comb stand" (Fig. 143). As will be seen this holds two frames. The platform is handy to receive tools, and the drawer serves well to hold scissors, knife, queen-cages, etc.

Mr. M. G. Young has invented an "Easel" (Fig. 143) for the same purpose. This will hold several combs. Of course it will not do to leave combs thus exposed, except when bees are busy in the field, or we will have great trouble with robber-bees. I have not found such devices of sufficient use to trouble with them.

TO GET OUR ITALIAN QUEENS.

At present the novice, and probably the honey-producer who prefers to purchase rather than rear his queens, better send to some reliable, experienced breeder, and procure "dollar queens." Unless these are impurely mated, which will rarely happen with first-class breeders, they are just as good as "tested queens." Testing only refers to the matter of pure mating.

I have felt, and still feel, that this cheap queen-traffic tends to haste, not care, in breeding, and that with "dollar queens" ruling in the market, there is lack of inducement for that careful, painstaking labor that is absolutely requisite to give us the best race of bees. It is justly claimed, however, in favor of the "dollar queen" business, that it has hastened the spread of Italian bees, gives those who rather buy than rear their queens a cheap market in which to purchase, and, best of all, weeds out of the business all but the most skillful, cautious, and honest breeders. Only skillful men can make it pay. Only cautious, honest men can find a market for their stock. We know that men are making a handsome profit in the business, and at the same time are giving excellent satisfaction. This is the best argument in favor of any business. I repeat, then, that the beginner better purchase "dollar queens" of some reliable breeder—one who has made queen-rearing a success for years, and given general satisfaction.

I have feared that this "cheap queen" traffic would crush the hard effort, requiring study, time, money, and the most cautious experiment and observation, necessary to give us a very superior race of bees. There is reason to hope now that it will, at most, only delay it. Enterprising apiarists see in this the greatest promise for improved apiculture, and already are moving forward. Enterprising bee-keepers will purchase and pay well for the bee of the future that gives sure evidence of superior excellence. One thing is certain, "dollar queens" are in the market, and are in demand; so, whether the busi-

FIG. 144.

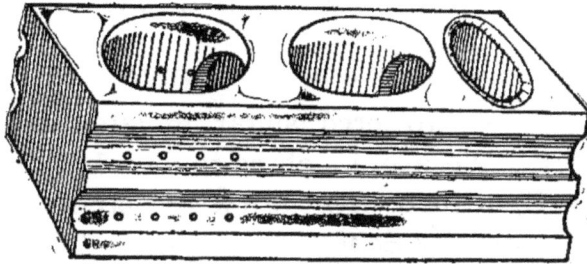

Benton Cage.—From Department of Agriculture.

ness tends to our good or evil, as rational men we must accept the situation and make the most of things as they exist.

Let me urge, however, upon the progressive apiarist, that there is no possible doubt but that the bees of the future will be immensely superior to those of to-day. Man can and will advance here as he has in breeding all other stock. If the obstacles in the way are greater because of the peculiar natural history of the bee, then the triumph, when it comes, will be greater, and the success more praiseworthy.

TO SHIP QUEENS.

For shipping queens the character of the shipping-cage and of the food are of the first importance. Nothing serves better for a cage than Benton's (Fig. 144), already mentioned.

As shown in the figure, the block from which the cage is made has three holes bored almost through it, which do not touch, but are connected by another smaller hole. The hole at one end is ventilated by small holes, as seen in the figure. The grooves prevent suffocation when the cage is wrapped or is snug in the mail-bags. At the other end the hole is waxed, and into this the candy is packed, and before the wire-gauze is added comb foundation is laid on to preserve the moisture. Over the wire-gauze, which covers the holes, wood is tacked. The candy or food apartment has a corked opening at the end, the others at the side. The food should never be honey. This may daub the queen and cause her death. If the food consists of hard candy, then the cage must contain a bottle of water, the cork of which has a small opening, through which is passed a small cotton string. These bottles are not satisfactory, and so our queen-breeders have discovered a moist candy which makes them unnecessary. Fig. 145 shows the cages ready for mailing.

THE "GOOD" CANDY.

This consists of powdered sugar moistened with the best extracted honey. We are indebted to Mr. I. R. Good for this cheap and excellent food, although a similar candy was recommended in Germany by Scholz years ago. The only caution required is to get it just moist enough to keep it soft, and not so moist that it will drip at all. The honey is heated, but only to thin it; then the sugar is stirred in and kneaded. It should stand two or three days. If too thin more sugar may be added. For spring feeding, as before stated, it has been suggested to stir in one-fourth of rye meal, to serve for pollen. In many sections this is unnecessary.

PREPARATIONS TO SHIP.

We have only to catch the queen and about ten workers and introduce them into the cage. We hold the cage in the left hand with the thumb over the hole, to keep the bees in, and with the right hand pick up the queen and eight or ten worker-bees—bright ones, neither very young nor old—by grasping the wings with thumb and index finger, and put them into the cage. Close the opening by inserting the cork, and our queen is ready to mail.

We should send queens by mail (Fig. 145). They go as safely as by express, and it costs but a cent or two. *No one should presume, on any account, to send a queen by mail, unless the queen-cage is covered by this double screen and is provisioned as before directed, instead of with honey.* If shippers neglect these precautions, so that the mails become daubed, or the mail agents stung, we shall again lose the privilege of sending queens by mail.

We have already learned how to introduce the queen. We have only to place her in the hive under the quilt or between

FIG. 145.

Mailing-Cage.—From A. I. Root Co.

two frames, and to withdraw the cork at the candy end. The bees will soon eat the candy, and the queen will be free. If we use this cage to introduce a virgin of some age, we may well paste paper over the holes to delay the exit.

TO MOVE COLONIES.

Should we desire to purchase Italians or other colonies, the only requisites to safe transport are : A wire-cloth cover for ventilation—in shipping by express in hot weather it is wise to put wire below as well as above—securely fastening the frames so they can not possibly move, and combs so old that they shall not break down and fall out. Of course the Hoffman (Fig. 99) close-fitting frames need little fastening.

They fasten themselves.   In spring, wire-gauze over the
entrance usually affords enough ventilation.   If the colony is
very large, and the weather very warm, the entire top of the
hive should be open and covered with gauze, or the bees may
smother.   The entrance ought also to be covered with gauze.
Dr. C. C. Miller, in his valuable book, "Forty Years Among
the Bees," offers a good suggestion.   It is to double a narrow
piece of wire-gauze, a little longer than the entrance to the
hive, and tack the cut edges to one side of a similar shaped
piece of soft wood, so that it will project one-half inch below.
By screwing or tacking this strip just above the entrance of a
hive, we quickly shut the bees in.   Several of these may be
made in advance. I find them very convenient.   If combs are
built from wired foundation they will not break down, even if
new.   Bees thus shut up should never be left where the sun
can shine on them.   I believe that comb partly filled with
water would be grateful to the bees in case of a long journey
in hot weather.   In the cars the frames should extend lengthwise of the cars.   In moving in a wagon, springs or a good
bed of straw should be used, and the frames should extend
crosswise of the wagon.   I would never advise moving bees in
winter in regions of cold winters, though it has often been
done with entire safety.   I should wish the bees to have a
flight very soon after such disturbance.   Of course this does
not apply to such localities as California.

CHAPTER XII.

EXTRACTING AND THE EXTRACTOR.

The brood-chamber is often so filled with honey that the queen has no room to lay her eggs, especially if there is any neglect to give other room for storing. Honey in brood-combs is unsalable, because the combs are dark, and the size undesirable. Comb is very valuable, and should never be taken from the bees, except when desired to render the honey more marketable. Hence, the apiarist finds a very efficient auxiliary in the

HONEY-EXTRACTOR.

No doubt some have expected and claimed too much for this machine. It is equally true that some have blundered quite as seriously in an opposite direction. For, since Mr. Langstroth gave the practical movable frame to the world, the apiarist has not been so deeply indebted to any inventor as to him who gave us, in 1865, the "Mel Extractor," Herr von Hruschka, of Germany.

The principle which makes this machine effective is that of centrifugal force, and it was suggested to Major von Hruschka by noticing that a piece of comb which was twirled by his boy at the end of a string was emptied of its honey. Herr von Hruschka's machine was essentially like those now so common, though in lightness and convenience there has been a marked improvement. His machine consisted of a wooden tub, with a vertical axle in the center, which revolved in a socket fastened to the bottom of the vessel, while from the top of the tub fastenings extended to the axle, which projected for a distance above. The axle was thus held exactly in the center of the tub. Attached to the axle was a frame or rack to hold the comb, whose outer face rested against a wire-cloth. The axle with its attached frame, which latter held the uncapped comb, was made to revolve by rapidly unwinding a string which had been previously wound about the axle, after the manner of top-spininng. Replace the wooden tub with one of tin, and the string with gearing, and it will be seen that we have essentially the neat extractor of to-day.

FIG. 146.

FIG. 147.

Comb-Basket.
—From C. F. Muth.

FIG. 148.

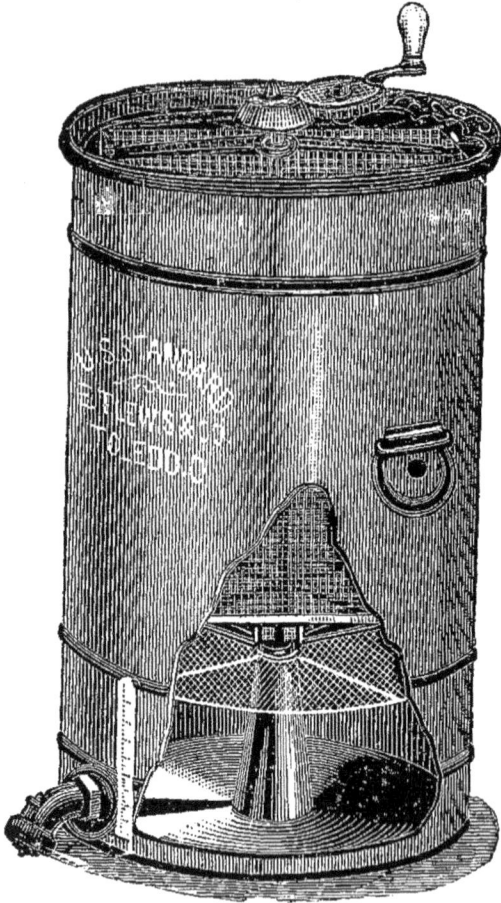

United States Extractor.
—From American Bee Journal.

Muth Extractor.
—From C. F. Muth.

The machine is of foreign invention, is not covered by a patent, and so may be made by any one who desires to do so.

The first American honey-extractor was that made by Messrs. Langstroth and Wagner (see American Bee Journal, Vol. III, No. 10), in the year 1867. As we should expect, our enterprising friends, A. I. Root (Novice), M. M. Baldridge, who, I think, was first, and others were soon in the field.

Some of these early extractors, like the Peabody, ran without gearing ; others, like Mr. Baldridge's, were of wood, while Mr. Langstroth's, if we may judge from the engraving, was very much like the ones of to-day.

DESIRABLE POINTS IN AN EXTRACTOR.

The machine (Fig. 146) should be as light as is consistent with strength. It is desirable that the can be made of tin, as it will be neater and more easily kept sweet and clean. The can should be stationary, so that only a light frame (Fig. 147) shall revolve with the comb. In some of the extractors (Fig. 147) the walls of this frame incline. This keeps the frames from falling in when the machine is at rest, but varies the centrifugal force at the top and bottom of the comb, which is urged as an objection. Of course this difference in force is very slight. Some of the extractors, like the United States (Fig. 146), are made so that the whole center can be removed in a moment, and with the central axis removed so that combs can be reversed without removal from the extractor, both of which are substantial improvements. But the most decided improvement is seen in the automatic extractor. This extractor (Fig. 149) is so made that the combs can be quickly reversed without removal from the extractor. This machine, although it costs. more than any other, will be especially prized in large apiaries. By simply reversing the motion the combs are also reversed.

It is desirable that the machine should run with gearing, not only for ease, but also to insure or allow an even motion, so that we need not throw even drone-larvæ from the brood-cells while in the act of extracting. In some machines the crank runs in a horizontal plane (Fig. 146), in others in a ver-. tical plane (Fig. 148). Both styles have their friends. I think there is little choice between them. The arrangement for exit of the honey should permit a speedy and perfect shut-off. A molasses-gate is excellent to serve for a faucet. I also prefer that the can should hold 30 or 40 pounds of honey before it would be necessary to let the honey flow from it. Large apiarists, like Mr. McIntyre, use power to run the extractor, and let the honey run continuously into a large tank below.

In case of small frames, I should prefer that the comb-basket might hold·four, or even more, frames. The comb-basket should be placed so low in the can that no honey will be thrown over the top to daub the person using the extractor. I think that a wire attachment with a tin bottom (Fig. 150, *a, b*) and made to hook on to the comb-basket, which will hold pieces of comb not in frames, is a desirable addition to an extractor. At present all our large apiarists use automatic reversing

FIG. 149.

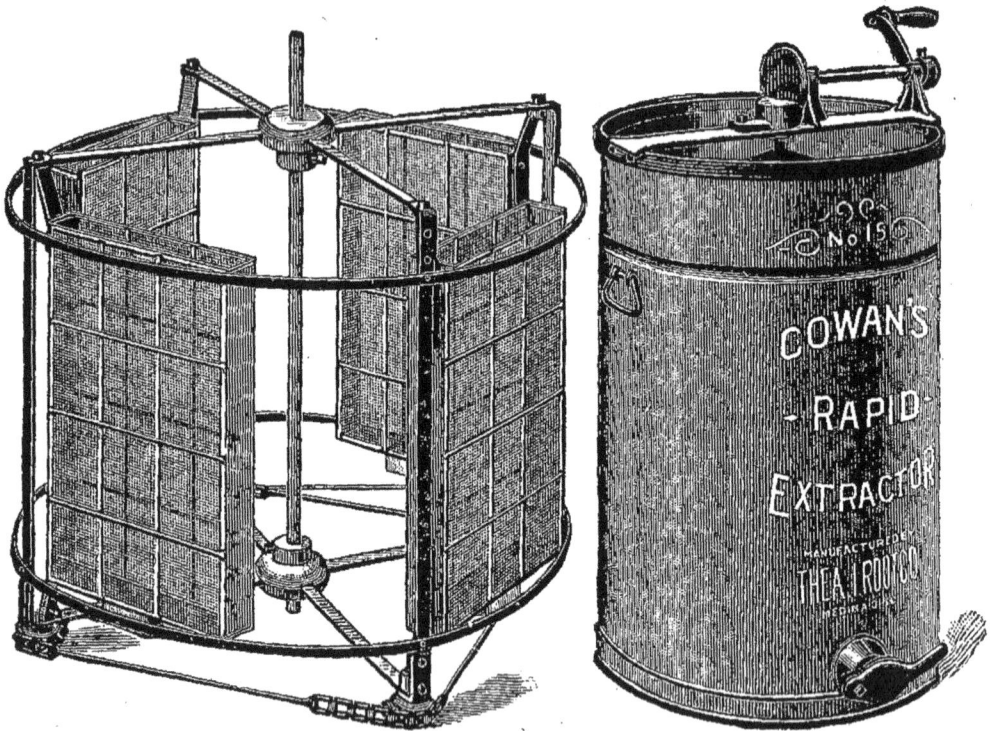

Inside Cowan's Automatic Extractor.
—From A. I. Root Co.

extractors, invented, I think, by Mr. Thos. W. Cowan. These reverse the combs automatically while in motion, and so are a great saving of time.

The can, if metal, which is lighter, and to be preferred to wood, as it does not sour nor absorb the honey, should be of tin, so as not to rust. A cover (Fig. 148) to protect the honey from dust, when not in use, is very desirable. The circular cloth cover, gathered around the edge by a string or a rubber, as made by Mr. A. I. Root, is excellent for this purpose. As no capped honey can be extracted, it is necessary to uncap it, which is done by shaving off the thin caps. To do this, nothing is comparable in excellence to the Bingham & Hetherington honey-knife (Fig. 151). After a thorough trial of this

FIG. 150.

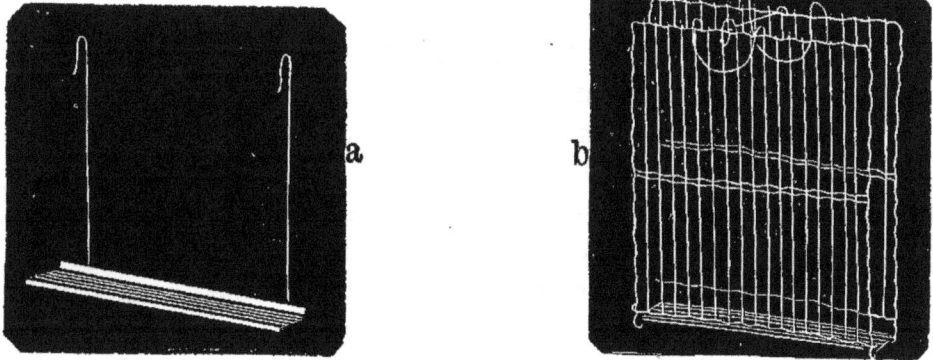

From American Bee Journal.

knife, here at the college, we pronounced it decidedly superior to any other that we have used, though we have several of the principal knives made in the United States. I do not think the bee-keeper can afford to use any other knife. This knife is peculiar for its thick blade, which is beveled on the edge.

USE OF THE EXTRACTOR.

Although some of our most experienced apiarists say nay, it is nevertheless a fact, that the queen sometimes remains idle,

FIG. 151.

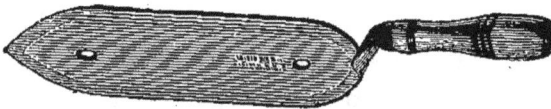

Bingham Knife.—From T. F. Bingham.

or extrudes her eggs only to be lost, simply because there are no empty cells. The honey-yield is so great that the workers occupy every available space, and sometimes they even become unwilling idlers simply because of necessity. It is true that the proper arrangement and best management of frames for surplus would prevent this. Yet in every apiary such a condi-

tion will occasionally occur ; at such times we should always
extract from the brood-chamber.

The extractor also enables the apiarist to secure honey—
extracted honey—in poor seasons, when he could get very little,
if any, in sections or boxes. By use of the extractor we can
largely avoid swarming, and thus work for honey instead of
increase of colonies.

By use of the extractor, at any time or season, the apiarist
—especially the beginner—can secure nearly, if not quite,
double the amount of honey that he could get in combs. It
requires much more skill to succeed in procuring comb honey
than is required to secure extracted. The beginner will
usually succeed far better if he work for extracted honey.

The extractor enables us to remove uncapped honey in the
fall, which, if left in the hive, may prove injurious to the bees.
It is usually better, however, to let the bees do this. By giving
many frames—hundreds at a time—these may be given to bees
in a box or in hives piled high above each other, right in the
apiary.

By use of the extractor, too, we can throw the honey from
our surplus brood-combs in the fall, and thus have a salable
article, and have the empty combs, which are invaluable for
use the next spring.

If the revolving racks of the extractor have a wire basket
attachment (Fig. 150), as I have suggested, the uncapped sections can be emptied in the fall and used the following spring
at a marked advantage. These, of course, may be cleaned by
the bees, as above described, or, if we have but few, by placing them in a super above any strong colony. Pieces of drone-
comb cut from the brood-chamber, which are so admirable for
starters in the sections, can be emptied of their honey at any
season.

By use of the extractor we can furnish at two-thirds the
price we ask for comb honey, an article which is equal, if not
superior, to the best comb honey, and, which, were it not for
appearance alone, would soon drive the latter from the market.
Extracted honey is also much more easily and safely shipped.

Indeed, extracted honey is gaining so rapidly in public

favor that even now its production is far in excess of that of comb honey.

Of course, extracted honey will never sell at a price equal to that of comb. Yet many bee-keepers will secure enough more to more than make up for this. Again, the extractor is ever a temptation to rob the bees, so that with winter will come starvation.

WHEN TO USE THE EXTRACTOR.

The novice should not extract unless the bees are working, else he will be very likely to induce robbing. Even the experienced bee-keeper must be very careful at such times. The bee-tent, soon to be described, is a great help then.

If extracted honey can be sold for half that secured for comb honey, the extractor may be used profitably the summer through ; otherwise it may be used as suggested by the principles stated above.

I would always extract just as the bees commence capping the honey. Then we avoid the labor of uncapping, and still have the honey thick and nearly ripe, as it is styled. I have proved over and over that honey may be extracted when quite thin, and artificially ripened or evaporated, and be equal to the very best. However, as there is danger of imperfect ripening it is wisest to leave it in the combs till the bees commence capping it. Many tier up and leave all in the hive till the busy season is over, then extract all, when the honey is of course thick and of the best quality. This is the method advised and practiced by such able authorities as the Dadants. This requires great care because of robbing. Unripe honey usually has a greenish tinge, and does not granulate as completely as does well-ripened honey. If the honey granulates, it can be reduced to the fluid state with no injury by heating, though the temperature should never rise above 200 degrees F. This can best be done by placing the vessel containing the honey in another containing water, though if the second vessel be set on a stove, a tin basin or pieces of wood should prevent the honey vessel from touching the bottom, else the honey will burn. As before stated, the best honey is pretty sure to crystallize, but it may often be prevented by keeping it in a temperature which is constantly above 80 degrees F. If canned honey is set on top of a furnace in which a fire is kept burning, it will remain liquid indefinitely. If honey is heated to 180 degrees F. and sealed, it will be uninjured, and generally remains ever after liquid. It may be cheaply canned in the usual fruit cans, or in bottles, if we dip the tops in melted wax after corking, to insure making all air-tight. Granulated honey, if reduced, will often remain permanently liquid. It is a curious fact that unripe honey is quicker to granulate than is honey that is thoroughly evaporated. If we drain the liquid from honey that is partially granulated, and melt the hard crystals, we secure a very superior product. If candied honey is but partially crystallized, the liquid part may color all as we melt the crystals, even though we do not burn it.

The fact that honey granulates is the best test of its purity. To be sure, some honey does not crystallize, but if honey does we may pretty safely decide that it is unadulterated.

To render the honey free from small pieces of comb or other impurities, it should either be passed through a cloth or wire sieve—I purposely refrain from the use of the word strainer, as we should neither use the word strained, nor allow it to be used, in connection with extracted honey—or else draw it off into a barrel, with a faucet or molasses-gate near the lower end, and after all particles of solid matter have risen to the top, draw off the clear honey from the bottom. In case of very thick honey, this method is not so satisfactory as the first. I hardly need say that honey, when heated, is thinner, and will of course pass more readily through common toweling or fine wire-cloth. If a sheet of queen-excluding zinc is used between the brood-chamber and upper story we shall have no brood above. This saves great loss in honey, for rearing drones is very expensive, and also saves a deal of vexation. The apiarist enjoys full frames of honey, and is annoyed at great patches of drone-brood in the extracting supers; neither does he enjoy cutting off the heads of drone-brood to rid the hives of these expensive hangers on.

Never allow the queen to be forced to idleness for want of empty cells. Extract all uncapped honey in the fall, and the honey from all the brood-combs not needed for winter, unless we allow the bees to clean the combs as above described. The honey should also be thrown from pieces of drone-comb which are cut from the brood-frames, and from the uncapped comb in sections at the close of the season.

FIG. 152.

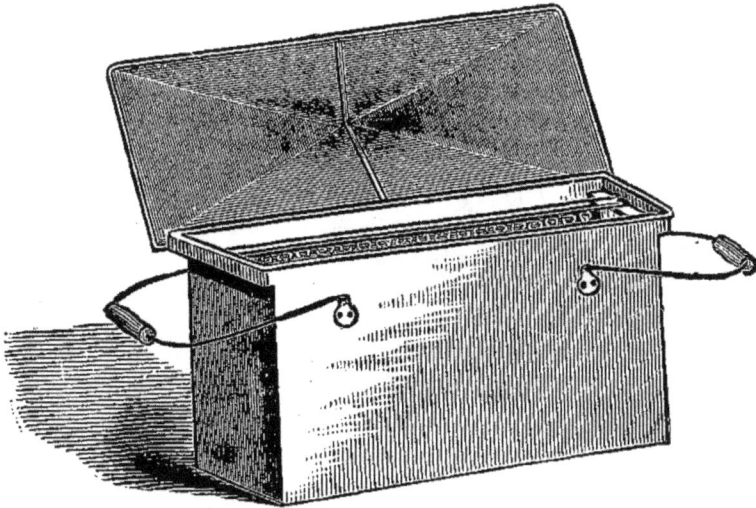

Comb-Box.—From A. I. Root Co.

The apiarist should possess one or two light comb-boxes or baskets (Fig. 152) of sufficient size to hold all the frames from a single hive. These should have convenient handles, and a close-fitting cover. Many large apiarists prefer a comb-cart

FIG. 153.

Osburn's Comb-Cart.—From A. I. Root Co.

(Fig. 153). The box of this is much like a hive, and to one end a cloth cover is tacked. Thus, the combs are easily covered and carried. The bees may be shaken off or brushed off with a large feather, pine twig, or other brush. A little experience makes it easy to shake bees—even Italians—from a comb. A quick, forcible, vertical jerk will always do it. We often find that a mild jar, quickly followed by an energetic one, will fell nearly every bee from the comb. The Davis brush (Fig.

FIG. 154.

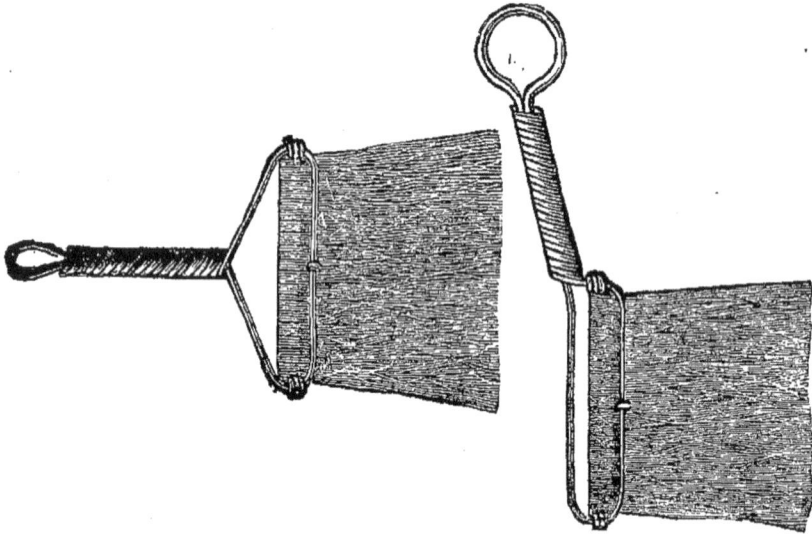

Davis Brush.—From A. I. Root Co.

154) is excellent for removing bees from the combs. It is kept for sale by supply dealers. A soft brush broom (Fig. 155) is excellent. It should be long and slim, and will be less harsh if partially thinned. If the bees are troublesome, close the

FIG. 155.

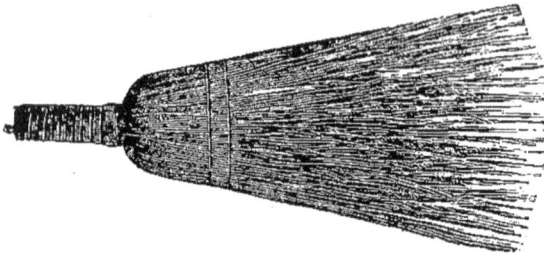

Coggshall Bee-Brush.—From A. I. Root Co.

box or cart cover as soon as each comb is placed inside. The Porter or other good bee-escape (Fig. 160) is a bonanza in extracting. We should have several honey-boards, each with an escape. One is placed under the extracting-combs of each colony at nightfall. In the morning the bees will be all below, and so each extracting-hive of combs can be carried at once to the extractor. By having extra sets of extracting-combs we may at once replace the ones removed, and thus have only to go to a hive once. Because of the cool nights these escapes do not always work as well in California. Extract the honey from these, using care not to turn so hard as to throw out the brood. If capped, and it always should be partially capped before extracting, with a thin knife pare off the caps, and after throwing the honey from one side turn the comb around

FIG. 156.

FIG. 157.

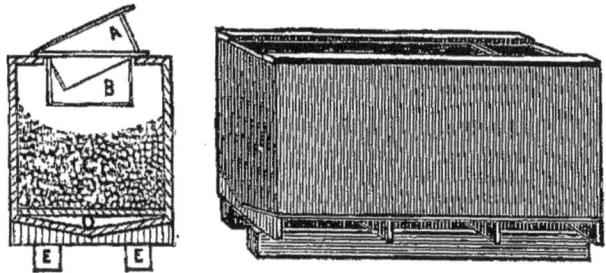

McIntyre's Uncapping-Box.—
From A. I. Root Co.

Dadant's Uncapping-Can.—
From A. I. Root Co.

and extract it from the other. The Dadant uncapping-can (Fig. 156) will be very convenient. It is formed like the extractor, and consists of two parts, about equal. The upper fits into the lower, which has a fine wire screen at the top, and a discharge gate at the bottom. A comb-rest holds the combs. This drains the cappings, and gives us a very superior quality of honey. Mr. McIntyre uses a large box (Fig. 157) six feet long, with a tin tray at the bottom. The comb-rest is on a pivot, so as to turn readily. The large size insures quick drainage, so that the cappings are soon ready for the solar wax-extractor. If the combs are of very different weights, it will be better for the extractor to use those of nearly equal weights on opposite sides, as the strain will be much less. Now take these combs to another colony, whose combs shall be replaced by them. Then close the hive, extract this second set of combs, and thus proceed till all the honey has been extracted. At the close, the one or two colonies from which the first combs were taken shall receive pay from the last set

FIG. 158.

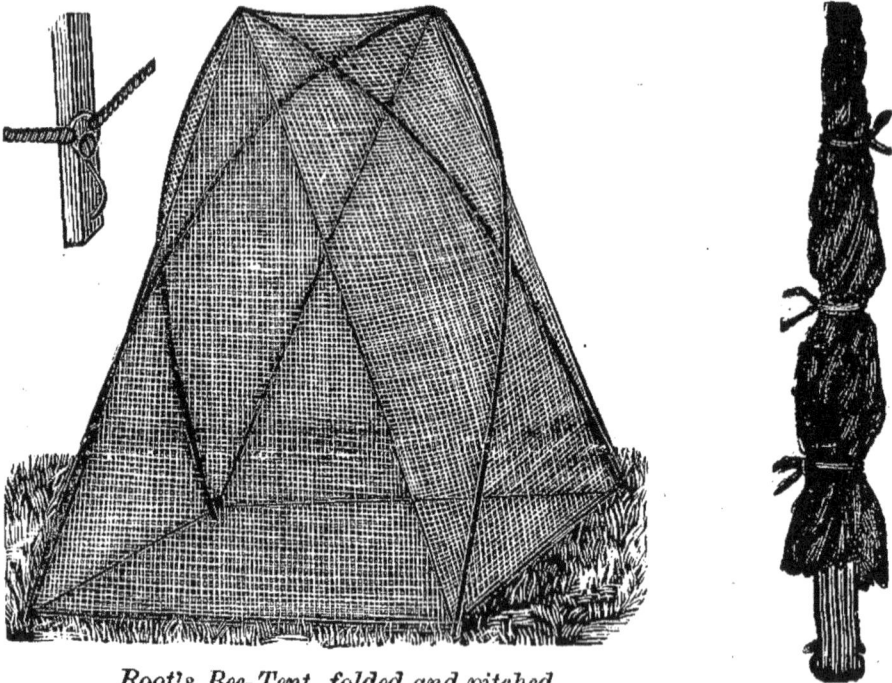

Root's Bee-Tent, folded and pitched.
—From A. I. Root Co.

extracted, and thus, with much saving of robbing, in case there is no gathering, we have gone rapidly through the apiary.

Some apiarists take the first set of combs from a single colony, and leave the colony without combs till they are through for the day. A better way is to keep an extra set of combs on hand. If the bee-keeper works for extracted honey, the extracting-combs should be kept separately in an upper story (Figs. 84 and 87), while the queen and breeding should be kept below in the lower story of the hive.

In case the bees are not gathering, we shall escape robbing and stings by the use of the tent (Fig. 158). This covers the hive and operator. The one figured is very ingenious in its construction, is light and cheap. Mr. Root sells it all made for use for $1.50.

TO KEEP EXTRACTED HONEY.

Extracted honey, if to be sold in cans or bottles, may be run into them from the extractor. The honey should be thick, and the vessels may be sealed or corked, and boxed at once.

If large quantities of honey are extracted, it may be most conveniently kept in barrels. These should be first-class, and ought to be waxed before using them, to make assurance doubly sure against any leakage. No rosin should be used with the wax, as it injures the honey. Good second-hand alcohol barrels are excellent, and cost but $1.00. These and whisky barrels need no waxing. They must be thoroughly cleaned, but must never be charred by burning inside. To wax the barrels, we may use beeswax, but paraffine is cheaper, and just as efficient. Three or four quarts of the hot paraffine or wax should be turned into the barrel, the bung driven in tight, the barrel twirled in every position, after which the bung is loosened by a blow with the hammer, and the residue of the wax turned out. Economy requires that the barrels be warm when waxed, so that only a thin coat will be appropriated. Barrels must be tight without soaking, though it is best to drive the hoops well before using them. We should also test them by use of a little hot water before use. If, when sealed, no steam escapes they are surely tight. Cypress kegs are much used for smaller vessels, but are more expensive, while the second-hand alcohol barrel holds about 500 pounds. Kegs that hold respectively 50, 100, and 175 pounds cost 40, 60, and 80 cents. Barrels or kegs should never be soaked, as the honey absorbs the water, and leaking will almost surely follow. If driving the hoops and waxing will not make them tight, then they are unfit for use.

Large tin cans, waxed and soldered at the openings after being filled, are cheap, and may be the most desirable receptacles for extracted honey. Tin cans are rapidly replacing barrels for honey. These are made of various sizes, and are shipped either in a wooden jacket (Fig. 189) or packed in barrels. In the dry climate of California these are absolutely necessary. Barrels are unsafe.

Extracted honey, unless sealed, should always be kept in dry apartments. If thin when extracted, it should be kept in open barrels or cans in a warm, dry room till it has thoroughly ripened. If quite thin it must be kept in a quite warm room, in very shallow vessels. In this way I have ripened very thin honey, so it was of excellent quality. In all such cases the vessels should be covered by cheese-cloth. To remove extracted honey from a barrel, etc., we may remove one head, or, if practicable, the vessel may be put in hot water, which should never be above 180° F., and soon the honey will run off.

CHAPTER XIII.

WORKING FOR COMB HONEY.

While extracted honey has so much to recommend it, and is rapidly growing in favor with American apiarists, still such reports as that of Dr. C. C. Miller, who in 1884 increased his 174 colonies to 202, and took 16,000 pounds of comb honey in one-pound sections, which netted him very nearly $3000; and that of Mr. Doolittle, who has secured nearly 100 pounds of comb honey per colony for a long series of years, may well lead us not to ignore this branch of our business. The showy horse, or the red short-horn, may not be intrinsically superior to the less attractive animals; but they will always win in the market. So comb honey, in the beautiful one-pound sections, will always attract buyers and secure the highest price. As more embark in the production of extracted honey, higher will be the price of the irresistible, incomparable comb honey. Well, then, may we study how to secure the most of this exquisite product of the bees, in a form that shall rival in attractiveness that of the product itself, for very likely the state of the market in some localities will make its production the most profitable feature of apiculture.

POINTS TO CONSIDER.

To secure abundance of comb honey the colonies must be very strong, and the brood-combs full of brood at the dawn of the honey harvest. The swarming-fever must be kept at bay or cured before the rapid storing commences, and the honey should be secured in the most attractive form.

TO SECURE STRONG COLONIES.

By feeding daily, whenever the bees are not storing, commencing as soon as the bees commence to store pollen, we shall most certainly, if the bees have been well wintered, secure this result. Yet bees are naturally active after their long winter's rest, and this stimulative feeding rarely pays. We should also use the division-board, and keep the bees crowded, especially if weak in the spring. Only give them the number of combs that they can cover. It is very important to keep all warm. Doolittle says this necessitates a telescope cover to the hive. Though this last may with proper management be unnecessary, it certainly does no harm; it may aid greatly. True, Mr. Heddon objects to this work of feeding and manipulating division-boards, and secures much honey and money. I have often wondered what his genius and skill would accomplish should he vary his method in this respect. Instead of feeding by use of the Smith (Fig. 127) or other feeder, we may uncap a comb of honey and with it separate combs of brood as the bees get two or three full frames of the latter. This will stimulate the bees, and as they will carry the honey from the uncapped cells the queen will be impelled to most rapid laying. We may also fill empty combs as already described, and place these in or close beside the brood-nest. By turning around the brood-combs, or separating them by adding combs with empty cells as the colonies gain in strength, we hasten brood-rearing to the utmost. This matter of separating the brood-combs must be very cautiously managed or brood will be chilled and much harm done. Most . bee-keepers do not take all this pains. Each one's experience must be guide.

TO AVOID THE SWARMING-FEVER.

This is not always possible by any method, and has ever been the obstacle in the way of successful comb-honey production. The swarming impulse and great yields of this delectable product are entirely antagonistic. Messrs. Heddon, Davis, and others, let the bees swarm. They hive these swarms on foundation, and hope to have this all done, and both colonies strong, in time for the honey harvest. Messrs. Hutchinson and Doolittle hive the swarm on empty frames, always, however, with starters, placing sections with their foundation, or better, comb, on the hive at once. It is specially desirable to have a few combs in the sections, to bait the bees and attract them to the supers. They also restrict the brood-chamber, either by filling the space with division-boards (Doolittle), or by using the lower half of a horizontally divided brood-chamber (Hutchinson), see new Heddon hive, p. 189. In this way the whole working force is put at once into the sections. Some of our best Michigan and New York bee-keepers, with Dr. Miller, let the bees swarm, and return them, either caging the queen or placing her in a nucleus for seven days, then return her to the bees, after cutting out the queen-cells. This takes nothing from the energy of the bees, and will doubtless work best of all methods in the hands of the beginner. In this case, as the full energy of the colony is turned to storing, the amount of honey would be theoretically greater. My practice sustains the theory. Such authorities as Messrs. Hetherington and Elwood practice this method. J. H. Robertson kills the old queen, and in seven days destroys all but the largest queen-cell, and so gains the same end, and requeens his apiary. If increase is desired, however, then Mr. Hutchinson's method should be followed. The yield of comb honey in this last case will not usually be so great, though in excellent seasons it may be greater.

Some very able bee-keepers manipulate so skillfully by adding empty combs to the hives, as to keep this swarming impulse in check, and still keep the bees- increasing most rapidly. Others divide the colonies, and so hold at bay the swarming-fever. All must practice as their own experience proves best, as the same method will not have equal value with different persons. We must work as best we can to secure strong colonies, and check or retard the swarming-fever, and, while learning by experience to do this, may well work the most of our bees for extracted honey, which is more easily secured, and is sure to be in demand, even though the price is less. The quantity may more than compensate for lower price.

ADJUSTMENT OF SECTIONS.

As before suggested, a wide space between bottom-bars of sections—three-eighths inch—is desirable. J. A. Green has half-story supers with frames only one-half as deep for extracting. These are put one on each hive at the dawn of the honey harvest. As soon as the bees commence to work in them, they are removed or raised and a section-case put in their place. As the bees commence in the sections these extracting half-story hives are used one above another with such colonies as are worked for extracted honey. The sections should be on at the very dawn of each honey harvest, as white clover, basswood, etc. At first the full set of sections better not be added, but as soon as the bees commence to work well in them, they all should be added, on side and top, if side-storing is practiced, and, if we wish to tier up, the case of sections first added should be raised and others added below. I like this practice of tiering up very much. As soon as the bees are working well in all the sections I raise the case and place another underneath. This is continued, often till there are three cases of sections on a single hive. Some think that if the unfilled case of sections is placed above instead of below, that less unfilled sections will remain at the close of the season, as the warmth higher up is grateful to the bees. As already stated, it is best not to have the sections too closely shut in. Slight ventilation is often desirable.

If the queen troubles by entering the sections, use may be made of the perforated zinc (Fig. 91), or, better still, the queen-excluding honey-board (Fig. 91), to keep her from them. As already suggested, we must arrange the form and size of sections as the market and our hives and apparatus make most desirable. We may vary the size and form of our sections so as to make them smaller, and yet use the same cases or frames that we used with larger sections. Small sections are most ready of sale, and safest to ship ; yet with their use we may secure less honey.

If we can get nice, straight combs by having them less thick without using separators in the sections, so that these latter can be readily placed side by side in shipping-cases, then we, by all means, better omit the separators. If we use separators, we can use wood or tin. Wood is cheapest, and I find that in practice it serves even better than tin. The plain sections with fence (page 241) give separators and wide connections, and are sure to grow in favor.

GETTING BEES INTO SECTIONS.

The crowded hive or brood-chamber, with no intent to swarm, the wide spaces between sections, and a rich harvest of nectar will usually send the bees into the sections with a rush. If they refuse to go, sections with comb, a little drone-brood, or the exchange of sections temporarily from above to the brood-nest, or the moving of a brood-frame up beside the sections for a short time, as before described, will frequently start the bees into the sections. Some apiarists will have their cases with sections so made that they can be placed between the brood-frames till the bees commence to work in the sections; others accomplish the same end by inverting the frames. Sections filled with foundation—only very thin foundation should be used in sections—are more attractive to the worker-bees. I find that a few sections full of comb in the section-case very greatly aid to tempt the bees to work in the sections. We often may gain our point by taking a case of sections, bees and all, from a hive whose bees are working in the sections, and giving them to the reluctant colony. Or we may gain the same end by giving the bees a one-half story or case of extracting-combs. The bees may enter these at once, when we may raise them and add our section-cases. Later these half-story extracting-combs may be used elsewhere, and may serve there to cut short unprofitable loafing, and to prevent swarming. I also have used the invertible frames to excellent purpose in obtaining the same result. I invert the frames and at the same time uncap the honey in them. The hives must always be shaded from the hot sun. With experience will come the skill which can accomplish this, and make comb-honey production the most fascinating feature of bee-keeping.

REMOVAL OF SECTIONS.

The three-eighths inch space between the upper as well as the lower bars of the sections enables us to see quickly the condition of each section just by removal of the cover. Each section should be removed as soon as capped, if we would have it very nice. Yet it is certainly true that the rich, delicate flavor will be increased if left on the hive even for a month or more. This, of course, can not be done unless we use separators. Any delay will make it dark and hurt its sale. During the harvest we should add other sections to take the place of those removed. Towards the close of the harvest we should not add other sections, for, by contracting the space, the last sections will be more surely filled and quickly capped. To

FIG. 159.

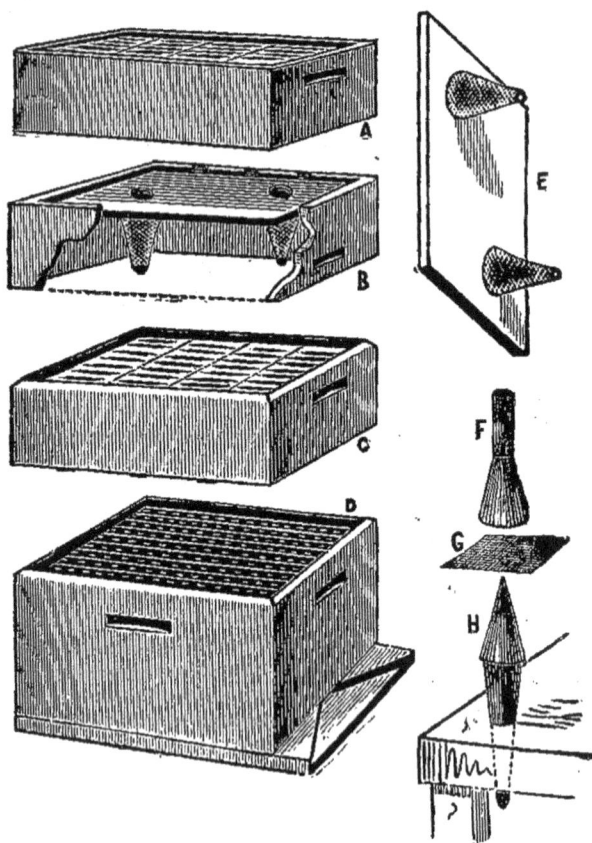

The Reese Cones.—From A. I. Root Co.

remove the bees from single sections taken from frame or case, we have only to brush them off.

Few bee-keepers will stop to remove single sections. In fact, the tiering-up process is, in my opinion, the key to the successful production of comb honey. If we remove a full case we can often shake a large portion of the bees from the sections, then by piling the cases in a box overspread by a sheet, or in a bee-tent, or even in the honey-house, the bees will all leave the sections. J. S. Reese, of Kentucky, invented double cones of wire-gauze, one smaller than and within the other, to remove the bees from sections. These are fastened with their bases (Fig. 159) just over an inch hole in a board just the size of a section-case. When it is desired to clear a

FIG. 160.

Porter Bee-Escape.—From A. I. Root Co.

case of sections of bees, the case is raised and an empty case with the board upon it, and the cones projecting downward, is placed beneath (Fig. 159). One need not try this to know that it would be practical.

The Porter bee-escape (Fig. 160) is much superior to the cones. It lies horizontal, and so requires no more space than the board (Fig. 161) which contains it.

FIG. 161.

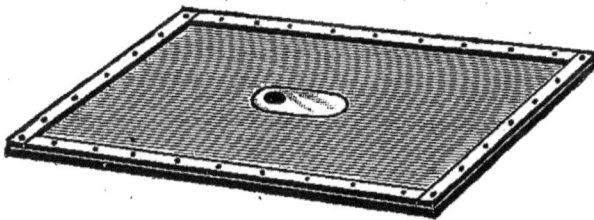

Porter Bee-Escape in Honey-Board.—From A. I. Root Co.

To remove the sections from the case, we invert the case and set it on a shallow box just the size of the case. This need not be more than two inches high. We now lay a block, which will cover a row, on the sections, when, by a smart blow from a mallet, a whole row of sections is loosened at once. Even with the best care and management there will be some unfilled sections at the close of the season. In large apiaries, where there are thousands of these, they may be safely placed in hives, one above the other, and fed to the bees right in the apiary. This will never do except on a very large scale, as it would cause robbing. If nearly full they may be sold in the local market. They may be extracted and the sections used as decoys the next year. Occasionally we can feed extracted honey, and have them filled. This is likely to cause robbing unless very carefully managed, and usually fails to pay.

Of course, all sections must be scraped, as any stain or show of propolis makes second-grade honey. Scraping requires much care, or the honey will be bruised, which would make a bad matter worse. Special boxes of convenient height, with shelves at ends to hold sections so that the edge of section may be flush with edge of the shelf, are used to advantage in cleaning sections. Some have used machinery, such as moving belts, wheels, or moving sandpaper, to accomplish this work. Most use the common case-knife, and usually, especially with the small bee-keeper, that is the best. The box, however, to catch litter, and with end shelves to bear the sections, first suggested by Mr. Boomhower, of New York, is a valuable feature.

If there is any possible danger of moths, the comb honey should be fumigated by use of burning sulphur (see Bee-Moth). Bisulphide of carbon may be used instead of sulphur. It is equally good, and requires less labor. As this last is thought to kill the eggs, it is much to be preferred. This is a wise precaution, even though the bee-keeper rarely sees one of these insects. A single moth can stock several cases of sections with the fatal eggs.

R. L. Taylor, one of Michigan's most successful bee-keepers, who produces large harvests of comb honey, gives the following points to be heeded in producing comb honey :

1. Bees must winter well.

2. There must be a goodly amount of honey in the hive in the spring. Bees never prosper on scant rations.

3. Keep colonies warm in spring.

4. Tier up and leave sections on the hive till just at the close of the season.

5. When removed, pile the cases of sections one upon another, fumigate, and keep in a warm room till sold.

The above are points well worthy consideration, and may be called the axioms of comb-honey production.

CHAPTER XIV.

HANDLING BEES.

But some one asks the question, Shall we not receive those merciless stings, or be introduced to what "Josh" calls the "business end of the bee?" Perhaps there is no more cause-less or more common dread in existence than this of bees' stings. When bees are gathering, they will never sting unless provoked. When at the hives—especially if Italians or Car-niolans—they will rarely make an attack. The common belief, too, that some persons are more liable to attack than others, is, I think, erroneous. With the best opportunity to judge, with our hundreds of students, I think I may safely say that one is almost always as liable to attack as another, except that he is more quiet, or does not greet the usually amiable passer-by with those terrific thrusts, which would vanquish even a practiced pugilist. Occasionally a person *may* have a peculiar odor about his person that angers bees and invites their darting tilts, with drawn swords, venom-tipped; yet, though I take my large classes each season, at frequent inter-vals, to see and handle the bees, each for himself, I still await the first proof of the fact that one person is more liable to be stung than another, providing each carries himself with that composed and dignified bearing that is so pleasing to the bees. True, some people, filled with dread, and the belief that bees regard them with special hate and malice, are so ready for the battle that they commence the strife with nervous head-shakes and beating of the air, and thus force the bees to battle, *nolentes volentes.* I believe that only such are regarded with special aversion by the bees. Hence, I believe that *no one* need be stung.

Bees should never be jarred, nor irritated by quick motions. It is always wise, also, to stand at one side and never in front of the hive. Those with nervous temperament—and I plead very guilty on this point—need not give up, but at first better protect their faces, and, perhaps, even their hands, till time and experience show them that fear is vain; then they will divest themselves of all such useless encumbrances. Bees are more cross when they are gathering no honey, and at such times, black bees and hybrids especially, are so irritable that even the experienced apiarist will wish a veil. Exposing honey about the apiary at such times will increase quickly this irritability. There are some bees that are chronically cross, and are ever about with their menacing cry. Doolittle advises killing these at once. He uses a small paddle-like board for this purpose. I have never minded these chronic grumblers. They usually respect indifference; at least they rarely sting me.

<h2 style="text-align:center">THE BEST BEE VEIL.</h2>

This should be made of black tarlatan, or, better, silk tulle, sewed up like a bag, a half yard long, without top or bottom, and with a diameter of the rim of a common straw-hat. Gather the top with braid, so that it will just slip over the crown of the hat—else, sew it to the edge of the rim of some cheap, cool hat; in fact, I prefer this style—and gather the bottom with rubber cord or rubber tape, so that it may be drawn over the hat rim, and then over the head, as we adjust the hat.

Some prefer to dispense with the rubber cord at the bottom (Fig. 162), and have the veil long so as to be gathered in by the coat or dress. If the black tarlatan troubles by coloring the shirt or collar, or if the silk tulle is thought too expensive, the lower part may be made of white netting; indeed, all may be made of white netting except a small square to be worn just in front of the eyes. When in use, the rubber cord draws the lower part close about the neck, or the lower part tucks within the coat or vest (Fig. 162), and we are safe. This kind of a veil is cool, does not impede vision at all, and can be made by any woman at a cost of less than 20 cents. Common buckskin or sheepskin gloves can be used, as it will scarcely pay to get special gloves for the purpose, for the most timid person— I speak from experience—will soon consider gloves as unnecessary and awkward.

Special rubber gloves are sold by those who keep on hand

apiarian supplies. It is reported that heavily starched linen is proof against the bee's sting, and so may be used for gloves or other clothing. Some apiarists think that dark clothing is specially obnoxious to bees. It is certainly true that fuzzy woolen, and even hairs on one's hands, are very irritating to them. Clothes with a heavy nap should be rejected by the bee-keeper, and the Esaus should singe the hair from their hands. .

For ladies, my friend, Mrs. Baker, recommends a dress which, by use of a rubber skirt-lift or other device, can be

FIG. 162.

Bee-Veil.—Original.

instantly raised or lowered. This will be convenient in the apiary, and tidy anywhere. The Gabrielle style is preferred, and of a length just to reach the floor. It should be belted at the waist, and cut down from the neck in front one-third the length of the waist, to permit the tucking in of the veil. The underwaist should fasten close about the neck. The sleeves should be quite long to allow free use of the arms, and gathered in with a rubber cord at the wrist, which will hug the rubber gauntlets or arm, and prevent bees from crawling up the sleeves. The pantalets should be straight and full, and should also have the rubber cord in the hem to draw them close about the top of the shoes.

Mrs. Baker also places great stress on the wet "head-cap," which she believes the men even would find a great comfort. This is a simple, close-fitting cap, made of two thicknesses of coarse toweling. The head is wet with cold water, and the cap wet in the same, wrung out, and placed on the head.

Mrs. Baker would have the dress neat and clean, and so trimmed that the lady apiarist would ever be ready to greet her brother or sister apiarists. In such a dress there is no danger of stings, and with it there is that show of neatness and taste, without which no pursuit could attract the attention, or at least the patronage, of our refined women.

TO QUIET BEES.

In harvest seasons the bees, especially if Italians, can almost always be handled without their showing resentment. Our college bees—hybrids, between Syrians and Carniolans—are so gentle that I go freely among them without protection each May and June, with my large classes. At first each student puts on the veil, but soon these are thrown aside, and it is rare indeed that any one gets a sting. Even Mr. Doolittle *always* uses a veil when steadily at work in the apiary. But at other times, and whenever they object to necessary familiarity, we have only to cause them to fill with honey—very likely it is the scare that quiets the bees—to render them harmless, unless we pinch them. This can be done by closing the hive so that the bees can not get out, and then rapping on the hive for a short time. Those within will fill with honey, those without will be tamed by surprise, and all will be quiet. Sprinkling the bees with sweetened water will also tend to render them amiable, and will make them more ready to unite to receive a queen, and less apt to sting. Still another method, more convenient, is to smoke the bees. A little smoke blown among the bees will scarcely ever fail to quiet them, though I have known black bees, in autumn, to be very slow to *yield*. It is always wise upon opening a hive to blow a little smoke in at the entrance.

The Syrian bees, when first imported, are maddened rather than quieted by use of smoke. I find, however, that with handling they soon become more like Italians. Deliberation is specially desirable when we first open the hive of Syrian bees. Dry cotton-cloth, closely wound and sewed or tied, or, better, pieces of dry, rotten wood are excellent for the purpose of smoking. These are easily handled, and will burn for a long time. But best of all is a

BELLOWS SMOKER.

This is a tin tube attached to a bellows. Cloth, corn-cobs, damp shavings, or rotten wood (that which has been attacked by dry-rot is the best) can be burned in the tube, and will remain burning a long time. The smoke can be directed at

FIG. 163.

FIG. 164.

Direct-Draft Perfect
BINGHAM
Bee Smoker

PATENTED 1878, 1882 and 1892.

Quinby Smoker.
—From L. C. Root.

Bingham Smoker.—From T. F. Bingham.

pleasure, the bellows easily worked, and the smoker used without any disagreeable effects or danger from fire.

THE QUINBY SMOKER.

This smoker (Fig. 163) was a gift to bee-keepers by the late Mr. Quinby, and was not patented. Though a similar device had been previously used in Europe, without doubt Mr. Quinby was not aware of the fact, and as he was the person to bring it to the notice of bee-keepers, and to make it so perfect

as to challenge the attention and win the favor of apiarists
instanter, he is certainly worthy of great praise, and deserving
of hearty gratitude.

Mr. Bingham was the first to improve the old Quinby
smoker in establishing a direct draft (Fig. 164). Later Mr.
Bingham added a wire fire-guard to the chimney, and hinged
the latter to the fire-tube. Mr. Clark next added the cold draft
(Fig. 165). This has a large fire-chamber, but it is awkward
in form, and the small cold-air tube soon chokes with soot.

FIG. 165.

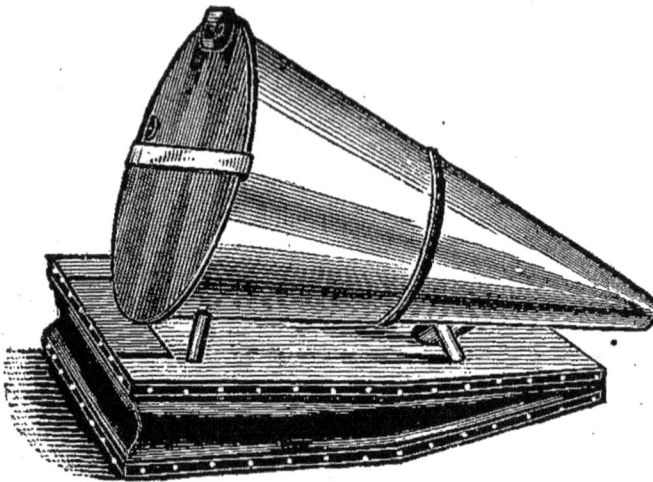

Clark Smoker.—From A. I. Root Co.

There are now several smokers on the market, each of
which has its merits and its friends. I have tried nearly all,
and, in my opinion, the Bingham is incomparably superior to
any other. I should have it at double or triple the price of any
other. Still, I know excellent bee-keepers who prefer the
Clark. No person who keeps even a single colony of bees can
afford to do without some one of them.

TO SMOKE BEES.

Approach the hive, blow a little smoke in at the entrance,
then open from above, and blow in smoke as required. If, at
any time, the bees seem irritable. a few puffs from the smoker
will subdue them. Thus, any person may handle his bees
with perfect freedom and safety. If, at any time, the fire-

chamber and escape-pipe of the smoker become filled with soot, they can easily be cleaned by revolving an iron or hardwood stick inside of them.

CHLOROFORM.

Mr. Jones finds that chloroform is very useful in quieting bees. He puts a dry sponge in the tube of the smoker, then a sponge wet in chloroform—it takes but a few drops—then puts in another dry sponge. These dry sponges prevent the escape of the chloroform, except when the bellows is worked. Mr. Jones finds that bees partially stupefied with chloroform receive queens without any show of ill-will. As soon as the bees begin to fall, the queen is put into the hive, and no more of the vapor added. I tried this one summer with perfect success. This was recommended years ago in Germany, but its use seems to have been abandoned. It is more than likely that Mr. Jones' method of applying the anæsthetic is what makes it more valuable. The smoker diffuses the vapor so that all bees receive it, and none get too much. I should use ether instead of chloroform, as with higher animals it is a little more mild and safe. Our British friends of late are recommending carbolic acid in lieu of smoke to quiet bees. By means of a feather the liquid is brushed about the entrance and along the top of the frames, or else a cloth dampened with the acid is placed over the frames. This is also used to fumigate the bees for the same purpose. Mr. Cheshire advises a little creosote placed in the common smoker, to make the smoke more effective. There is no question but that this obnoxious substance will quiet the bees ; but it seems to me, from a brief experience, that it is far less convenient than the smoker. With fuller experience I say unhesitatingly that for convenience and effectiveness, smoke is quite superior to any of these substances.

TO CURE STINGS.

In case a person is stung, he should step back a little for a moment, as the pungent odor of the venom is likely to anger the bees and induce further stinging. By forcing a little smoke from the smoker on the part stung, we will obscure this odor. The sting should be rubbed off at once. I say rubbed, for we should not grasp it with the finger-nails, as that crowds more poison into the wound. If the pain is such as to prove troublesome, apply a little ammonia. The venom is an acid, and is neutralized by the alkali. A strong solution of saltpeter I have found nearly as good to relieve pain as the ammonia. Ice-cold water drives the blood from any part of our body to which it is applied, and so it often gives relief to quickly immerse the part stung in very cold water. In case horses are badly stung, as sometimes happens, they should be taken as speedily as possible into a barn (a man, too, may escape angry bees by entering a building), where the bees will seldom follow, then wash the horses in soda water, and cover with blankets wet in cold water. Cows picketed many rods from the apiary, in the line of flight to a spring, have been stung to death. Unlike horses, cows will not run off. This fact surely suggests caution.

A wash or lotion, "Apifuge," is praised in England as a preventive of stings. The hands and face are simply washed in it. I have tried it, but could see no advantage. The substances used are oil of wintergreen or methyl salicylate.

THE SWEAT THEORY.

It is often stated that sweaty horses and people are obnoxious to the bees, and hence almost sure targets for their barbed arrows. In warm weather I perspire most profusely, yet am scarcely ever stung, since I have learned to control my nerves. I once kept my bees in the front yard—they looked beautiful on the green lawn—within two rods of a main thoroughfare, and not infrequently let my horse, covered with sweat upon my return from a drive, crop the grass while cooling off, right in the same yard. Of course, there was some danger, though less as I always kept careful watch, but I never knew my horse to get stung. Why, then, the theory? May not the more frequent stings be consequent upon the warm, nervous condition of the individual? The man is more ready to strike and jerk, the horse to stamp and switch. The switching of the horse's tail, like the whisker trap of a full beard, will anger even a good-natured bee. I should dread the motions more than the sweat.

Often when there is no honey to gather, as when we take the last honey in autumn, or prepare the bees for winter, the bees are inordinately cross. This is especially true of black

FIG. 166.

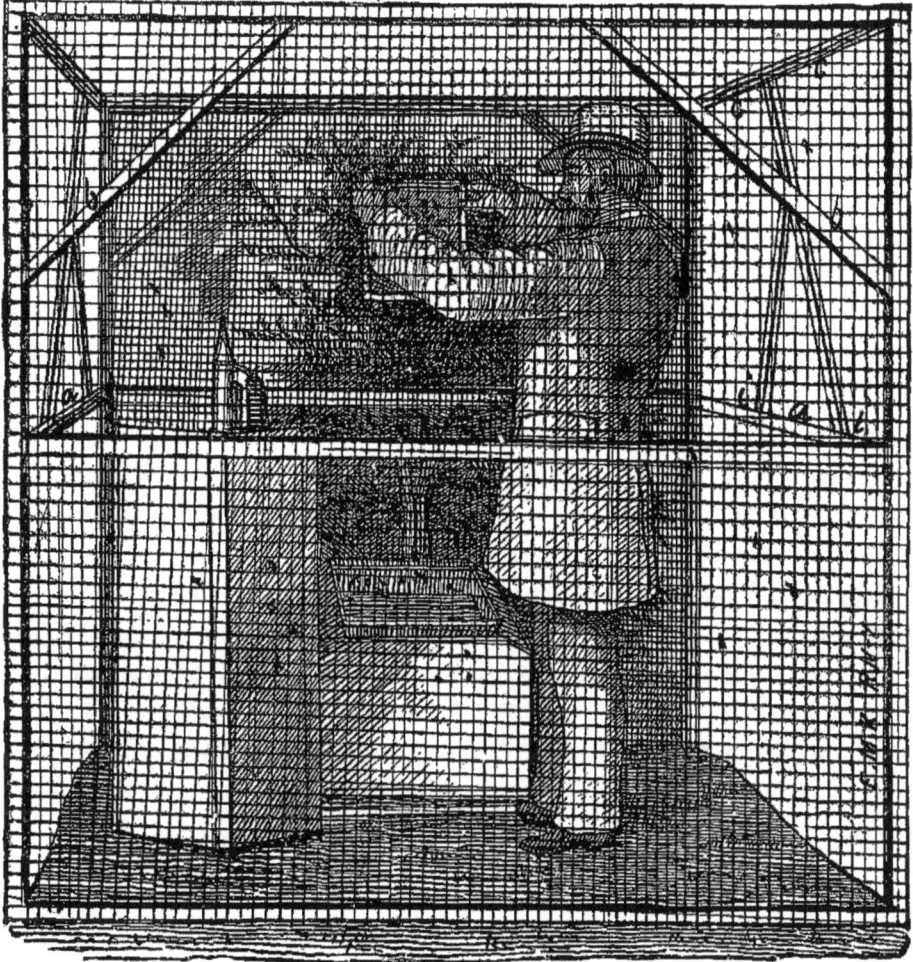

Bee-Tent.—Original.

bees and hybrids. At such times I have found an invaluable aid in

THE BEE-TENT.

This also keeps all robbers from mischief. It is simply a tent which entirely covers the hives, bees, bee-keeper and all. The one I use (Fig. 166) is light, large, and easily moved, or folded up if we wish to put it in the house. The sides are rectangulur frames made of light pine strips, well braced (Fig. 166, *b*, *b*), and covered with wire-cloth. The top and ends are covered with factory-cloth, firmly tacked except at one end, where it is fastened, at will, by rings which hook over screws. The two sides have no permanent connection of wood, except at the ends (Fig. 166, c, c). The small strips which connect at these places are double, and hinged to the side frames, and the two parts of each hinged together. Thus these may drop, and so permit the side frames to come close together where we wish to "fold our tent." The sides are kept apart by center cross-strips at the ends (Fig. 166, a, a), from which braces (Fig. 166, i, i) extend to the double cross-strips above. These center strips, with their braces hinged to them, are separate from the rest of the frame, except when hooked on as we spread the tent. I have since made a similar tent, and for end-pieces used simply four round sticks, the ends of which fitted very closely into holes bored into the uprights of the side frames, one into the top and one into the middle of each. These end-pieces are as long as can be crowded in. This is very simple and excellent.

After use of this tent several years, I can not praise it too highly. It is also admirable in aiding to get bees out of sections—in which case cones, like the Reese cones (Fig. 159), will permit the bees to escape, and to use at fairs, when bees can be manipulated in the tent. I have so used it. The tent should always be used, if we must handle bees when no gathering is being done. There no robbing will be caused. I have already referred to a cheap tent made by A. I. Root (Fig. 158). That, however, is not as convenient as this one.

CHAPTER XV.

COMB FOUNDATION.

Every apiarist of experience knows that empty combs in frames, comb-guides in the sections, to tempt the bees and to insure the proper position of the full combs, in fact, combs of almost any kind or shape, are of great importance. So every skillful apiarist is very careful to save all drone-comb that is cut out of the brood-chamber—where it is worse than useless, as it brings with it myriads of those useless gourmands, the drones—to kill the eggs, remove the brood, or extract the honey, and transfer it to the sections. He is equally careful to keep all his worker-comb, so long as the cells are of proper size to domicile full-sized larvæ, and never to sell any comb, or even comb honey, unless a greater price makes it desirable.

FIG. 167.

Comb Foundation.—From American Bee Journal.

No wonder, then, if comb is so desirable, that German thought and Yankee ingenuity have devised means of giving the bees at least a start in this important yet expensive work of comb-building, and hence the origin of another great aid to the apiarist—comb foundation (Fig, 167).

HISTORY.

For more than forty years the Germans have used impressed sheets of wax as a foundation for comb, as it was first made by Herr Mehring, in 1857. These sheets are several times as thick as the partition at the center of natural comb. This is pressed between metal plates so accurately formed that the wax receives rhomboidal impressions which are a *fac simile* of the basal wall or partition between the opposite cells of natural comb. The thickness of this sheet is an objection, as it is found that the bees do not thin it down to the natural thickness, though they may thin it much, and they use the shavings to form the walls. Prof. C. P. Gillette (Bulletin 54, Colorado Experiment Station), by mixing lampblack with wax, proved what we have long known, that bees extend the midrib and foundation to complete the cells. As we have seen, the bees form comb in the same way, when they make their own foundation.

AMERICAN FOUNDATION.

Mr. Wagner secured a patent on foundation in 1861, but as the article was already in use in Germany, the patent was, as we understand, of no legal value, and, certainly, as it did nothing to bring this desirable article into use, it had no virtual value. Mr. Wagner was also the first to suggest the idea of rollers. In Langstroth's work, edition of 1859, page 373, occurs the following in reference to printing or stamping combs: "Mr. Wagner suggests forming these outlines with a simple instrument somewhat like a wheel cake-cutter. When a large number are to be made, a machine might easily be constructed which would stamp them with great rapidity." In 1866, the King Brothers, of New York, in accordance with the above suggestion, made the first machine with rollers, the *product* of which they tried to get patented, but failed. These stamped rollers were less than two inches long. This machine was useless, and failed to bring foundation into general use.

In 1874, Mr. Frederick Weiss, a poor German, invented the machine which brought the foundation into general use. This was the machine on which was made the beautiful and practical foundation sent out by " John Long," in 1874 and 1875, and which proved to the American apiarists that foundation machines, and foundation, were to be a success.

In 1876, A. I. Root commenced in his energetic, enthusiastic way, and soon brought the roller machine (Fig. 168) and foundation into general use. These machines, though a great aid to apiculture, were still imperfect, and though sold at an extravagantly high price—through no fault of Mr. Root, as he informs me—were in great demand. Next, Mrs. F. Dunham greatly improved the machine by so making the rolls that the foundation would have a very thin base and high, thick walls, which, in the manufacture, were not greatly pressed. These three points are very desirable in all foundation—thin base and thick, high walls, which shall not be compactly pressed.

FIG. 168.

Roller Comb Foundation Machine.—From American Bee Journal.

Mr. Chas. Ohlm invented a machine for cutting the plates, which greatly cheapened the machines. This was purchased by Mr. Root, and he says that ninety percent of the foundation made in the United States has been made on machines, the rollers of which were embossed by this Ohlm machine.

Mrs. Dunham is not only entitled to gratitude for the superior excellence of the machines she manufactured, but by putting so excellent a machine on the market at a lower price, all roller machines had to be sold more reasonably. Mr. Vandervort also improved the rollers, so that his machine secures the same results as does Mrs. Dunham's, while the form of the foundation is somewhat more natural, though not preferred by the bees, I think. Another form of foundation—that with flat bottom—is made by the Van Deusen mill. This has a very

FIG. 169.

Given Press.—From American Bee Journal.

thin base, and is very handsome. It was made to use with wires. This can be made very thin, and many bee-keepers praise it very highly. Mr. P. H. Elwood, I think, still prefers it for use in sections. Mr. Root has kept his machine abreast with the latest improvements. Mr. A. B. Weed has shown great inventive genius in manufacturing very complete comb with natural base and cells nearly complete, so that it is very like natural comb. The bees, however, seem to prefer that with less length of cell, and the greater cost and more difficult transportation makes its use undesirable. At least, it has made no hit in practical bee-keeping. Mr. Weed uses types for the cells, and so the cells must be exactly alike.

THE PRESS FOR FOUNDATION.

Mr. D. A. Given, of Illinois, has made a press (Fig. 169) that stamps the sheets by plates and not by rolls, which, for a time gave nearly, if not quite, as good satisfaction as the improved roller machines. This shuts up like a book, and the wax sheets, instead of passing between curved metal rollers, are stamped by a press after being placed in position.. The advantages of this press, as claimed by its friends, are that the foundation has the requisites already referred to, par excellence, that it is easily and rapidly worked, and that foundation can at once be pressed into the wired frames. Rubber plates have also been made, but as yet have not won general favor or acceptance. Plaster of Paris molds made directly from the foundation are made and used satisfactorily by some excellent bee-keepers. At present I think the press is little used. The roller machine seems to have quite displaced it. Mr. Root says this is because it is slow. Yet he thinks the press gives the most perfect foundation. All of the improved machines give us foundation of exquisite mold, and with such rapidity that it can be made cheap and practical. As Mr. Heddon says, the bees in two days, with foundation, will do more than they would in eight days without it. Every one who wishes the best success must use foundation, often in the brood-chamber, and always in the sections, unless nice white comb is at hand. Whoever has 100 colonies of bees may well own a machine for himself, though it usually pays better to purchase. The specialist can make nicer foundation than the mere amateur.

HOW FOUNDATION IS MADE.

The process of making the foundation is very simple. Thin sheets of wax, of the desired thickness, are pressed between the plates or passed between the rolls, which are made so as to stamp either drone or worker foundation, as desired. Worker is best, I think, even for sections. The only difficulty in the way of very rapid work is that from sticking of the wax sheets to the dies. Mr. Heddon finds that by wetting the dies with concentrated lye the wax is not injured, and sticking is prevented. Mr. Jones uses soapsuds with excellent success for the same purpose. Think of two men running through fifty pounds of foundation in an hour! That is what I saw two men do at Mr. Jones', with a Dunham machine, by use of soapsuds. The man who put in the wax sheets was not delayed at all. The kind of soap should be selected with care. Mr. Root prefers common starch to either lye or soapsuds. New machines are more liable to trouble with sticking than are those that have been used for some time. It is said that dipping the sheets in salt brine also prevents this troublesome sticking. Mr. Baldridge gives this hint, but conceals the name of the discoverer. Mr. Weed now secures the wax in continuous sheets, wound on a spool, and these are fed con tinuously. So the old, sticking trouble is done away with. Mr. Root says three-fourths of our foundation and one-half of that of the world is now made by this new Weed process.

TO SECURE THE WAX SHEETS.

The wax should be melted in a double-walled tin vessel, with water between the walls, so that in no case would it be burned or overheated.

To form the sheets, a dipping-board of the width and length of the desired sheets is the best. It should be made of pine, and should be true and very smooth. This is first dipped into cold water—salt in the water makes it easier to remove the sheets—then one end is dipped quickly into the melted wax, then raised till dripping ceases—only a second—this end dipped into the cold water, grasped by means of a dextrous toss with the hands, and the other end treated the same way. The thing is repeated, if necessary, till the sheet is thick enough. Twice dipping is enough for brood-combs, once for sections. We now only have to shave the edges with a sharp knife, and we can peel off two fine sheets of wax. As the Weed machine forms continuous sheets which can be readily fed into a roller machine, and the sheets of foundation accurately cut and all perfect and automatic, of course, the dipping of wax sheets will soon be entirely a thing of the past.

For cutting foundation nothing is so admirable as the

Carlin cutter (Fig. 170, *a*), which is like the wheel glass-cutters sold in the shops, except that a larger wheel of tin takes the place of the one of hardened steel. Mr. A. I. Root has suggested a grooved board (Fig. 170, *b*) to go with the above, the distance between the grooves being equal to the desired width of the strips of comb foundation to be cut.

For cutting smaller sheets for the sections the same device

FIG. 170.

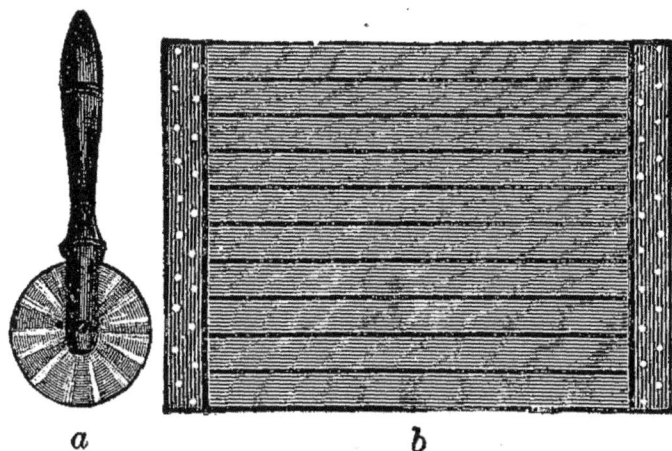

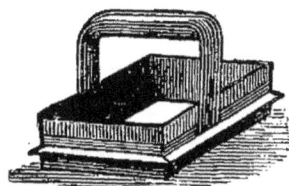

FIG. 171.

a   *b*

From A. I. Root Co.

may be used. I saw Mr. Jones cut these as fast as a boy would cut circular wads for his shot-gun, by use of a sort of modified cake-cutter (Fig. 171).

USE OF FOUNDATION.

Unless to force the bees into sections, when, as we have seen, it is better to hive swarms on empty frames, with mere starters, we better always use foundation in brood-frames. It is astonishing to see how rapidly the bees will extend the cells, and how readily the queen will stock them with eggs. *The foundation should always be the right size for worker-comb.* Even for surplus comb honey the small cells are best. The honey evaporates more quickly, and so will be sooner capped, and it looks better. For brood-combs I prefer wired frames. The sheet of foundation should not quite fill the frame. The advantage of foundation is, first, to insure worker-comb, and thus worker-brood; and, second, to furnish straight, nice combs. We have proved in our apiary repeatedly, that by use of foundation and a little care in pruning out the drone-comb, we could limit or even exclude drones from our hives, and we have but to examine the capacious and constantly crowded stomachs of these idlers to appreciate the advantage of such a course. Bees may occasionally tear down worker-cells, and build drone-cells in their place; but such action, I believe, is not sufficiently extensive ever to cause anxiety. I am also certain that bees that have to secrete wax to form comb do less gathering. Wax-secretion seems voluntary, and when rapid seems to require quiet and great consumption of food. As before suggested, may this not be due to greater or less activity of the bees? If we make two artificial colonies equally strong, supply the one with combs, and withhold them from the other, we will find that this last sends less bees to the fields, while all the bees are more or less engaged in wax-secretion. Thus, the other colony gains much more rapidly in honey; first, because more bees are storing; second, because less food is consumed. This is undoubtedly the reason why extracted honey can be secured in greater abundance than can comb honey.

It also pays remarkably well to use foundation in the sections. If we use very thin foundation—eleven or twelve feet to the pound—all talk about "the fish-bone" need not frighten any one. Foundation for the sections should be twelve or thirteen feet to the pound, while that for the brood-chamber is better at seven or eight feet. Prof. Gillette's experiments and measurements show that the thickness of midrib of natural comb varies from .003 to .006 of an inch in worker, and from .0048 to .008 in drone, and is thickest towards the top. The cell-walls were found as thin where foundation was used as were the natural walls. The walls vary in thickness from .0018 to .0028 of an inch. Bees always thin the base if thicker than natural, but never thin it to equal the natural base. Prof. Gillette found drone-comb weighed 4.32 feet to the pound, worker 5.40, and that from thin foundation 4.23. As comb honey is generally in drone or store comb, we see we get but little more in wax honey from thin foundation. The foundation may or may not fill these sections. It is recommended by Dr. Miller and our Canadian friends, to put two pieces of foundation in each section—an inch strip from the bottom, and a piece from the top to reach within one-eighth of an inch of the lower strip. Of course, this takes time and care. When only one piece is used, I have had best success leaving one-eighth inch space on sides and bottom. Many prefer to fasten to both top and bottom. Of course, foundation for the sections—in fact, all foundation—should be made only of nicest, cleanest wax. *Only pure, clean, unbleached wax should be used in making foundation.* We should be *very careful* not to put on the market any comb honey where the foundation has not

FIG. 172.

Parker Foundation Fastener.—From American Bee Journal.

been properly thinned by the bees. If we always use **thin** foundation there will be no trouble.

Foundation can be fastened into the sections by means of melted wax. This method, however, is too slow; though my friend, R. L. Taylor, has an ingenious arrangement whereby he melts the edges of the foundation and fastens it in the sections with great accuracy and rapidity.

The Parker foundation fastener (Fig. 172) for pressing starters or full sheets of foundation into sections, is prized very highly by most who have used it. The figure shows how it is used,

The Daisy fastener (Fig. 173) uses heat from a lamp, and so fixes by melting rather than pressure. It is preferred, as it is quicker, neater, and saves wax. In the Parker the pressed portion is of course lost.

Still other machines for the same purpose are in the market. Our British friends recommend grooving the sections on all sides in the center to receive the foundation, as we often groove the top. They also recommend splitting the top in

FIG. 173.

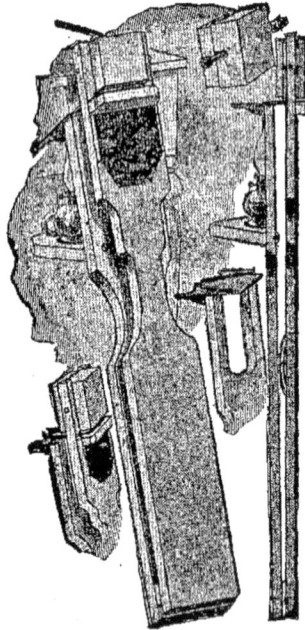

Daisy Foundation Fastener.—From A. I. Root Co.

the middle and in placing together, after adjusting one-half, add the foundation, and then crowd down the other side, thus holding the foundation in place. These methods may be easily tried.

Foundation can be fastened in the brood-frames rapidly and very securely by simply pressing it against the rectangular projection from the top-bar already described. This may be done by use of a case-knife, dipped in honey to prevent its sticking. In this case a block (Fig. 174, a) should reach up into the frame from the side which is nearest to the rectangular projection—it will be remembered that the projection is a little to one side of the center of the top-bar, so that the foundation shall hang exactly in the center—so far that its upper surface would be exactly level with the upper surface of the rectangular projection. This block has shoul-ders (Fig. 174, c), so that it will always reach just the proper distance into the frame. It is also rabbeted at the edge where the projection of the top-bar of the frame will rest (Fig. 174, b), so that the projection has a solid support, and will not split off with pressure. We now set the frame on this block, lay on the foundation, cut the size we desire, which will be as long as the frame, and nearly as wide. The foundation will rest firmly on the projection and block, and touch the top-bar at every point. We now take a board as thick as the projection

FIG. 174.

FIG. 175.

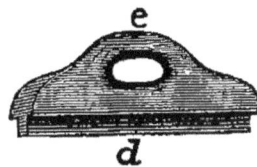

Original.

is deep, and as wide (Fig. 175, d) as the frame is long, which may be trimmed off, so as to have a convenient handle (Fig. 175, e), and by wetting the edge of this (Fig. 175, d) either in water, or better, starch-water, and pressing with it on the foundation above the projection, the foundation will be made to adhere firmly to the latter, when the frame may be raised *with the block*, taken off, and another fastened as before. I have practiced this plan for years, and have had admirable success. I have very rarely known the foundation to drop if made of good wax, though it must be remembered that our hives are shaded, and our frames small. If the top-bar of the frame has not the projection, the comb can be pressed directly on the top-bar and then bent at right angles, as with the Parker foundation fastener. To make this more secure a narrow strip may be tacked to the top-bar, pressing the foundation. Our English friends use a double top-bar which is dovetailed to the uprights of the frame. Thus, in putting .together the frame the foundation is pressed between the two

FIG. 176.

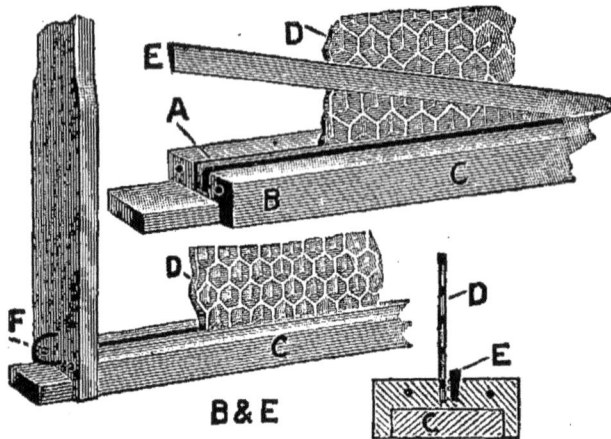

From A. I. Root Co.

halves of the top-bar, and so firmly held in place. Sometimes a groove is cut into the top-bar, which may receive the edge of the foundation, which is held by a wedge (Fig. 176), which is pressed in beside it.

The above methods are successful, but probably will receive valuable modifications at the hands of the ingenious apiarists of our land. If we have frames with the V-shaped top-bar (Fig. 96), we may easily break the foundation and press it on, as shown in Fig. 177.

WIRED FRAMES.

But as foundation does sometimes fall or sag, so that many cells are changed to drone-cells, or warp in awkward shapes, especially if the hive is unshaded, or receives a full colony of bees with all its frames full of foundation, and as the wax is sometimes so brittle that it will not hold together, however well fastened, wired frames (Figs. 178 and 70) are rapidly coming into use. Another point strongly in favor of such frames is that they can be handled or shipped, and there is not the least danger of their combs falling from the frames. Mr. Jones states that with wired frames we may use thinner

FIG. 177.

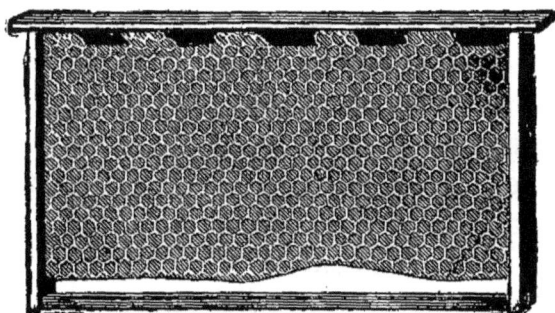

From American Bee Journal.

foundation, and thus save one-third the expense. The wires should be two inches apart, and the extreme wires not more than one-half inch from the side of the frame. They may be fastened by passing through holes in the top and bottom bars of the frames, which must be exactly in the center, or they may be hooked over little hooks, such as may be made by driv-

FIG. 178.

Root's Wired Frame. End-wires are too far from End-bars.
—From A. I. Root Co.

ing a staple into the frame after we have cut one limb of the staple off near the curve. If holes are not made through the top-bar of the frame, they can be easily formed by use of sharp awls. If these are set in a strong block, like an iron rake, each bar can be pierced at one stroke by use of a lever press. If the foundation is to be stamped in the frame by the Given press, then the wire should be No. 36; if it is to be put on by hand, then No. 30 must be used. Tinned wire should be used. To cut wire the right length for frames and not have it tangle, it may be wound lengthwise about a board of the right length, so that one round of wire will be just enough for a frame. Then tie two or three strings tightly around board, wire and all. The strings extend at right angles to the wire. We now cut across all the wires at one end of the board. Thus, the wires are all the proper length, and are held firmly ready for use. Some, even with the Given press, prefer to put the foundation on the wires by hand. In this case the foundation should be warmed till quite soft, then laid on a board and the frame placed over all so that the wires rest on the foundation. Then by use of a shoe-buttoner, with a longitudinal groove cut into the convex side of the curve, the wires are pressed into the foundation. This work is easily and rapidly performed. A tin

FIG. 179.

Wire-Imbedder.—From A. I. Root Co.

wire-imbedder (Fig. 179) works admirably and costs very little. Mr. Cheshire states that the brood dies over the wires. There is no such trouble in my apiary. In Germany it is recommended to press the foundation for extracting-combs on a board, and so have the cells built out only on one side and elongated so as to hold much honey. This gives strong combs and saves turning the frames when extracting. But wired combs are strong, and our improved extractors make turning very easy and rapid. Again, evaporation or ripening in deep cells is very slow. I have also found that bees object to foundation on a board, and often bite it off.

SAVE THE WAX.

As foundation is becoming so popular, it behooves us all to be very careful that no old comb goes to waste. Even now the supply of wax in the country is scarce equal to the demand. Soiled drone-comb, old, worthless worker-comb, all the comb in the old hives, if we use Mr. Heddon's method of transfer-

FIG. 180.

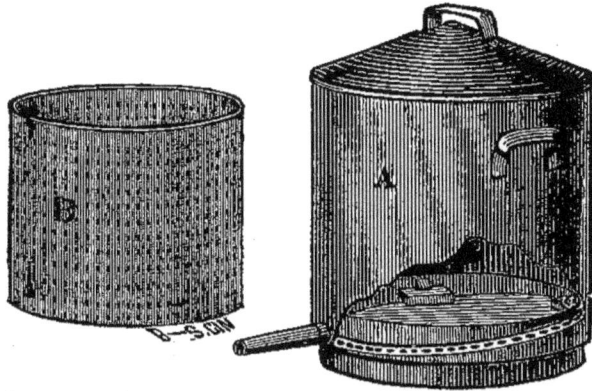

Swiss Wax-Extractor.—From American Bee Journal.

ring, and all fragments that can not be used in the hives, together with cappings, after the honey is drained out, should be melted, cleansed and molded into cakes of wax, soon to be again stamped, not by the bees, but by wondrous art.

METHODS.

A slow and primitive method is to melt in a vessel of heated water, and to purify by turning off the top, or allowing it to cool, when the impurities at the bottom are scraped off, and the process repeated till all impurities are eliminated.

A better method to separate the wax is to put it into a strong, rather coarse bag, then sink this in water and boil. At intervals the comb in the bag should be pressed and stirred. The wax will collect on top of the water.

To prevent the wax from burning, the bag should be kept from touching the bottom of the vessel by inverting a basin in the bottom of the latter, or else by using a double-walled vessel with hot water between the walls. The process should be repeated till the wax is perfectly cleansed.

But as wax is to become so important, and as the above methods are slow, wasteful, and apt to give a poor quality of wax, specialists, and even amateurs who keep ten or twenty colonies of bees, may well procure a wax-extractor (Fig. 180). This is also a foreign invention, the first being made by Prof. Gerster, of Berne, Switzerland. These cost from five to seven

FIG. 181.

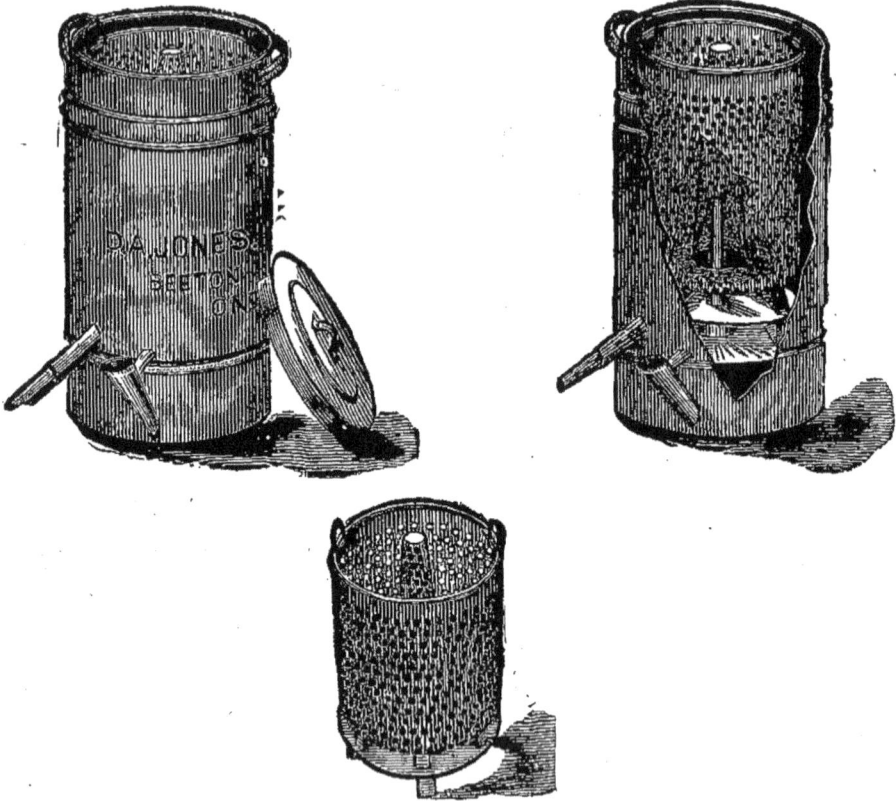

Jones Wax-Extractor.—From D. A. Jones.

dollars, are made of tin, are very convenient and admirable, and can be procured of any dealer in apiarian supplies.

The comb is placed in the perforated vessel, and this in the larger can, which is set on a kettle of boiling water. The clean, pure wax passes out the spout. Mr. Jones has improved the common wax-extractor (Fig. 181). This is what he says of it:

"Put the extractor on the stove in the same manner as an ordinary pot, having beforehand filled the lower tank with water, and the perforated basket above the tank with broken comb or whatever material you wish to extract wax from. The steam passes through the perforated metal walls of the basket, melting every particle of wax from the crude material: the wax runs out of a spout for the purpose, turned downwards; under this spout have a receptacle, which have slightly oiled, to keep the wax from adhering to its walls. The tube turned upwards serves two very important purposes, viz.: To fill water into the lower tank, and to see if the tank requires replenishing, without taking out the basket above. Keep everything but the spout closed, in order to lose no steam and give it full force. When not in use as an extractor it is excellent as an uncapping-can; the cappings drop into basket,

FIG. 182.

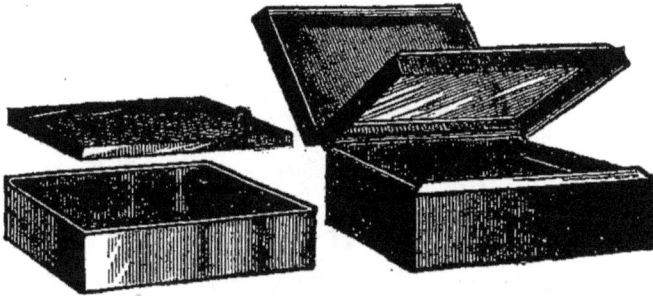

Solar Wax-Extractor.—From A. I. Root Co.

the honey drains off, leaving the cappings just where you want them to extract from."

Still better than the above is the solar wax-extractor (Fig. 182). This is cheap, and can be easily made at small cost. A box lined with tin has hinged to its top, first, a glass cover, and then to the top of this glass cover, a wooden cover lined with tin, or a glass mirror.

A perforated tin wax-pan is made to set just under the glass cover. This is placed conveniently where the sun can strike it, and is always ready for pieces of wax. By raising the upper cover the reflector hastens the work. I value the solar wax-extractor very highly. It is always ready for pieces of comb. The Boardman extractor (Fig. 183) has only the glass cover, and is on rockers to give proper incline to catch the sun. The solar wax-extractor, indeed, all the methods thus far described, fail to secure all the wax from old, black combs, hence the

WAX-PRESS.

This valuable invention was given us by Mr. Wm. W. Cary. Mr. C. A. Hatch says it will pay if one has one hundred pounds of wax to render. In old combs it will save ten percent or more. Mr. Hatch says two men will render three hundred pounds in one day. It is also a neat way. Neatness, despatch, thoroughness—surely, a grand trio.

The press is used much as we use a cider-press to express apple-juice. Mr. Hatch uses a large kettle in which to melt

F ɪ ɢ. 183.

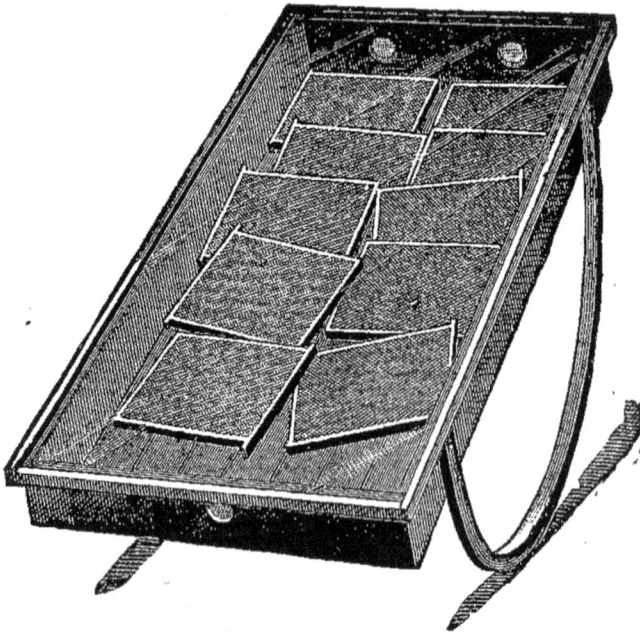

Boardman Solar Wax-Extractor.—From A. I. Root Co.

the wax. This is done out-of-doors. About eight gallons of water are kept in the kettle. Only four other parts are needed. A frame with screw (Fig. 184), which may be turned down as seen in figure; a tray about eighteen inches square with lip, a form fifteen inches square and four inches high, and the slotted rack, which Mr. Hatch makes of triangular pieces. such as he uses for the top-bars of his frames. These may be one-eighth of an inch apart. Of course, a good quality of burlap and a square board follower are required. The cheese is made by dipping the melted comb and wax into the form, which has the slotted rack below, and the burlaps laid over all. Mr. Hatch pins the burlaps over the cheese with long, slim wire nails. Of course the form is in the tray, and it is easy to see that the press will do the work.

By these last inventions all the wax, even of the oldest combs, can be secured, in beautiful condition, and as it is perfectly neat, there is no danger of provoking the " best woman

FIG. 184.

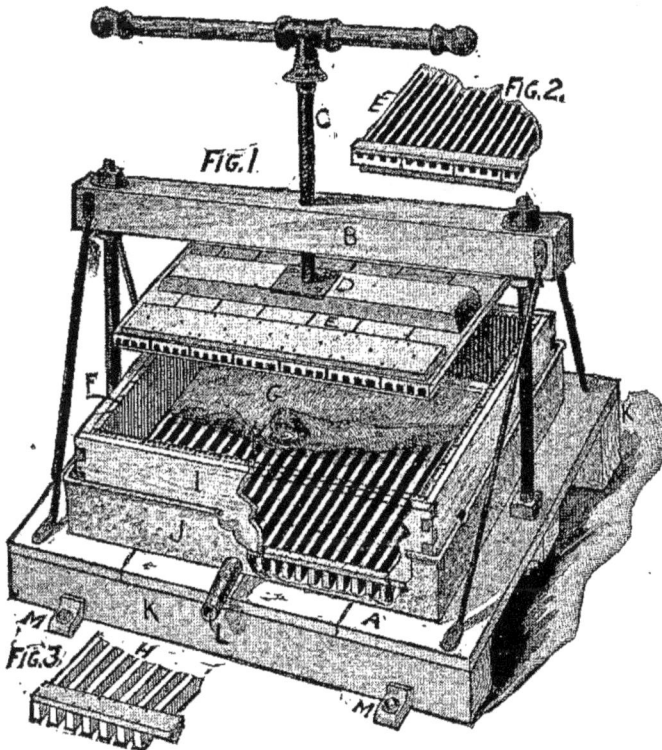

Wax-Press.—From A. I. Root Co.

in the world," as we are in danger of doing by use of either of the first-named methods—for what is more untidy and perplexing than to have wax boil over on the stove, and perhaps get on the floor, and be generally scattered about ?

All pieces of comb should be put into a close box, or in the solar wax-extractor if we have one, and if any larvæ are in it, the comb should be melted so frequently that it will not smell badly. It will often pay to use the press on comb that has been melted in the solar wax-extractor—nearly always in case of very old comb. By taking pains, both in collecting and melting, the apiarist will be surprised at the close of the season, as he views his numerous and beautiful cakes of wax, and rejoice as he thinks how little trouble it has all cost.

Beeswax as bought on the market is of all colors, and often full of impurities. If this is melted in water containing sulphuric acid—one pound to over 100 gallons of water—it may be entirely cleansed, and made uniform. In very dirty comb the acid should be doubled. If the comb is quite clean, not more than half as much is required. Mr. Doolittle uses vinegar and water, half and half. One pint of vinegar answers for ten pounds of wax. This is more expensive than is the sulphuric acid. This is usually melted in a wooden vessel—a barrel serves well. It is melted by steam, and so there is no danger of burning. Care is necessary that it does not boil over, and that all the wax is melted. Thus, after it seems melted it should simmer for a time. When cooled down to near the point of solidification, it is dipped out, down to any foreign matter, then cooled, and any remaining wax scraped off. Wax thus cleansed makes the finest foundation.

Wax is readily bleached by placing thin sheets or ribbons in the sun. Unbleached wax is better for foundation, and in use is practically as beautiful.

Wax is adulterated with tallow, paraffine and ceresin. We can usually detect tallow by the odor and taste. The latter is betrayed by chewing. Wax is brittle, while wax adulterated with these coal-oil products is salvy, and so chews up like gum. As stated on page 176, these petroleum products are lighter than wax, so if we add alcohol to water until a specimen of wax of known purity just sinks, we have a sure detection of this latter kind of adulteration. Mr. Root says he can nearly always detect adulteration in these ways.

Hot water and benzine are excellent to clean wax from vessels, etc. We must not melt wax in galvanized-iron vessels, as it will injure the wax.

CHAPTER XVI.

MARKETING HONEY.

No subject merits more attention by the apiarist than that of marketing honey. There is no question but that the supply is going to increase continually; hence, to sustain the price we must stimulate the demand, and by doing this we shall not only supply the people with a food element which is necessary to health, but we shall also supersede in part the commercial syrups, which are so often adulterated as not only to be crowded with filth the most revolting, but are often even teeming with poison. (Report of Michigan Board of Health for 1874, pages 75–79.) To bring, then, to our neighbor's table the pure, wholesome, delicious nectar, right from the hive, is philanthropy, whether he realizes it or not.

Nor is it difficult to stimulate the demand. I have given special attention to this topic for the last few years, and am free to say that not a tithe of the honey is consumed in our country that might and should be.

HOW TO INVIGORATE THE MARKET.

First. See that no honey goes to market from your apiary that is not in the most inviting form possible. Grade *all the honey thoroughly*, and expect prices to correspond with the grade. If, as estimated by two of our most successful bee-keepers, it costs from five to eight cents to produce extracted honey, and from seven to thirteen cents to produce comb honey, we see that all should labor that prices for first-class honey should never fall below eight cents for extracted and twelve cents for comb. The best grades ought always to sell for ten cents for extracted and fifteen cents for comb. See that every package and vessel is not only attractive, but so arranged as not to make the dealer any trouble, or cause him any vexation. One leaky can or case may do great injury.

Second. See that every grocer in your vicinity has honey

constantly on hand. Do all you can to build up a home market. The advice to sell to only one or two dealers is, I think, wrong. Whether we are to buy or sell, we shall find almost always that it will be most satisfactory to deal with men whom

FIG. 185.

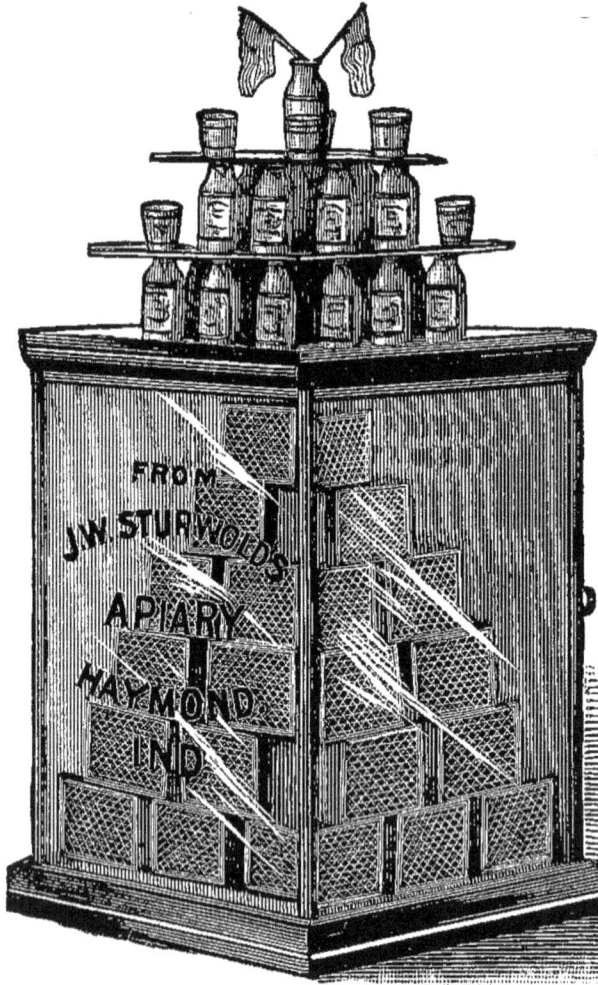

Show-Case.—From A. I. Root Co.

we know, and who are close at hand. Only when you outgrow your home market should you ship to distant places. This course will limit the supply in large cities, and thus raise the prices in the great marts, whose prices fix those in the country. Be sure to keep honey constantly in the markets.

Third. Insist that each grocer make the honey *very* conspicuous. If necessary, supply large, fine labels, with your own name almost as prominent as that of the article.

Fourth. Deliver the honey in small lots, so that it will be sure to be kept in inviting form, and, if possible, attend to the delivery yourself, that you may know that all is done "decently and in order."

Fifth. Instruct your grocers that they may make the honey show to the best effect (Fig. 185), and thus captivate the purchaser through the sight alone.

Sixth. Never send honey to a commission man of whose standing you are not assured. Your banker may be able to secure this for you. The fact that a commission man advertises in the bee-papers is a pretty safe guarantee of his honesty. It is for the interest of the journals to protect the bee-keepers in this regard.

Seventh. Call local and general conventions, that all in the community may know and practice the best methods, so that the markets may not be demoralized by poor, unsalable honey.

Eighth. There should be a Bee-Keepers' Exchange which should be modeled after the very successful Citrus Fruit Exchange, of Southern California. Such co-operation in every State would remove all uncertainty. It is sure to come. All bee-keepers should do all in their power to hasten the day of its coming.

It is of the greatest importance to encourage State, inter-State, and National Associations. Happily, our civilization makes every person affected by the acts of each person. Selfishness, not less than Christianity, urges us all to be interested in each other. The honey-traffic reaches from State to State. Bee-keeping will never be perfect as an art till all bee-keepers act as one man. He is short-sighted that decries conventions. It is the experience of the world that they are valuable in other arts. Bee-keeping is no exception. Let us all urge that the associations act in unison, from the local to the general; that all other apiarian interests no less than the markets shall be in the highest degree fostered. Each association, from the most local to the most general, has its special mission which no other can perform. Such associations will usually promote general co-operation,

PREPARATIONS FOR MARKETS.

Of course, the method of preparation will depend largely upon the style of honey to be sold, so we will consider the kinds separately.

EXTRACTED HONEY.

As before intimated, extracted honey has all the flavor, and is in every way equal, if not superior—comb itself is innutritious and very indigestible—to comb honey. As Dr.

FIG. 186.

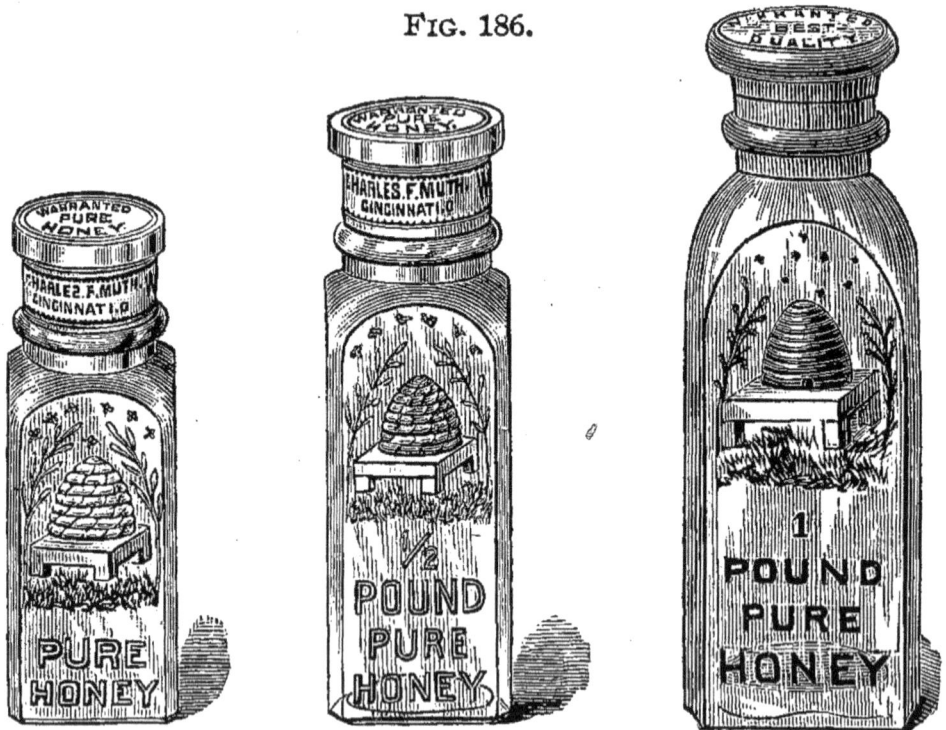

Miller has pointed out, granulated honey, thoroughly drained and then melted, gives a most delicious article. When people once know its excellence—know that it is not "strained"— then the demand for extracted honey will be vastly increased, to the advantage both of the consumer and the apiarist.

Explain to each grocer what we mean by the word "extracted," and ask him to spread wide the name and character of the honey. Leave cups of honey with the editors and men of influence, and get them to discuss its origin and merits.

I speak from experience, when I say that in these ways the reputation and demand for extracted honey can be increased to a surprising degree, and with astonishing rapidity.

HOW TO TEMPT THE CONSUMER.

First. Have it chiefly in small cups or pails. Many persons will pay twenty-five cents for an article, when, if it cost fifty cents, they would not think of purchasing.

Second. Study the kinds of receptacles that will take best with the buyers. Some persons will prefer such vessels as jelly-cups or glass fruit-jars, etc., that will be useful in every household when the honey is gone. As Dr. Mason and Mr. Cutting have shown, jelly-cups, by simply dipping the upper edge in melted wax, then quickly filled and covered, are quite securely sealed. Mr. Root recommends that the honey be covered with a paper dipped in white of egg, which further seals the vessel. Others will prefer more showy vessels, like the Muth one-pound and two-pound jars (Fig. 186), even

FIG. 187.

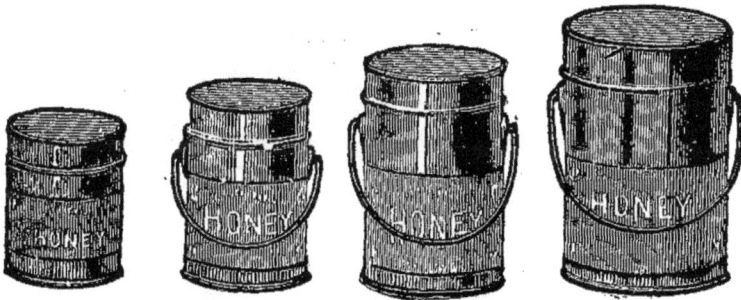

From American Bee Journal.

though they cost more. At present the neat tin pails (Fig. 187), holding from one-half pound to twelve pounds, are very popular in the markets. The covers shut inside, and if the honey is granulated they are very excellent. The bails make them more convenient and salable. Mr. Jones has a pail that is easily sealed with wax strings, and is beautifully decorated with chromoed labels. Such pails are cheap, convenient, and leave little to be desired. Their beauty aids the sale. Mr. A. I. Root pronounces them the best receptacle for extracted honey.

If the honey is to be sent to a distant market it may be put in soft wood—spruce, pine or hemlock—kegs (Fig. 188). These are light, and if we carefully drive the hoops, and test by use of boiling water, we need not wax them. Hard wood barrels

FIG. 188.

From American Bee Journal.

must be waxed, then if the honey granulates the hoops must be loosened to take out the head. This cracks the wax and a leak results. As before stated on page 333, alcohol barrels are cheap, and safe even without waxing. At present large tin

FIG. 189.

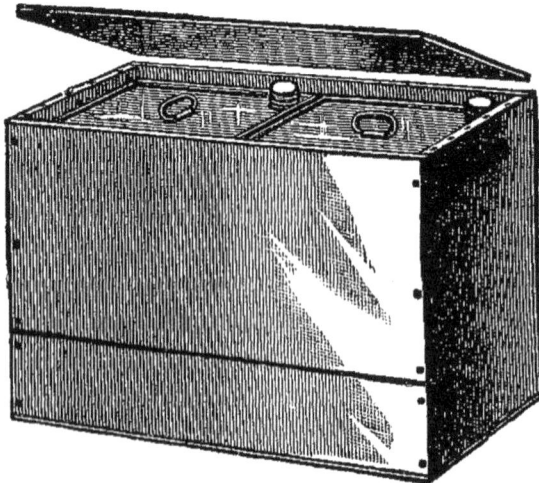

Cans for Extracted Honey in Jacket.—From A. I. Root Co.

vessels in wooden jackets (Fig. 189) are rapidly gaining in favor. These are absolutely necessary in such dry climates as California. Even small tin vessels of honey can be safely and cheaply shipped as freight by packing in barrels, using straw to make all close and secure. Mr. Doolittle has even boxed thoroughly candied honey and shipped it safely for long distances. He has quite a trade in such packages.

Third. Explain to the grocer that if kept above the temperature of 70° or 80° F., it will not granulate; that granulation is a pledge of purity and superiority, and show him how easy it is to reduce the crystals, and ask him to explain this to his customers. If necessary, liquefy some of the granulated honey in his presence. Put on the labels directions for reliquefying candied honey. Honey, like many other substances, will not granulate if heated to 180° F., and then sealed while hot. This does no injury to the honey, but it is trouble, and makes the honey less convenient to ship, though at times it may pay until we can educate our patrons in reference to the excellence of granulated honey.

Lastly. If you do not deliver the honey yourself, be sure that the vessels will not leak in transit. It is best, in case jelly-cups are used, that they be filled at the grocery, and sealed as already described. Do not forget the large label, which gives the kind of honey, grade, and producer's name.

If the honey is extracted before it is fully ripened—before the bees cap it—it should always be kept in an open can or barrel, covered with cloth, and in a *dry*, warm room. Thus arranged it will thicken as well as in the hive. *No honey should ever be kept in a cool, damp room.*

The admirable work of the late Mr. C. F. Muth, in Cincinnati, educating people in reference to extracted honey, fighting all adulteration, pushing it into the candy, tobacco, and confectionery establishments, deserves our hearty gratitude. Mr. Muth's market became stupendous, and graphically shows what this trade is to be in the near future, when all our cities have a Muth to work for us. I would also recommend to all the very valuable little pamphlet of Mr. Chas. Dadant, on the production and sale of extracted honey. It is most interesting reading to the honey-producer, and shows what energy and thought may accomplish in this direction. Every bee-keeper should watch the markets, and so must have one, or, better, two of the best bee-periodicals. He should also circulate honey leaflets to encourage sales.

COMB HONEY.

This, from its wondrous beauty, especially when light-colored and immaculate, will always be a coveted article for the table, and will ever, with proper care, bring the highest price paid for honey. So it will always be best to work for this, even though we may not be able to procure it in such ample profusion as we may the extracted. He who has all kinds will be able to satisfy every demand, and will most surely meet with success.

RULES TO BE OBSERVED.

This should be chiefly in small sections (Fig. 108), for, as before stated, such are the packages that surely sell. Sections from three to six inches square will just fill a plate nicely, and look very tempting to the proud housewife, especially if some epicurean friends are to be entertained.

The sections should surely be in place at the dawn of the white clover season, so that the apiarist may secure the most of this irresistible nectar, chaste as if capped by the very snow itself. They should be taken away as soon as all are capped, or at least as soon as the harvest is over, as delay makes them highways of travel for the bees, which always mars their beauty.

In case old combs are near by, the bees incorporate chippings from it in the cappings, much to the injury of the comb honey. Thus sections should be always produced in supers above the brood-combs, or distant from old, dark combs.

When removed, if demanded, glass the sections, but before this we should place them in hives one upon another, or special boxes made tight, with a close cover, in which to store either brood-frames in winter or sections at any season, and fume them with burning sulphur. This is quickly and easily done by use of the smoker. Get the fire in the smoker well to burning, add the sulphur, then place this in the top hive, or top of the special box. The sulphurous fumes will descend and deal out death to all moth-larvæ. *This should always be done* before shipping the honey, if we regard our reputations as precious. It is well to do this within two weeks after removal, and also two weeks later, so as to destroy the moth-larvæ not hatched when the sections are removed. Bisulphide of carbon is more easily used than is sulphur, and is quite as effective. This needs only to be turned into the close box holding the sections. Sections may be treated in a close barrel covered with oil-cloth. The vapors form very quickly, and are deadly to all insects. It is used in mills to kill flour insects; in special houses or barrels to kill pea and bean weevils; in their runs to kill squirrels and gophers; in holes, or in their hills, to kill ants. In all such use great care must be exercised, as it is as inflammable as is gasoline, and it vaporizes even more quickly. The quick vaporization is what makes it so effective. An inferior article, which is as good for all these purposes, sells very cheaply.

If one-pound sections are used with separators bees will seldom enter them to store pollen, and, with no pollen at all in

FIG. 190.

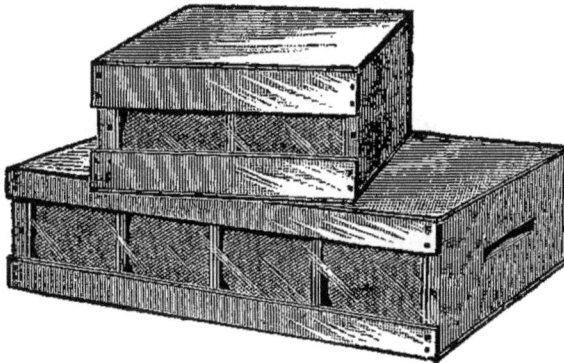

12-lb. and 24-lb. Shipping-Cases.—From A. I. Root Co.

the combs, moths are not likely to be troublesome. If separators have been used, these sections are in good condition to ship, as they may stand side by side and not mar the comb.

The shipping-case (Fig. 190) should be strong, neat and cheap, with handles—such handles are also convenient in the ends of the hives (Fig. 159), and can be cut in an instant by having the circular saw set to wabble. With handles the case is more convenient, and is more sure to be set on its bottom. The case should also be glassed, as the sight of the comb will say: "Handle with care." It is always wiser to buy shipping-cases in the "knock-down." They are neater, and usually cheaper than home-made ones. Strong paper trays should be placed in them. The sections should rest on cleats, which are nailed to hold the paper. We must do all possible to prevent leaking.

Mr. Heddon makes a larger case (Fig. 191), which is neat and cheap. It is best to have single-tier cases (Fig. 190), and when full they should not weigh more than twenty-four pounds.

FIG. 191.

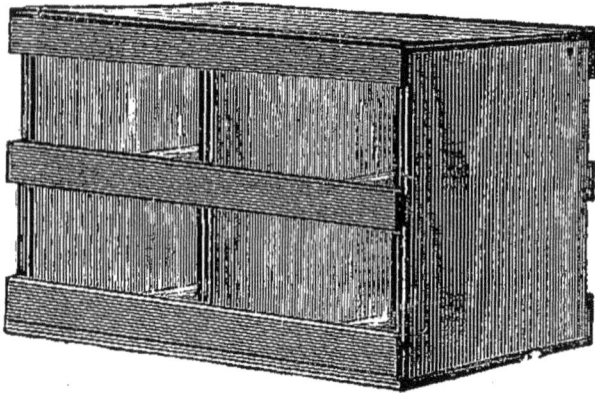

Shipping-Case.—From James Heddon.

However, some prefer forty-eight pound cases. These are double (Fig. 191). Even twelve-pound cases are preferred by many.

FIG. 192.

Carton for Comb-Honey.—From A. I. Root Co.

It may be well to wrap the sections in paper, as thus breakage of one will not mean general ruin. A carton (Fig. 192) is often very helpful. These are neat and convenient, and with neat label cost less than one cent. Mr. Crane, of

Vermont, praises these very highly. Grocerymen may well be urged to use them.

In shipping in freight cars, it is desirable that the sections be set lengthwise of the cars, as the danger from the shocks of starting and stopping will be much less. Always ship a car-

FIG. 193

Fancy.   No. 1.   No. 2.

Comb-Honey.—From A. I. Root Co.

load, if possible, so as to avoid re-shipping. When moving honey in a wagon the combs should extend crosswise of the wagon.

In groceries, where the apiarist keeps honey for sale, it will pay him, unless the groceryman will use a fine exhibition case, to furnish his own boxes. These should be made of white-wood, very neat, and glassed in front to show the honey, and the cover so fixed that unglassed sections—and these, probably will soon become the most popular—can not be punched or fingered. Be sure, too, that the label, with kind of honey, grade, and name of apiarist (Fig. 185) be so plain that "he who runs may read."

The grading of the honey can not be too carefully and honestly done. One or more inferior sections in the middle of a case may, and ought to, do the packer great harm. "An honest pack" should be the motto and pride of every man who has honey or any other commodity to sell. All sections well filled should be called "fancy" (Fig. 193), and all filled wholly, "extra fancy" (Fig. 194). If not quite filled out at the corners

FIG. 194.

Comb Honey, Extra Fancy, in Plain Sections.
—From A. I. Root Co.

it may be No. 1; when quite a space is empty, No. 2. (See Fig. 193.) These four grades will be enough. The kind of honey should be on the label, as "Buckwheat, Extra Fancy," "Clover, Fancy," etc. All honey below No. 2 should be kept, and after being cleaned out as before described, retained for baits the next season.

Every bee-keeper should encourage the sale of honey by broadly circulating the honey leaflets, showing how honey can be used in cookery, etc. The following recipes are used in making gems and jumbles, which are largely sold in the markets:

GEMS.—2 quarts flour, 3 tablespoonfuls melted lard, ½ pint honey, ½ pint molasses, 4 heaping tablespoonsfuls brown sugar, 1½ even tablespoonfuls soda, one even teaspoonful salt, ½ pint water, ½ teaspoonful vanilla extract.

JUMBLES.—2 quarts flour, 3 tablespoonfuls melted lard, 1 pint honey, ¼ pint molasses. 1½ even teaspoonfuls soda, 1 even teaspoonful salt, ¼ pint water, ½ teaspoonful vanilla extract.

Mr. Root, in the "A B C of Bee-Culture," gives many recipes, besides the above, which call for honey.

Comb honey that is to be kept in the cool weather of autumn, or the cold of winter, must be kept in warm rooms, or the comb will break from the sections when handled. By keeping it quite warm for some days previous to shipment, it may be sent to market even in winter, but must be handled very carefully, and must make a quick transit.

Above all, *let "taste and neatness" ever be your motto.*

MARKETING BEES.

Before leaving this subject, let me say a word about selling bees.

SELLING QUEENS.

As queen-rearing and shipping have already been sufficiently described, it only remains to be said that the vender of queens can not be too prompt, or fair, or cautious. Success, no less than morality, demands the most perfect honesty. If, for any reason, queens can not be sent promptly, the money should be returned at once, explanation made, and, if reasonable, delay may be requested. The breeder, who, by careful selection and care in following the rules of breeding, shall secure a type of bees pronounced in excellence, will surely win in the race. There is no reason why the capable, persistent breeder of bees should not equal in success the best breeders of cattle and horses.

I have described shipping bees. The rules just given should guide also in selling whole colonies.

SELLING BEES BY THE POUND.

This has been quite a business, and originated, I think, with Mr. A. I. Root. The bees are put, by use of a large tin funnel, into a cage (Fig. 195) made of sections, as shown in the figure. The handle makes it easy to carry them, and they get careful handling without any special request. It is said that a pound of bees can be prepared for shipment in five minutes. The cages are provisioned with "Good candy." It is always

FIG. 195.

From A. I. Root Co.

safe to get a pound of bees in June or July, with a queen, expecting to have a good colony by winter. It is reported that from such a start, even five good colonies have been secured, all of which wintered. In this case they were fed.

VINEGAR FROM HONEY.

Mr. T. F. Bingham utilizes the cappings secured while extracting to produce wax and a most excellent quality of vinegar. The honey is drained from the cappings, which are then covered for an hour or two with water. The cappings from 1000 pounds of honey will sweeten enough water for forty-five gallons of vinegar. The water is now drained into an open barrel, which should be kept covered with cloth. The scum should be removed as it rises. In about a year the change to first-class vinegar will have been accomplished.

After the water is drained from the cappings they can be converted into pure wax, as already described.

The poorer grades of honey and rinsings from cleaning barrels and honey from utensils may also be profitably used in the same way. One and one-half pounds of honey will make one gallon of the best vinegar. Mr. E. France adds honey to water until an egg sinks so as to expose a surface about the size of a ten-cent piece. It should be put in a close barrel with a one-inch auger-hole to permit escape of gases. Some good vinegar or yeast should be added to start fermentation. After working or fermentation commences draw off a pailful occasionally and turn it back. If one or two kegs or barrels are working at the same time, turn from one into the other. It is well to turn old vinegar into new or unripe, but the reverse should never be done, as it injures the keeping qualities. By using old, sour barrels and old vinegar to start fermentation, vinegar may be made in one year, else it will take two.

FAIRS AND THE MARKET.

Our English friends have demonstrated that large honey exhibitions are most powerful aids in developing the honey market.

Till within a few years our American honey exhibits have been a disgrace and a hindrance, and they are largely so to-day. A little second-rate honey sandwiched in with sugar and syrups, and supplemented by a cake or two of black, dirty wax, describes the honey exhibit of most of our fairs to-day. The premiums range from twenty-five cents to fifty cents.

WHAT SHOULD WE HAVE?

Our industry demands a separate building, filled with tons, not pounds, of honey, and exhibiting everything that is valuable in modern apiculture. Bees may be exhibited in hives covered by wire-gauze, and if it is desired to manipulate them, this can be readily done in a bee-tent, to the great satisfaction and pleasure of many who know nothing of such matters. I have proved this by actual trial.

It can be arranged with the managers that sales of honey

and all apparatus be made at any time at this building, on conditions that the exhibit should be in nowise interfered with. The premiums should range from one dollar to twenty, and the total should reach to the hundreds.

We have found in Michigan that all that is necessary to effect this grand and invaluable transformation is a little life and energy on the part of the bee-keepers. Through the enterprise of H. D. Cutting and others, the bee-keeping exhibit of our State fairs, in a separate building, leaves little to be desired, and is a credit to the industry.

EFFECTS OF SUCH EXHIBITS.

They show that apiculture is no second-rate business. They attract attention and educate as nothing else can. They go hand in hand with local conventions in instructing bee-keepers so that no inferior honey will go on the markets. They enable bee-keepers to see and buy just what they need in the more intelligent prosecution of their business. They scatter the little pint, half-pint, and gill pails of honey into thousands of homes, and develop a knowledge and taste that stimulate the honey market most powerfully. Tons of honey have been sold at the Toronto fairs, the influence of which has been a lasting surprise even to the most enterprising producers. I believe that the great quartet that is to advance apiculture is fairs, associations, co-operative organizations, and improved bees.

CHAPTER XVII.

HONEY-PLANTS.

As bees are dependent mainly upon flowers for honey, it of course follows that the apiarist's success will depend largely upon the abundance of honey-secreting plants in the vicinity of his apiary. True it is that certain bark and plant lice

FIG. 196.

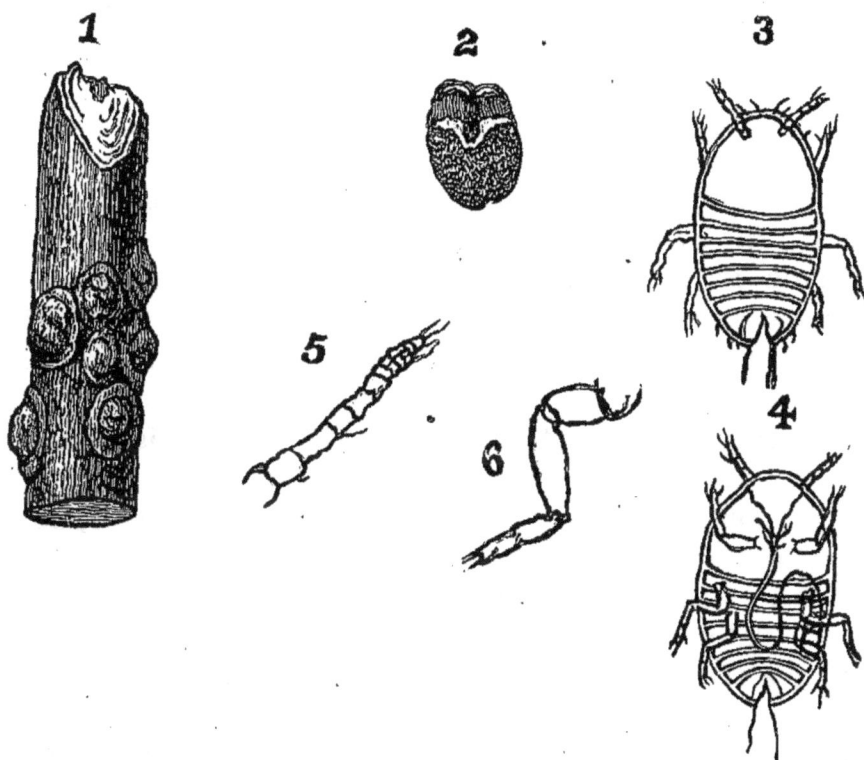

Tulip-Tree Bark-Louse, 3, 4, 5, and 6, Greatly Magnified.—Original.

1   Scale on Twig.   2 Underside of scale.
3, 4 Young Lice.   6 Leg.
5   Antenna.

secrete a kind of liquid sweet, which, in the dearth of any-thing better, the bees seem glad to appropriate. I have thus seen the bees thick about a large bark-louse which attacks the tulip-tree, and thus often destroys one of our best honey-trees.

I have described this insect (Fig. 196) under the name of

Lecanium tulipifera. In 1870 it did no small injury to our tulip-trees at the Michigan Agricultural College. It has seriously injured this tree in the States bordering the Ohio River. The tulip is often called poplar, which is quite incorrect. The poplar belongs to the willow family, the tulip to the magnolia. This louse is of double interest to bee-keepers. It ruins one of our best honey-trees, and supplies a poor substitute for plant nectar to the bees. All bark-lice, which include the orange-tree scale and the San Jose scale, are best destroyed by use of kerosene oil. This latter is best applied in the form of an emulsion, with soap. To make the kerosene and soap emulsion I make a very strong suds, using one-eighth of a pound of whale-oil soap, or one quart of soft soap, and two quarts of water. To this is added one quart of kerosene oil, and all churned by use of a force-pump, pumping it back into itself till it is thoroughly and permanently mixed. I then dilute with water till the kerosene oil forms one-twelfth of the whole. In California it is found that a distillate emulsion is more effective than kerosene emulsion. One-fourth pound of whale-oil soap is dissolved in one gallon of water. Then one gallon of untreated distillate is added and all is violently stirred. This is then diluted with water one to ten. It is cheap and effective. It is found that spraying can not be done thoroughly enough for evergreens like the orange-tree, and so fumigation by aid of tents with cyanide of potassium is adopted by most of the progressive citrus fruit-men of California. This emulsion often spots the fruit.

I have also seen the bees thick about several species of plant-lice. One—the Erisoma imbricator, Fitch—works on beech-tree. Its abdomen is thickly covered with long wool, and it makes a comical show as it wags this up and down upon the least disturbance. The leaves of trees attacked by this louse, as also those beneath the trees, are fairly gummed with a sweetish substance. I have found that the bees avoid this substance, except at times of extreme drouth and long-protracted absence of honeyed bloom.

Another species, Thalaxes ulmicola, gives rise to certain solitary galls, which appear on the upper surface of the leaves of the red elm. These galls are hollow, with a thin skin, and within the hollows are the lice, which secrete an abundant sweet that often attracts the bees to a feast of fat things, as the gall is torn apart, or cracks open, so that the sweet exudes. This sweet is anything but disagreeable, and may not be un-

FIG. 197.

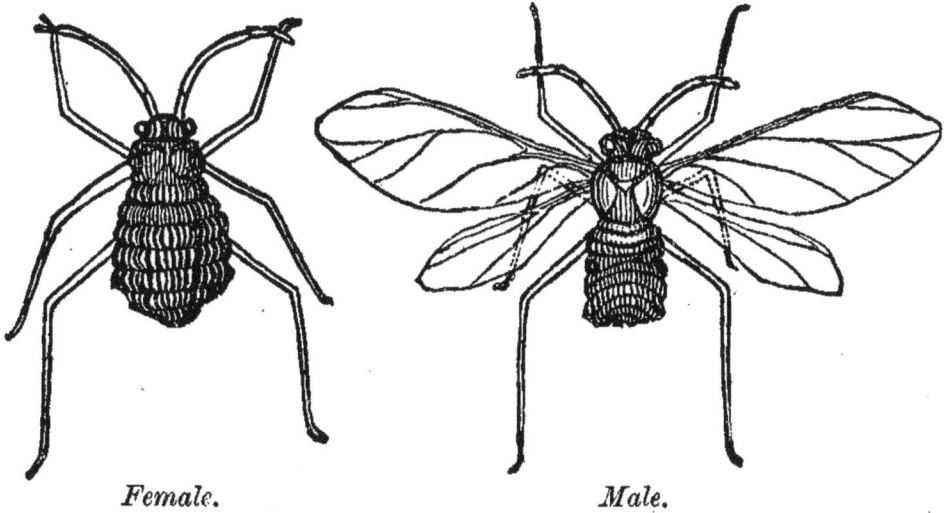

Female.   *Male.*

Sycamore Plant-Louse, much enlarged.—Original.

FIG. 198.

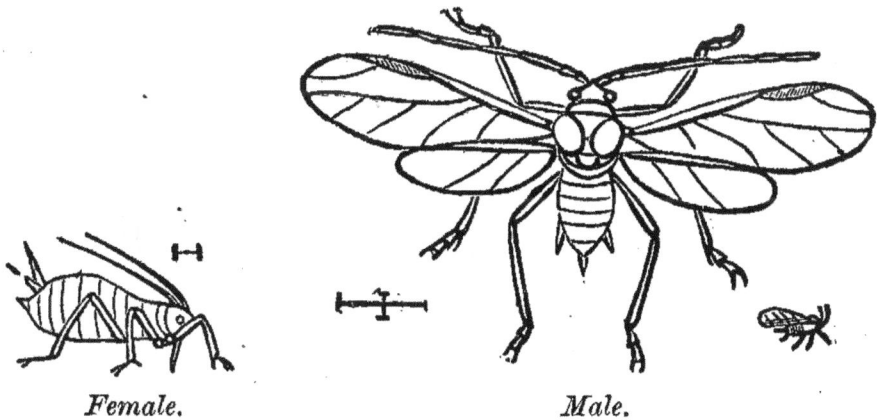

Female.   *Male.*

Apple-Tree Aphis, much magnified.—Original.

wholesome to the bees. The larch-louse, Lachnus laricis, secretes a liquid that is greedily taken by the bees. The honey is very excellent.

Another of the aphides, of a black hue, works on the

branches of our willows, which they often entirely cover, and thus greatly damage another tree valuable for both honey and pollen. Were it not that they are seldom numerous two years in succession, they would certainly banish from among us one of our most ornamental and valuable honey-producing trees. These are fairly thronged in September and October, and not infrequently in spring and summer, if the lice are abundant, by bees, wasps, ants, and various two-winged flies, all eager to lap up the oozing sweets. This louse is the Lachnus dentatus of Le Baron, and the Aphis salicti of Harris.

I have received from apiarists of Indiana and Ohio a large, dark-gray plant-louse, which worked on the sycamore, and is reported from both States as keeping the bees actively employed for some weeks. This louse is one-fourth of an inch long. The winged lice measure three-eighths of an inch to the tips of their wings. The veins of the wings, as also the short nectaries—the tubes at the posterior part of the abdomen —show that this louse (Fig. 197) belongs to the genus Lachnus. The lice of the genus Aphis—of which there are innumerable species—have longer nectaries (Fig. 198), from which ooze large drops of nectar. This is much relished by the ants, which often care for these lice as tenderly as for their own young.

Doubtless many have supposed that the bees were gathering secretion from the plants, when closer inspection would have shown that some species of plant-lice was wholly responsible. Honey-dew may not always be a secretion from insects; but that it is almost always is certainly true. We can see how it serves the insects. It attracts the bees and wasps, which repel the birds, which else would devour the insects. If plants do secrete nectar (?) from their leaves, it surely serves them some valuable purpose. It would seem, in attracting the black fungus—smut—that it harmed the foliage. Is honey-dew ever a product of the foliage ? This nectar from plant-lice is very often entirely wholesome and unobjectionable. I would, however, never consider it a safe food for bees in winter, unless it was agreeable to my taste, and fit for my own table.

REAL HONEY-DEW.

Many plants, like the cotton and cow-pea (Fig. 199) of the South, have extra floral-glands which secrete nectar. In case of the cow-pea these glands are on the peduncles or flower-

FIG. 199.

Cow-Pea—Original.

a, a Glands. *b* Flower. *c* Pods.

stems, just at the base of the flowers (Fig. 199, *a, a*). Prof. Trelese thinks that this nectar serves the plant by attracting bees, wasps, etc., which keep injurious insects from attacking it. If honey-dew is secreted from the general foliage, as so many believe, then surely, as stated above, it serves the plants in some such way.

SWEET SAP AND JUICES.

Bees often gather much nectar from the stubble of wheat that is cut early, while the straw is yet green. The sap from the maple and other trees and plants also furnishes them sweets. They gather juices of questionable repute from about cider-mills, some from grapes and other fruit which have been crushed or eaten and torn by wasps and other insects. Bees in gathering from cider-mills annoy the cider-maker, and store a product unfit for winter use. They are also often drowned in great numbers. It is wise, then, to screen them from the room where the juice is being expressed. By use of mosquito-netting this may be quickly and cheaply done. That bees ever tear grapes or other fruit is a question of which I have failed to receive any personal proof, though for years I have been carefully seeking it. I have lived among the vine-yards of California, and have often watched bees about vines in Michigan, but never saw bees tear open the grapes. I have laid crushed grapes in the apiary, when the bees were not gathering, and when they were ravenous for stores, which, when covered with sipping bees, were replaced with sound grape-clusters, which, in no instance, were mutilated. I have even shut bees in empty hives on warm days and closed the entrance with grape-clusters, which even then were not cut. I have thus been led to doubt if bees ever attack sound fruit, though quick to improve the opportunities which the oriole's beak and the stronger jaws of wasps offer them. Mr. Root finds that the Cape May warbler is even more ready than the oriole to pierce the grapes. Such habit is exceptional with the warblers, which are usually wholly insectivorous. My friend, Prof. Prentiss, suggests that when the weather is very warm and damp, and the grapes very ripe, the juice may ooze through small openings of the grapes and so attract the bees. It is at just such times that attacks are observed. I feel very certain that bees never attack sound grapes. I judge not only from observation and inquiry, but from the habits of the bee.

Bees never bore for nectar, but seek, or even know only of that which is fully exposed. Bees are, however, a tremendous aid to the fruit-grower in the great work of cross-pollination, which is imperatively necessary to his success, as has been so well shown by Dr. Asa Gray and Mr. Chas. Darwin. It is true that cross-pollination of the flowers, which can only be accomplished by insects, and early in the season by the honey-bee, is often, if not always, necessary to a full yield of fruit and vegetables. In diœcious plants, like the willows and many nut-bearing trees, the stamens that bear the pollen or male element, are on one plant or flower (Fig. 202), and the

FIG. 200.

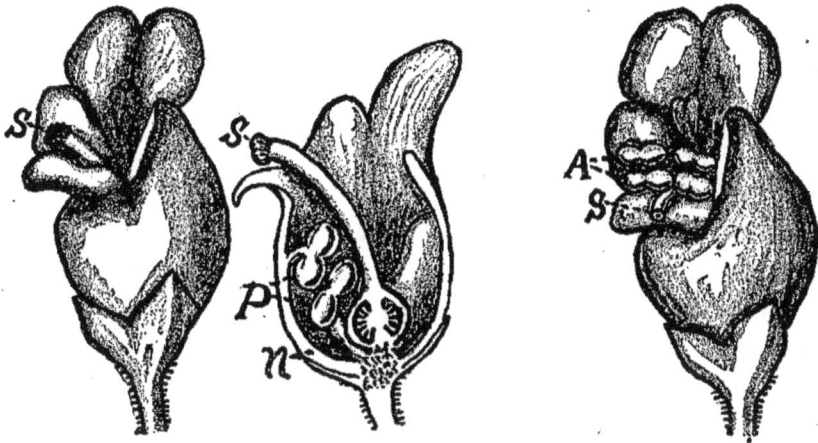

Blossoms of Figwort, after Gray.

A Developed stamens and pollen.
S In right-hand flower unripe stigma.
n Nectar.

S In two left-hand flowers ripe stigma.
P Unripe stamens.

pistils that grow the ovules—the female element—on another. Here, then, insects must act as "marriage priests" that fructification may be accomplished at all. In other plants, where the organs are all in the same flower, pollination is wholly dependent upon insects. The pollen-grains must reach the stigma. Often this is, from the very structure of the flower, entirely dependent upon insects. Often, as in the willow-herb (Fig. 252) and figwort (Fig. 200), as my colleague and esteemed friend, Dr. Beal, was first to discover, the pollen and stigma are not ripe simultaneously, and so pollen must be brought from one flower to the stigma of another, and this must be done by insects—chiefly bees. Nature thus makes close pollination impossible. Indeed, color and odor in flowers are solely to attract insects for the good of the flowers. In cases like red clover, where pollination is possible without aid, my colleague, Prof. Beal, has shown that, unless insects are present, the yield of seed is meager indeed. The seeds in the uncovered blossoms were to those in the covered as 236:5. Prof. Waite, of the Department of Agriculture, has shown that many varieties of pears, apples, etc., will fruit very scantily unless cross-pollinated by insects. I have proved the same in California with pears, plums, olives, and citrus fruits. The navel orange is an exception. It fruits just as fully without pollination, and so, of course, is usually seedless. It bears no pollen. Thus many fruit-growers keep bees to do this very important work, which they find they can not afford to neglect. I suspect in very favorable years, or in very favored localities, fruits like the Bartlett pear may be fertile to their own pollen, when at other times they will be wholly sterile. The fruit-men, then, must see that bees are abundant hard by their orchards. There is then entire reciprocity between the bees and flowers. The bees are as necessary to the plants as are the plants to the bees. I am informed by Prof. W. W. Tracy, that the gardeners in the vicinity of Boston keep bees that they may perform this duty. Mr. Root found in New York a greenhouse, where bees were kept at work all winter, to save the otherwise necessary hand-pollination, which was very laborious and expensive. That bees ever injure blossoms and thus effect damage to the fruitage of such plants as buckwheat—or to any plants, as is sometimes claimed—is utterly absurd and without foundation. It is now contended by able authorities, like Profs. Waite and Pierce, that bees carry the germs of pear-blight. Very likely this is true. Yet other insects are sufficiently abundant to do this, and yet too few to do the work of pollination. A few inoculations will scatter the blight, while pollination must be done in wholesale fashion.

But the principal source of honey is still from the flowers.

WHAT ARE THE VALUABLE HONEY-PLANTS ?

In the northeastern part of our country the chief reliance, for May, is the fruit-blossoms, willows, and sugar maples. In June, white clover, alsike clover, and raspberries yield largely of the most attractive honey, both as to appearance and flavor. In July, the incomparable basswood and sweet clover make both bees and apiarist jubilant. In August, buckwheat offers a tribute, which we welcome, though it be dark and pungent in flavor, while in Michigan, August and September give a profusion of bloom which yields to no other in the richness of its capacity to secrete nectar, and is not cut off till the autumn frosts—usually about Sept. 15.

Thousands of acres of willow-herb (Fig. 252), goldenrod, boneset, asters, and other autumn flowers of northern Michigan as yet have blushed unseen, with fragrance wasted. This unoccupied territory, unsurpassed in its capability for fruit-production, covered with grand forests of maple and basswood, and spread with the richest of autumn bloom, offers opportunities to the practical apiarist rarely equaled except in Texas and the Pacific States.

In the following table will be found a list of valuable honey-plants. Those mentioned first are annual, biennial or perennial; the annual being enclosed in a parenthesis thus: ( ); the biennial enclosed in brackets thus: [ ] ; while those mentioned later are shrubs or trees; the names of shrubs being enclosed in a parenthesis. The date of the commencement of bloom is, of course, not invariable. The one appended, in case of plants which grow in that State, is about average for Central Michigan. Those plants whose names appear in small capitals yield very superior honey. Those with (a) are useful for other purposes than honey-secretion. All but those with a * are native or very common in Michigan. Those written in the plural refer to more than one species. Those followed by a † are very numerous in species. Of course I have not named all, as that would include some hundreds which have been observed at the college, taking nearly all of the two great orders, Compositæ and Rosaceæ. I have only aimed to give the most important, omitting many foreign plants of notoriety, as I have had no personal knowledge of them.

HERBACEOUS HONEY-PLANTS, TREES AND SHRUBS.

DATE OF BLOOMING.	ANNUALS AND PERENNIALS.
February to July.......	*Gilias—California—Blue Pollen.
February and March....	*Gallberry—South.
March to fall	ALFALFA OR LUCERNE (a)—Calif., Colo.
April....................	Skunk Cabbage.
April....................	Crocus.
April and May..........	Dandelion.
April and May..........	Crowfoot.
April and May..........	Strawberry. (a)
April and May..........	Crimson Clover (a)—South—Not California.
May.	(Seven-Top Turnip.)
May and June..........	*Hoarhound—California.
May and June..........	*Sumac—California.
May and June..........	*Coffee-Berry—California.
May and June..........	*HORSEMINT—Texas.
May and June..........	False Indigo.
May and June..........	Lupine.
May to August	Ground Ivy or Gill.
May to August	Cow-Pea. (a)
May to fall	(Cow-Pea.) (a)—South.
June	*Stone Crop—South.
June	Mammoth Red Clover. (a)
June	*California Figwort—California.
June	(Hemp.) (a)
June	Gumbo or Okra.
June to July..........	WHITE CLOVER. (a)
June to July..........	Beans (a)—California.
June to July..........	ALSIKE CLOVER. (a)
June to July..........	RED CLOVER. (a)
June to July..........	CRIMSON CLOVER. (a)
June to July..........	*[SWEET CLOVER.]
June to July..........	Hoarhound.
June to July..........	Ox-Eyed Daisy—Bad Weed.
June to July..........	Bush Honeysuckle.
June to July..........	*(Partridge Pea.)
June to July..........	Burdock—White Pollen.
June to August........	Matrimony Vine.
June to August........	*Sage.
June to August........	Motherwort.
June to frost............	*(Borage.)
June to frost............	*(Cotton.) (a)
June to frost............	Pleurisy Root.
June to frost............	Silk or Milk Weeds.
June to frost............	[Cabbage.] (a)
June to frost............	(Mustard) †
June to frost............	*(Rape.) (a)
June to frost............	St. John's Wort.
June to frost............	(MIGNONETTE.) (a)
July	(Corn.) (a)
July	*(Teasel.) (a)
July to August	Basils or Mountain Mint.
July to August	Catnip. (a)

July to August	Chamomile.
July to August	*Asparagus. (a)
July to August	(Cucumber,Squash, Pumpkins,Melons,etc) (a)
July to August	*(Rocky Mountain Bee-Plant)—Colorado.
July to August	*Viper's Bugloss (Blue Thistle.)
July to August	Blue Vervain or Verbena.
July to August	White Vervain or Verbena.
July to August	*Fog-fruit, Lippia—Texas.
July to August	Marsh Milkweed.
July to frost............	Boneset.
July to frost............	Bergamot.
July to frost	Chicory.
July to frost............	Figwort or Carpenter's Square.
July to frost............	Giant Hyssop.
July to frost............	Malva.
July to frost............	Ironweed.
July to frost............	Fireweed.
July to frost............	Culver's Root.
July to frost............	Indian Plantains.
July to frost............	*SPIDER FLOWER.
August.................	(Buckwheat.) (a)
August.................	(Snapdragon.)
August.................	(Touch-me-not or Swamp Balsam.)
August.................	(GREAT WILLOW-HERB, Fireweed.)
August to September ...	Golden Honey-Plant.
August to September ...	*Heartsease, or Smartweed, or Knotweed—
August to September ...	Mississippi Valley.
August to September ...	Large Smartweed.
August to frost........	(GOLDENROD.) †
August to frost........	ASTERS. †
August to frost........	Marsh Sunflower. †
August to frost........	Tick-Seed. †
August to frost........	Beggar-Ticks or Bur Marigold. †
August to frost........	Spanish Needles. † Coreopsis.
August to frost........	Rattlesnake Root or Tall White Lettuce.

TREES AND SHRUBS.

January to January.....	Lemon—South and California.
January to January.....	Eucalyptus, many species—California.
January to May	*Manzanita—California.
January to May.........	*Rattan—South.
January to May	*(Willow) †—California.
January to May	*Chaparall—California Lilac.
February to June.......	*(Gall Berry)—South.
March..................	*Orange—South and California.
March..................	Madrona—California.
April..................	Box-Elder or Ash-Leaf Maple.
April..................	Red or Soft Maple. (a)
April..................	Elm.
April..................	Poplar or Aspen.
April..................	Silver Maple.
April and May..........	(Willows) † also Trees.
April and May..........	*Black Gum—South.
April and May..........	*Judas Tree—South.
April to June...........	(Chinkapin, Chinquapin Dwarf Chestnut.)
April to July	Mesquite—Texas and West.

April to July	*WHITE SAGE—California.
April to July	*BALL, BLACK, or BUTTON SAGE.—California.
April to July	*Rhododendron—South and West.
April to July	Honeysuckle.
May	(Shadbush, or June-berry, or Service-berry.)
May	(Alder.)
May	Maples—Sugar Maple. (a)
May	Crab-Apple.
May	Apricot.
May	(Hawthorns.)
May	Fruit Trees—Apple,Plum,Cherry,Pear,etc(a)
May	(Currant and Gooseberry. (a)
May	*(Wistaria Vine—South.)
May	*(Chinese Wistaria Vine—South).
May	*(Japan Privet—South.)
May	*Varnish Tree—South.
May	*Acacia—South and California.
May	(Bladder Nut.)
May	*Persimmon (a)—South.
May	*SAW PALMETTO—South.
May	Mountain Laurel—South.
May	Buckeye.
May	Horse-Chestnut.
May and June	(Barberry.)
May and June	(Grape Vine.) (a)
May and June	Poplar, Tulip.
May and June	Tulip-tree or Whitewood, Poplar—South.
May and June	(Sumac.)
May and June	*Buckthorns—South. California Feb. to July.
May and June	*BLACK MANGROVE—Florida.
May and June	Locust or Black Locust.
May to September	*Banana—South.
May to September	Cobœa scandens. Vine.
June	Catalpa.
June	*Magnolias—South. California.
June	Honey-Locust.
June	Wild Plum.
June	(Black Raspberry.) (a)
June	Locusts.
June	(RED RASPBERRY.) (a)
June to July	(Blackberry.)
June to July	Chestnut. (a)
June to July	*Sourwood—South.
June to frost	Wild Buckwheat—California.
July	(Button Bush.)
July	BASSWOOD. (a)
July	(Virginia Creeper.) (a) Vine.
June and July	*CABBAGE PALMETTO.—South.
July	*Blue-Gum—California.
July	Catalpa. (a)
July to August	*Pepper-tree—California.
July to September	*(St. John's Worts.)
August	(Late Sumac.)
August to September ...	Indian Currant, Coralberry, or Buckbush, or Snowdrop.
August to September ...	
August to frost	*Red-Gum—California.
August to December	*Japan Plum—South.
August to January	(Germander or Wood-Sage.)

DESCRIPTION, WITH PRACTICAL REMARKS.

As this subject of bee-pasturage is of such prime importance, and as the interest in the subject is so great and widespread, I feel that details with illustrations will be more than warranted.

We have abundant experience to show that forty or fifty colonies of bees, take the seasons as they average (except in such very highly-favored localities as Southern California, where in good seasons two or three hundred are profitably kept in a single apiary, even six hundred having been proved in the best seasons to do well), are all that a single place will sustain to the greatest advantage. Then how significant the fact that when the season is the best full three times that number of colonies will find ample resources to keep all employed. So this subject of artificial pasturage becomes one well worthy of close study and observation. The subject, too, is a very important one in reference to the location of the apiary.

It is well to remember in this connection, that while bees do sometimes go from five to seven miles for nectar, two or three miles should be regarded as the limit of profitable gathering. That is, apiaries of from fifty to one hundred or more colonies should not be nearer than four or five miles of each other.

MARCH PLANTS.

In Florida the orange gives early bloom, and the thousands of trees in that land, not only of flowers but of honey, will have no small influence in building up the colonies for the grand harvest of mangrove and palmetto soon to follow.

The gall-berry of the South commences to bloom even in February, and yields abundant nectar. In Florida this shrub gives the main supply of honey during the swarming season.

APRIL PLANTS.

As we have already seen, the apiarist does not secure the best results, even in the early spring, unless the bees are encouraged by the increase of their stores of pollen and honey ; hence, in case we do not practice stimulative feeding—and

many will not—it becomes very desirable to have some early
bloom. Happily, in all sections of the United States our
desires are not in vain.

Early in spring there are many scattering wild-flowers, as
skunk-cabbage (Symplocarpus fœtidus), which supplies abundant pollen and some honey; the blood-root (Sanguinaria
canadensis), liver-leaf (Hepatica acutiloba), and various others

FIG. 201.

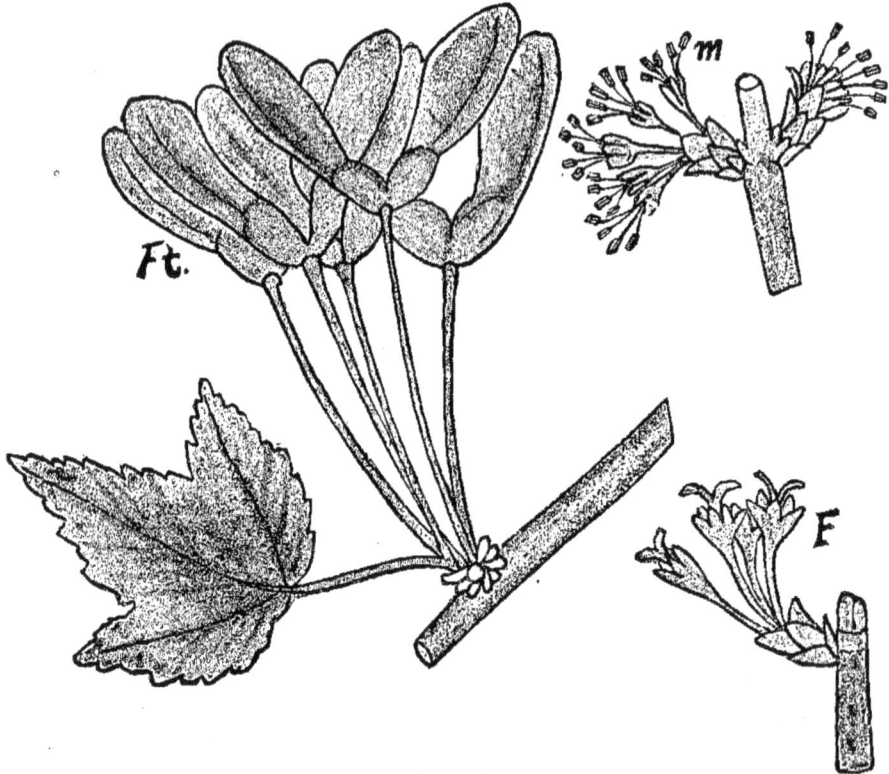

Red Maple.—Original.

M Male blossoms.   *F* Female blossoms.
Ft Fruit.

of the crow-foot family, as also many species of cress, which
belong to the mustard family, and the gay dandelion (Taraxa-
cum dens-leonis), which keeps on blooming for weeks, etc., all
of which are valuable and important. The dandelion affords
nectar for excellent honey. Were it not so concurrent with
fruit-bloom, it would be more valuable, yet it anticipates and
succeeds the orchard bloom.

The maples, which are all valuable honey-plants, also contribute to the early stores. Especially valuable are the silver maples (Acer dasycarpum), the red or soft maples (Acer rubrum) (Fig. 201), and the box-elder or ash-leaf maple (Negundo aceroides), as they bloom so very early, long before the leaves appear. The bees work on these in Michigan the first week of April, and often in March. They are also magnificent shade-trees, especially those that have the weeping habit. Their early bloom is very pleasing, their summer form

FIG. 202.

Willow.—Original.

and foliage beautiful, while their flaming tints in autumn are indescribable. The foreign maples, sycamore (Acer pseudo-platanus), and Norway (Acer platanoides), are also very beautiful. Whether superior to ours as honey-plants I am unable to say.

The willows, too (Fig. 202), rival the maples in the early period of bloom. Some are very early, blossoming in March, while others, like the white willow (Salix alba) (Fig. 202), bloom in May. The flowers on one tree or bush of the willow are all pistillate—that is, have pistils but no stamens—while on others they are all staminate, having no pistils. On the former, bees can gather only honey, on the latter only pollen. That the willow furnishes both honey and pollen is attested by the fact that I saw both kinds of trees, the pistillate and the staminate, thronged with bees the past season. Indeed, the willow furnishes abundant honey nearly every spring.

FIG. 203.

FIG. 204.

Mesquite.—*From Dept. of Agriculture.*   *Judas Tree.—Original.*

Were bees numerous thus early, and were the weather propitious, the honey from willow would be very important. It is in stimulating breeding. The willow, too, from its elegant form and silvery foliage, is one of our finest shade-trees. It grows everywhere in the United States. The mesquite (Prosopis juliflora), a shrub or tree of the bean family (Fig. 203), is exceedingly valuable for honey from Texas to Arizona. The honey is excellent in quality and very abundant. It blooms from April on to July.

In the south of Michigan, and thence southward to Kentucky, and even beyond, the Judas tree, or red-bud (Cercis canadensis), (Fig. 204), is not only worthy of cultivation as a honey-plant, but is also very attractive, and well deserving of attention for its ornamental qualities alone. The red flowers precede the foliage, and are very striking. This blooms from March to May, according to the latitude.

FIG. 205.

Eucalyptus Robusta.—From University of California.

The poplars—not the tulip—also bloom in April, and are freely visited by the bees. The wood is immaculate, and is used for toothpicks and sections for comb honey.

In California, the unique and exquisite Manzanitas (spe-

cies of Arctostaphylos), together with the willows and many
other flowering plants, keep the bees busy from January till
March.   The Gilias, many of which furnish blue pollen, bloom
in January and on till June.   The wondrously fragrant orange
and lemon hang out their profuse bloom even in January, and
the lemon is always in bloom.   Some species of gum eucalyp-

FIG. 206.

Pepper.—Original.

tus (Fig. 205) is in bloom nearly every month of the year. The
beautiful peppers (Fig. 206), so exquisite in grace, have a very
wide period of bloom.   The eucalyptus honey is agreeable, as
I have surely proved.   I think the pepper only furnishes pollen
extensively.   The tree is dioecious.   The female trees are the
more attractive for the red fruit of winter.   They are tran-
scendently beautiful for roadside planting, as are many of the
gums.

MAY PLANTS.

In May we have the grand sugar maple (Acer saccha-rinum), (Fig. 207), incomparable for beauty, also all our vari-

FIG. 207.

Sugar-Maple.—Original.

ous fruit-trees, peach, cherry, plum, apple, etc.; in fact all the Rosaceæ family. Despite adverse criticism, I have found fruit honey excellent. Our beautiful American wistaria (Wistaria frutescens), (Fig. 208), the very ornamental climber, or the still more lovely Chinese wistaria (Wistaria sinensis), (Fig. 209), which has longer racemes than the native, and often blossoms twice in the season. These are the woody twiners for the apiarist. I regret to say that neither one is hardy in Michigan.

The barberry, too (Berberis vulgaris), (Fig. 210), comes after fruit-blossoms, and is thronged with bees in search of nectar in spring, as with children in winter, in quest of the beautiful scarlet berries, so pleasingly tart.

. FIG. 208.

American Wistaria.—Original.

In California, the sumac, the hoarhound, the famous ball or black sages (Fig. 211), (Ramona stachyoides and R. palmeri), with their most beautiful and delicious honey, and the more common and equally excellent white sage (Ramona poly-stachya), (Fig. 212), keep the bees roaring with activity, in favorable seasons, from April even unto June. It is charac-teristic of California bloom to continue for weeks. The long racemes of white sage may open in April and continue in blossom away to June.

In the South, as I learn from that able apiarist, Dr. J. P,

H. Brown, they are no less favored. The Japan privet, the varnish tree, the acacia, the black-gum or sour-gum (Nyssa aquatica), and the persimmon (Diospyros virginiana) stir the bees up to their best endeavor in May. The banana (Musa sapientum) blooms not only in May, but, as Mr. W. S. Hart, of Florida, writes me, it is in blossom the year around. So rich are the flower-tubes in nectar that Mr. Hart says he could

FIG. 209.

FIG. 210.

Chinese Wistaria.—Original.   *Barberry.—Original.*

soon gather a teacupful, by hand, of clear, beautiful nectar of good flavor. Chinquapin (Castanea pumila) is an excellent honey-plant in the Carolinas.

The horse-mint (Fig. 213), (Monarda aristata), especially in Texas, is sending the bees loaded to their hives with its peculiar, aromatic nectar. This, with the buckthorns, species of Rhamnus and Ceonothus, yield honey into June. These plants often cover acres in Wisconsin and Minnesota. Mr.

FIG. 211.

Ball or Black Sage.—Original.

Freeborn, of Wisconsin, has often secured a large harvest from this source when all else failed. The buckthorns are very common in California. Some of the blooms are delicate

FIG. 212.

White Sage.—From A. I. Root Co.

blue, and are known as the "California lilacs." They are in bloom—different species—from March to July.

The saw palmetto (Serenoa serrulata) forms a dense growth

and makes clearing the land no small expense in Florida. The slim trunk creeps along the ground for twenty feet, and sends roots beneath for nourishment. The leaves arise from this stem, and are from four to six feet long. The clusters of

FIG. 213.

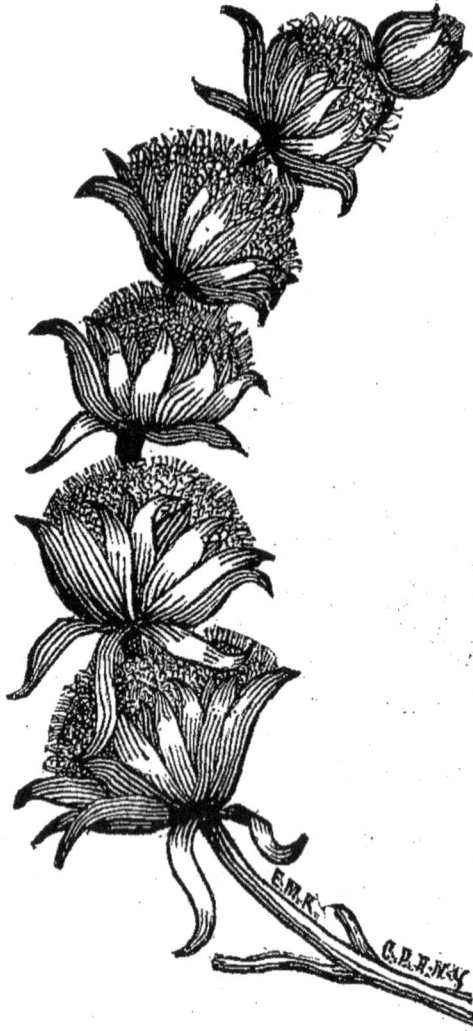

Horsemint.—Original.

small, yellowish-white blossoms are immense in size. The blossoms last from the middle of April till June. The honey is yellow, thick and fine. The fruit of this palm is about twice the size of the Concord grape, and from *October till Christmas* the oozing nectar keeps the bees at work. This is dark honey, but very good for stimulative feeding. The date-palms (species of Phœnix), and many others, grow magnifi-cently in California, and are valuable aids to the bee-keeper. The growing of date-palms promises a rich harvest in Califor-nia and Arizona.

JUNE PLANTS.

With June comes the incomparable white or Dutch clover (Trifolium repens), (Fig. 214), whose chaste and modest bloom betokens the beautiful, luscious, and unrivaled sweets which

FIG. 214.

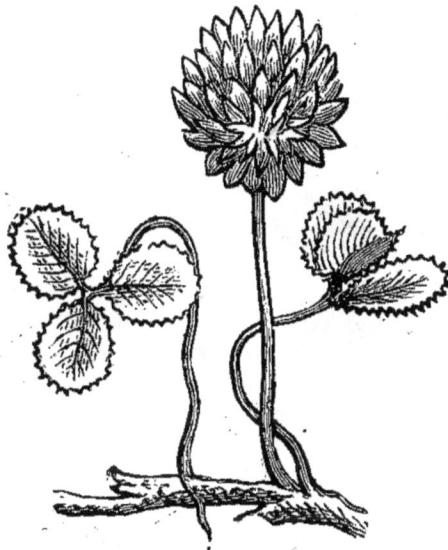

White or Dutch Clover.

are hidden in its corolla-tube; also its sister, Alsike or Swedish clover (Trifolium hybridum), (Fig. 215), which seems to resemble both the white and red clover, is now beautiful and fragrant. This is not a hybrid, as its name would suggest. It is a stronger grower than the white, and has a whitish blossom tinged with pink. Mr. Doolittle says the honey is often a little off-color, and its presence may so tinge basswood honey as to make it second grade. Messrs. Doolittle and Root think that white clover furnishes about fifty pounds of honey to the acre during the season. I am sure that Alsike may furnish much more than this, and I believe the same is true of

FIG. 215.

Alsike Clover.—From American Bee Journal.

white. This forms excellent pasture and hay for cattle, sheep, etc., and may well be sown by the apiarist. When pastured the bloom is much prolonged. It will often pay apiarists to furnish neighboring farmers with seed as an inducement to grow this excellent honey-plant. It will be easy to get all farmers within two miles of the apiary to sow this seed, if we sell it to them for six dollars per bushel of sixty pounds, when the price is eight or nine dollars. This would be a wise plan.

FIG. 216.

Mammoth Red Clover.—From A. I. Root Co.

Like white clover, it blooms all through June and into July. Both of these should be sown early in spring with timothy, four or five pounds of seed to the acre, in the same manner that red clover seed is sown. As Alsike seeds itself each year, and so lasts much longer than red clover, I think it pays well to mix the seed, using about three pounds of Alsike clover seed and five or six of red clover. As Alsike clover is visited freely by honey-bees the first growth of the season, unlike red clover, it seeds bountifully. By cutting Alsike clover

FIG. 217.

Crimson Clover.— From A. I. Root Co.

NATURAL SIZE
LEAF

just as it commences to bloom, it may be made to come into blossom the second time, so as just to fill the vacant space in August. This is a very important fact, and may well be acted upon. I have known Alsike clover to give a good harvest of nectar during a dry year, when white clover utterly failed. Red clover (Trifolium pretense), especially mammoth (Fig. 216), is a wondrous honey-plant, but its long flower-tubes place the nectar beyond the reach of black bees, and of most

FIG. 218.

Melilot Clover.—Original.

Italians. Can we breed longer tongues in our bees, or shorter tubes in the clover ? I see no reason why we may not do both.

Crimson clover (Trifolium incarnatum), (Fig. 217), is popular in some sections. It is a failure in Southern California. The blossoms are large and fine, and they are visited freely by the bees. It is used in the East and South for green manuring. I do not think it will compare with white or Alsike clover as a honey-plant.

Sweet clover, yellow and white (Melilotus officinalis and Melilotus alba), (Fig. 218), are well named. They bloom from the middle of June to the first of October. Their perfume scents the air for long distances, and the hum of bees that throng their flowers is like music to the apiarist's ear. The honey, too, is just exquisite. These clovers are biennial, not

FIG. 219.

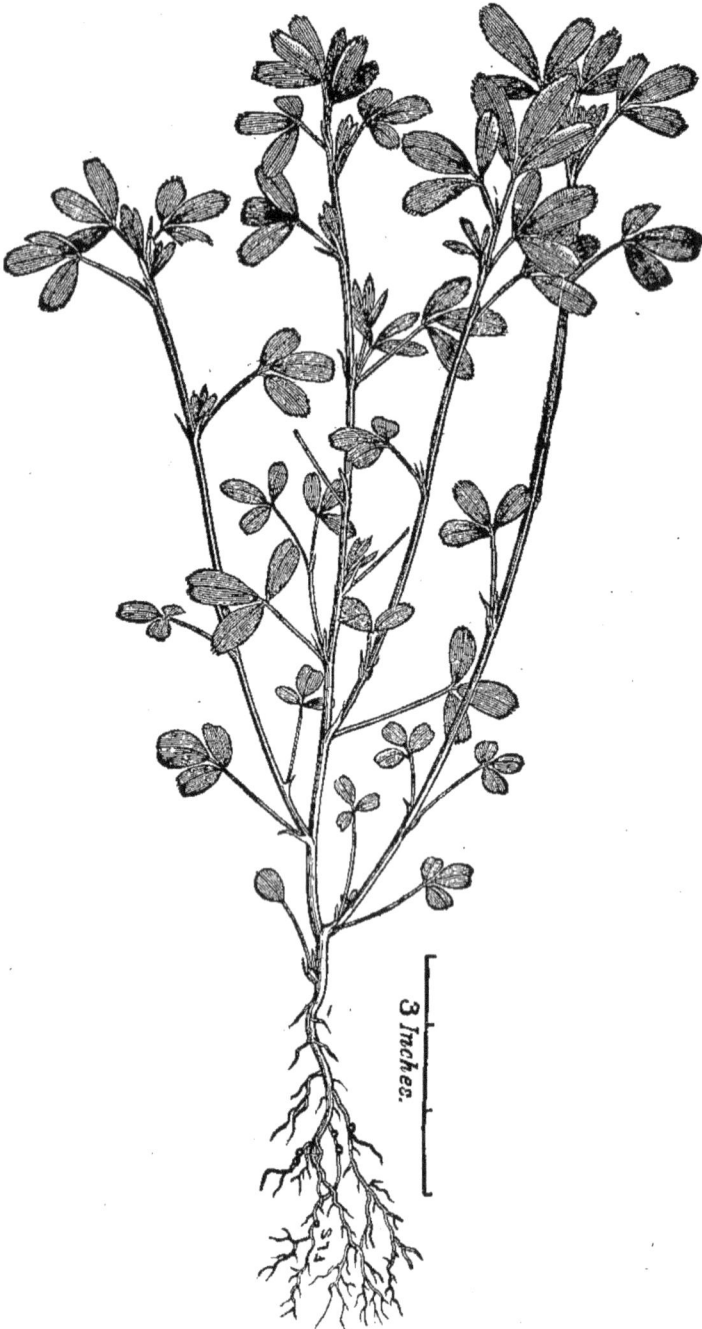

Alfalfa Honey-Plant.—From A. I. Root Co.

FIG. 220.

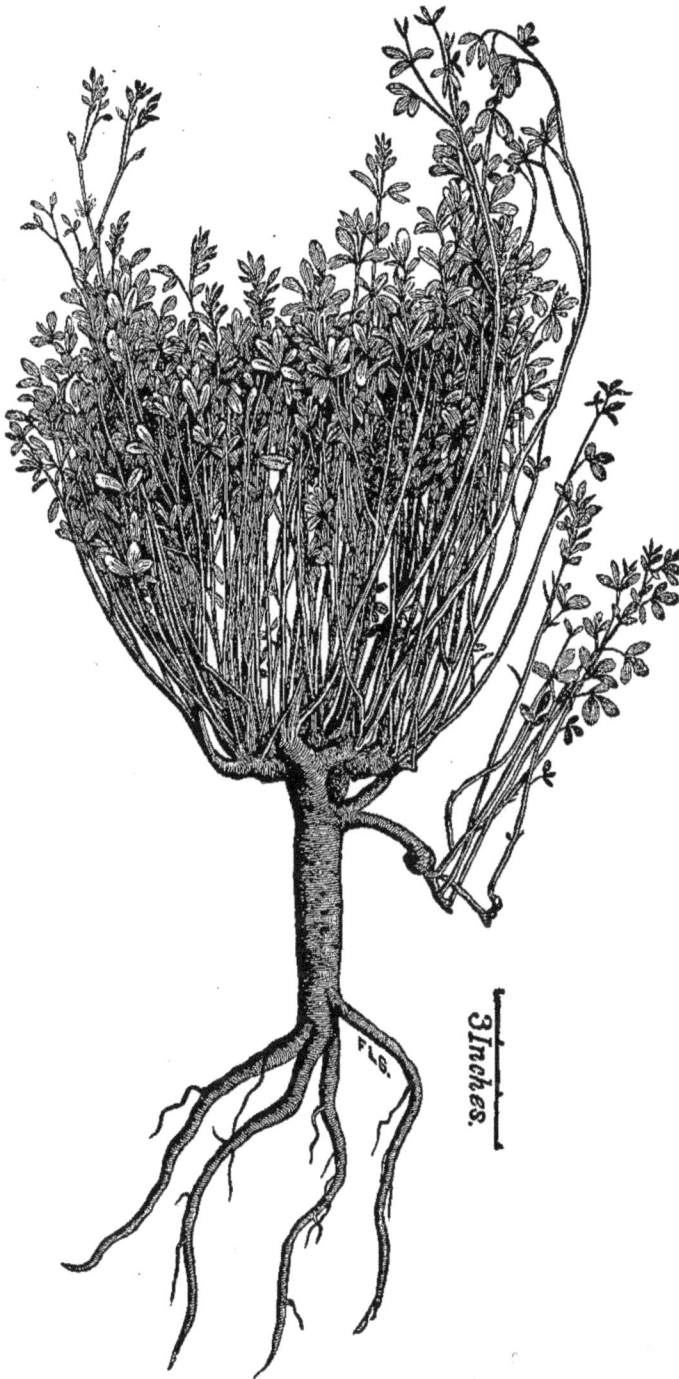

Alfalfa.—From A. I. Root Co.

blooming the second season. They perpetuate themselves, however, through the seed, so as really to become perennial. A disagreeable fact is that they have little value except for honey. Some bee-keepers claim to have found them valuable for pasturage and for hay. I have wished and tried to use them, but my horses and cow would not co-operate. The Bokhara clover is only a variety of the above.

FIG. 221.

Alfalfa in Bloom.—From A. I. Root Co.

a, b Seed pods.   c Seed.

The other clovers—lucerne, yellow trefoil, scarlet trefoil, and alfalfa—(Figs. 219, 220 and 221) have not proved of any value in Michigan, perhaps owing to locality. The alfalfa is valued highly for bees in the Western States. In California, Nevada, Arizona and Colorado it is a marvelous plant for both hay and honey. Eight cuttings a season have been made, though five are the average in Southern California. A yield of three tons per cutting per acre is not exceptional. The hay is the very best. It is pre-eminently a crop for irrigation, and so is not stayed by the drouth. Bee-keepers in central California and Arizona report two hundred pounds per colony from alfalfa, even with very large apiaries. While most prefer to cut before it is in full bloom, as it is eaten better, yet there

FIG. 222.

FIG. 223.

Borage.—Original.

Mignonette.

is always much bloom where it is grown extensively. It is a sure honey-producer in the famous San Joaquin Valley of California.

Borage (Borago officinalis), (Fig. 222), an excellent bee-plant, blooms from June till frost, and is visited by bees even in very rainy weather. It seems not to be a favorite, but is eagerly visited when all others fail to yield nectar.

Mignonette (Reseda odorata), (Fig. 223) blooms from the middle of June till frost, is unparalleled for its sweet odor, furnishes nectar in profusion, and is well worthy cultivation. It does not secrete well in wet weather, but in favorable weather it is hardly equaled. It will never be grown in quantities to give any large returns.

Okra, or gumbo (Hibiscus esculentus), (Fig. 224), also blooms in June. It is as much sought after by the bees in quest of honey, as by the cook in search of a savory vegetable, or one to give tone to soup.

Sage (Salvia officinalis), hoarhound (Marrubium vulgare), motherwort (Leonurus cardiaca), and catnip (Nepeta cataria), which latter does not commence to bloom till July, all furnish nice, white honey, remain in bloom a long time, and are very

FIG. 224.   FIG. 225.

Okra.—Original.   *Mint.—Original.*

desirable, as they are in bloom in the honey-dearth of July and August. They, like many others of the mint family (Fig. 225), are thronged with bees during the season of bloom. The first and last are of commercial importance, while very few of our native plants afford so much nectar, are such favorites with the bees, and are so independent of weather as motherwort (Fig. 226). It is crowded with bees from the dawn of its bloom till the last flower withers. By cutting it back in May it can be made to blossom just at the dearth of nectar-secreting bloom; otherwise it comes in June and early July, just when

linden is yielding its precious harvest. Few plants are more
desirable to sow in waste-places. Mr. Doolittle says truly,
"If any plant will pay to grow solely for honey, it is this
one." He is correct in the opinion that none will pay.

FIG. 226.

Motherwort.—Original.

The silk or milkweed furnishes abundant nectar from
June to frost, as there are several species of the genus Asclepias, which is wide-spread in our country. Indeed, pleurisy-
root or butterfly weed (Asclepias tuberosa) is the bee-plant
that Mr. Heddon has praised so highly. He thinks it one of

our best indigenous honey-plants. These are the plants which have large pollen-masses which often adhere to the legs of the bees (Fig. 227), and sometimes so entrap them as to cause their death. Prof. Riley once very graciously advised planting them to kill bees! I say graciously, as I have watched these very closely, and am sure they do little harm, and are rich in nectar. Seldom a bee gets caught so as to hold it long, and when these awkward masses are carried away with the bee, they are usually left at the door of the hive,

FIG. 228.

FIG. 227.

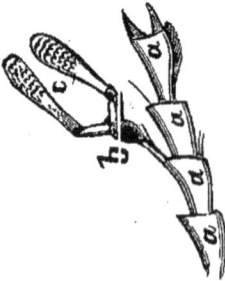

Pollen of Milk-Weed on Bee's Foot.
—Original.   *Black Mustard.—Original.*

where I have often seen them in considerable numbers. The river bank, hard by our apiary, is lined with these sweet-smelling herbs, and we would like even more. Occasionally, however, the bees are held to the plant by them, and more often become so burdened with these pollen-masses that the other bees drag or drive them from the hive, as no longer fit for labor or worthy to live. Bees are veritable Hottentots—they kill, though they do not eat the old and the feeble.

Black mustard (Sinapis nigra), (Fig. 228), white mustard

(Sinapis alba) and rape (Brassica campestris), (Fig. 229), all
look much alike—all are species of the great family Cruciferæ—
and are all admirable bee-plants, as they furnish much and
beautiful honey. The first, if self-sown, blooms in Michigan
July 1st, the others June 1st; the first about eight weeks after
sowing, the others three or four. The mustards bloom for
four weeks, rape for three. These are all specially commend-
able, as they may be made to bloom during the honey-dearth
of July and August, secrete honey late as well as early in

FIG. 229.

Rape.

the season, and are valuable plants to raise for seed. The
mustards were grown in Southern California for seed during
the Civil War, and have run wild in parts of San Diego County
where they grow very extensively. Though the hills are
yellow with this plant for miles, I have heard no special commendation of it for honey. Very likely our scant rainfall is
not favorable for nectar-secretion. Rape seems to be very
attractive to insects, as the flea-beetles and the blister-beetles
are often quite too much for it, though they do not usually
destroy the plants till after they have blossomed. Rape is

now grown extensively in Michigan and contiguous States for
sheep, etc.   It pleases the stockmen and the bee-keepers alike.
Three pounds per acre of seed is the amount to sow.   I have

FIG. 230.

Tulip.—Original.

several times purchased what purported to be Chinese mustard, dwarf and tall, but Prof. Beal, than whom there is no
better authority, tells me they are only the white and black,
and certainly they are no whit better as bee-plants.   These

plants, with buckwheat, the mints, borage and mignonette, are specially interesting, as they cover, or may be made to cover, the honey-dearth from about July 20th to August 20th.

The mustards may be planted in drills about eight inches apart, any time from May 1st to July 15th. Four quarts will plant an acre.

In this month (though I have known it to bloom in Michigan in May, while South it blossoms in April) blooms the

FIG. 231.

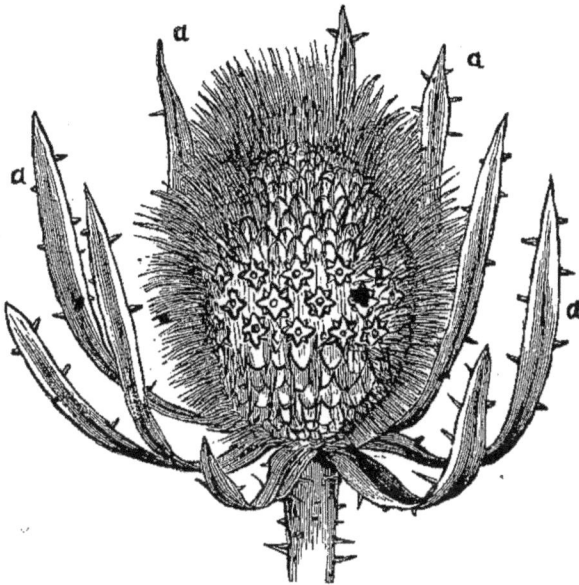

Teasel.—Original.

tulip-tree (Liriodendron tulipifera), (Fig. 230)—often called poplar in the South, which is not only an excellent honey-producer, but is one of our most stately and admirable shade-trees. It is also very valuable for its lumber, which is known as whitewood. It would be of more worth did it not shrink so much. Dr. Brown, of Georgia, says this is the great dependence—the basswood of the South. He says that along rivers especially the bloom is so prolonged, being earlier on the up-lands, that the harvest is long as bountiful. Now bloom the sumacs, though one species blooms in May, the wild plum, the raspberries, whose nectar is unsurpassed in color and flavor, and the blackberry. The raspberry is specially to be praised.

It blooms between the usual fruit-blossoms and clover. It yields nectar in wet weather, which most plants fail to do, and the honey is unexcelled. Bees sometimes gather the juice from very ripe raspberries. This colors the honey red. The blackberry comes quite late, some days after the raspberry. I think it is far less valuable as a honey-plant. Corn yields largely of honey as well as pollen, and the teasel (Dipsacus fullonum), (Fig. 231), is said, not only by Mr. Doolittle, but by English and German apiarists, to yield richly of beautiful

FIG. 232.

Common Locust.—From American Bee Journal.

honey. It blossoms at the same time with basswood, and the honey is much thinner at first. This last has commercial importance. In central New York it is raised in large quantities. The spinous fruit-heads are used in preparing woolen cloth. Machinery is now taking the place of teasel, and as no plant can be profitably grown for honey alone, this plant will be of little importance in the future. The fragrant locust (Robinia pseudacacia), (Fig. 232), opens its petals in June opportunely, for it comes between fruit-bloom and clover. Unfortunately, it furnishes nectar only occasionally. The honey is fine, as we should expect, as it belongs with the clovers, to the great family of Leguminosæ. From its rapid growth, beautiful form, and handsome foliage, it would rank among our first shade-trees, were it not that it is so tardy in spreading its canopy of green, and so liable to ruinous attack by the borers, which last peculiarity it shares with the incom-

FIG. 233.

Partridge Pea.—Original.

parable maples. Washing the trunks of the trees in June and July with soft soap will in great part remove this trouble.

In June, mammoth red clover (Trifolium pratense), (Fig. 216) comes out in one mass of crimson. This, unlike common red clover, has flower-tubes short enough for even the ligula of the black bee. It is pretty coarse for hay, but excellent for pasture and for green manuring. The partridge-pea (Cassia chamæcrista), (Fig. 233), furnishes abundant nectar, and, like

the cow-pea (Vigna sinensis) of the South, (Fig. 199) has extra
floral as well as floral glands.   Lupine (Lupinus perennis) and
gill or ground-ivy (Nepeta glechoma) began blooming in May,
and now are fully out.   This last is a mint, a near relative of
catnip.   I find there are foreign mints which are excellent
honey-plants, and very likely would pay well to sow in waste-
places.   The matrimony vine (Lycium vulgare), and the beautiful honey-locust (Gleditschia triacanthos), (Fig. 234) are now

FIG. 234.

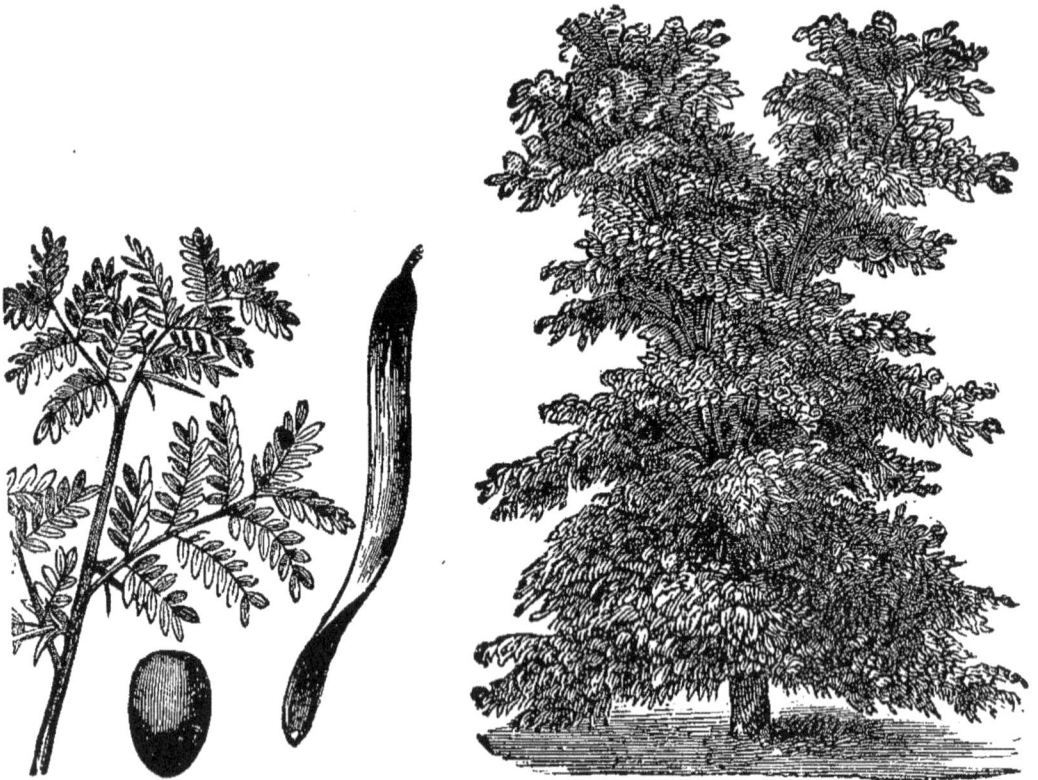

Honey-Locust.—From American Bee Journal.

full of life, as the bees come and go full-loaded with nectar.
In California, the figwort (Scrophularia californica) contributes
to the honey-supply.   The wild buckwheat (Fig. 235) blooms
profusely in all parts of Southern California from June to
frost.   It yields much amber honey of excellent quality,
though from its color it is second grade.   Next to the sages it
is the best wild honey-plant of the section.   Our brothers of
the South reap a rich harvest from the great staple, cotton

(Gossypium herbaceum), (Fig. 236), which commences to bloom early in June, and remains in blossom even to October. This belongs to the same family—Mallow—as the hollyhock, and, like it, blooms and fruits through the season.

FIG. 235.

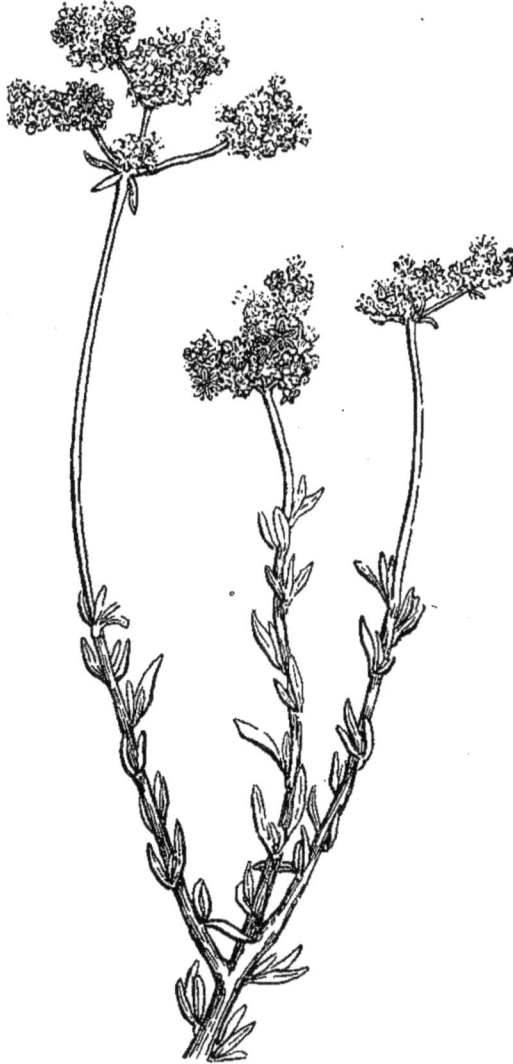

Wild Buckwheat.—Original.

The cow-pea, which blooms from April to August (Fig. 199), is not only good for bees, but for feed, and to enrich the soil. The stone-crop (Sedum pulchellum) is another valuable honey-plant of the South. In June the magnolias (Fig. 237)—

there are several species in the South—are in bloom. In many parts they commence to blossom in May. One of the finest of these is the Magnolia glauca (Fig. 237). One would suspect at once that it was a near relative of the tulip tree. This is also common in Southern California.

FIG. 236.

Cotton.—Original.

JULY PLANTS.

Early in this month opens the far-famed basswood or linden (Tilia Americana), (Fig. 238), which, for the profusion and quality of its honey, has no superior. Mr. Doolittle got 66 pounds of linden honey from a single colony in three days. It is what has given Wisconsin its proud place as a bee-section. There is rarely a year that it does not give us some of its incomparable nectar. It has been estimated that one linden tree would furnish, in a favorable year, fifty pounds of honey.

The tree, too, from its great, spreading top and fine foliage, is magnificent for shade. Five of these trees were within two rods of my study window, and their grateful fragrance and

FIG. 237

Magnolia.—Original.

beautiful form and shade were often the subject of remark by visitors. This tree is par excellence for roadside planting. It bears transplanting admirably, and is very little disturbed by

insects.   We have only to keep stock away from it, and they
are death to any tree.   Maples, and even elms in many parts

FIG. 238.

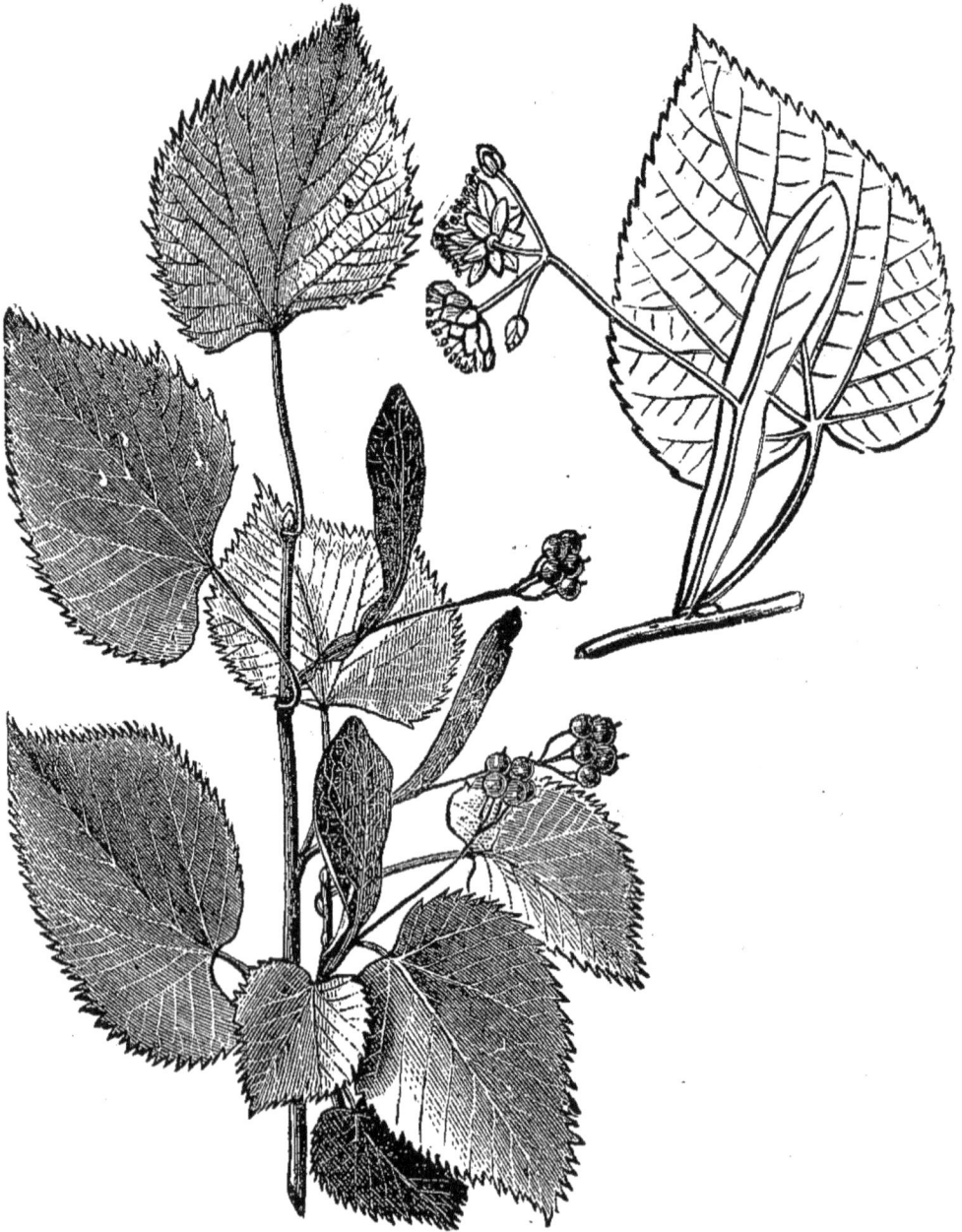

Basswood.—From A. I. Root Co.

of the United States, may well give place to the linden.   The
beautiful white lumber, and its rapidly growing use for boxes,

sections, and furniture, as also for pulp for paper, threaten the continuance of this incomparable honey-tree. Yet the fact

FIG. 239.

Figwort.—From A. I. Root Co.

that it will grow to a large tree in fifteen years, and will commence to bloom in five years from setting, is full of promise.

Figwort (Scrophularia nodosa), (Fig. 239), often called

Rattleweed, as the seeds will rattle in the pod, and Carpenter's Square, as it has a square stalk, is an insignificant looking weed, with inconspicuous flowers, that afford abundant nectar from the middle of July till frost. I have received almost as many for identification as I have of the asters and goldenrods. Prof. Beal remarked to me, years since, that it hardly seemed possible that it could be so valuable. We can not always rightly estimate by appearances alone. It is a very valuable

FIG. 240.

Chapman Honey-Plant.—From A. I. Root Co.

plant to be scattered in waste-places. The Chapman honey-plant (Echiops sphærocephalus), (Fig.240), commences to bloom late in July and continues till in August. We have many better native plants, and as no plant can be profitably grown for honey alone, we have no use for this foreigner. It takes its first name from its spines, and the second from its round flower-head.

FIG. 241.

Rocky Mountain Bee-Plant.—Original.

FIG. 242.

Boneset.—Original.

That beautiful and valuable honey-plant from Minnesota, Colorado, and the Rocky Mountains, Cleome, or the Rocky Mountain bee-plant (Cleome serrulata), (Fig. 241), if self-sown, or sown in the fall, blooms by the middle of July and lasts for long weeks. Nor can anything be more gay than these brilliant flowers, alive with bees all through the long fall. While this is a very valuable honey-plant in its native Colorado, it gives little or no promise East.

FIG. 243.

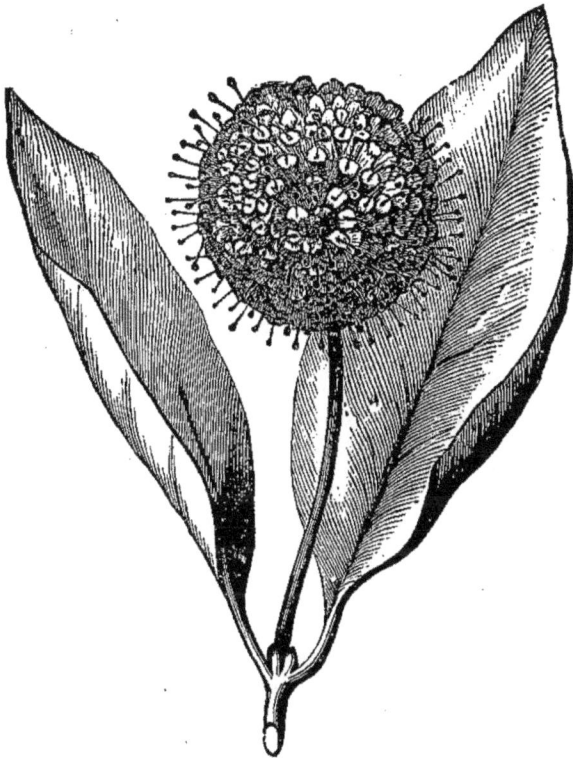

Button-Ball.—Original.

Now commence to bloom the numerous Eupatoriums, or bonesets, or thoroughworts (Fig. 242), which fill the marshes of our country, and the hives as well, with their rich, golden nectar. These are precursors of that profusion of this composite order, whose many species are even now budding, in preparation for the sea of flowers which will deck the marshlands of August and September. Wild bergamot (Monarda fistulosa), which, like the thistles, is of importance to the api-

arist, also blooms in July.   As before remarked, this is one of
the plants whose long flower-tubes are pierced by the Xylocopa
bees.   Then the honey-bees help to gather the abundant nectar.   This is a near relative of the horsemint which, as will be

FIG. 244.

Sour-Wood.—Original.

seen, it closely resembles.   The golden honey-plant (Actino-
meris squarrosa), so praised by Dr. Tinker, and rattlesnake
root (Nabalus altissimus), which swarms with bees all the day
long, are also composite plants.

The little shrub of our marshes, appropriately named but-
ton-bush (Cephalanthus occidentalis), (Fig. 243), also shares

the attention of the bees with the linden ; while apiarists of the South find sourwood, or sorrel tree (Oxydendrum arboreum), (Fig. 244), a valuable honey-tree. It yields much very excellent honey. The honey is not quite as light-colored as sage, clover and basswood. It is slow to granulate. This plant is grown at the Michigan Agricultural College, but it is not

FIG. 245.

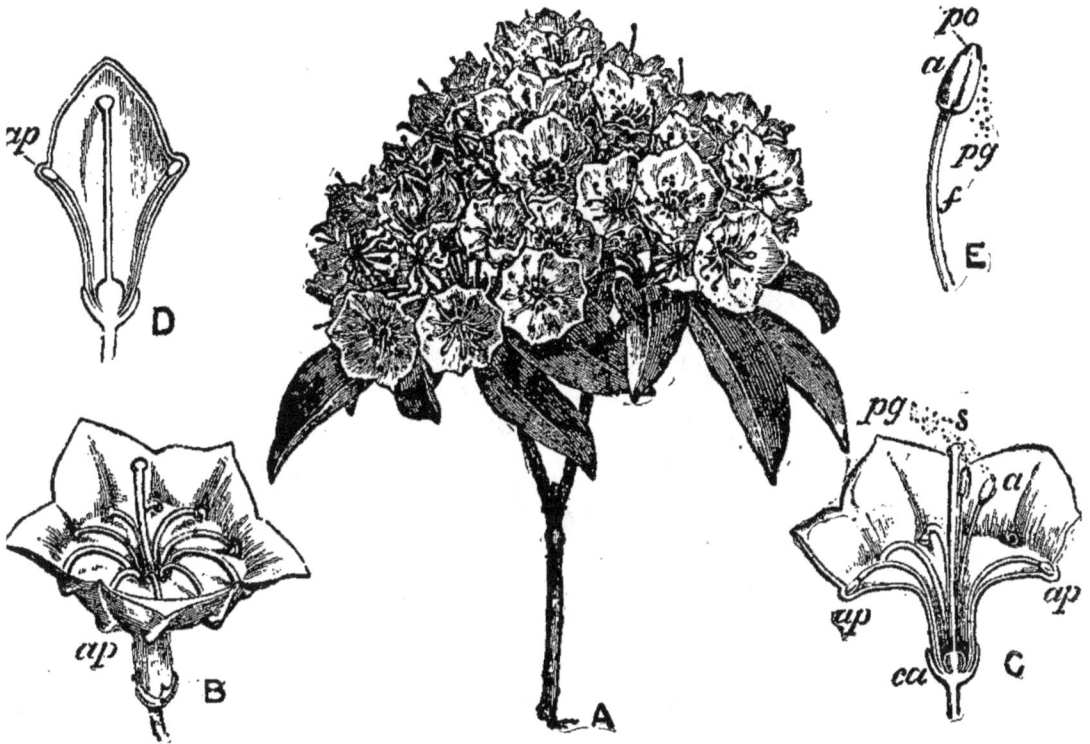

Mountain Laurel.—From Department of Agriculture

A Flowering branch.	*f* Filaments.
B C Expanded flower.	*p g* Shower of pollen-grains.
ap, ap Antler pockets.	*a* Free anthers.
s Stigma.	*c a* Calyx.
e Enlarged stamens.	*d* Section flower-bud.

hardy, as it kills back nearly every winter. It belongs to the Heath family, which includes the far-famed heather-bloom of England. It also includes our whortleberry, cranberry, blueberry, and one plant which has no enviable reputation, as furnishing honey which is very poisonous, even fatal to those who eat, the mountain laurel (Kalmia latifolia), (Fig. 245). There is good reason to question these reports as to poisonous honey. We can easily see how mistakes could occur. It is not easy to understand, if these plants furnish poisonous nectar, why poisonous honey (?) is so very rare an occurrence. A near relative of K. latifolia, which grows at the South (Andromeda nitida), is said to furnish beautiful and wholesome honey in

FIG. 246.

Yellow Jessamine

From A. I. Root Co.

great quantities. The yellow jasmine (Gelsemium *semper-virens*), (Fig. 246), is also said to furnish honey that is poisonous to both people and bees. It blooms in Georgia in February and March. Like Kalmia, it is a poisonous plant, which possibly accounts for the evil reputation of the honey. I have eaten freely of several samples of this so-called poisonous honey with not the slightest inconvenience. The Virginia creeper also blooms in July. I wish that I could say that this beautiful vine, transplendent in autumn, is a favorite with the honey-bee. Though it often, nay always, swarms with wild bees when in blossom, yet I have rarely seen honey-bees visit the ample bloom amidst its rich, green, vigorous foliage.

The St. John's wort (Hypericum), with its many species, both shrubby and herbaceous, offers bountiful contributions

FIG. 247.

Cabbage Palmetto.—Original.

to the delicious stores of the honey-bee. The catnip (Nepeta cataria), which Quinby said he believed would pay better than any other plant for special planting, blooms at this time; also asparagus—which, if uncut in spring, will bloom in June—so delectable for the table, and so elegant for trimming table meats and for banquets in autumn, come now to offer their nectarian gifts, and beautiful, orange pollen.

Basil, or mountain mint (Pycnanthemum lanceolatum), we might almost include all the mints; the blue and white ver-

vains or verbenas (Verbena hastata and V. stricta); also fog-
fruit (Lippia lyceroides), another of this family, is valued very
highly in Texas—it grows ten feet high, and bears beautiful

FIG. 248.

True Mangrove.—Original.

white flowers; the ironweeds (Vernonias), the malvas, Cul-
ver's root, Veronica Virginica—another of the figwort family;
Indian plantains, Cacalias, and viper's bugloss—the so-called

blue thistle—all contribute to the apiary in July ; the viper's bugloss (Echium vulgare), so-called blue thistle, though most common South, is very abundant at Beeton, Canada. Mr. Jones has it growing all about his apiaries. It is a near relative of borage, and does not belong even to the family—Compositæ—of the thistles. Like the borage, it is not a troublesome weed.

In California, the blue-gum and the red-gum (Eucalyptus globulus and E. rostrata), introduced from Australia, furnish

FIG. 249.

Buckwheat.—Original

honey from July and August till December. There are over one hundred species of gum-trees (Fig. 205). Some are very beautiful in habit, foliage and blossom. They blossom at nearly all seasons, summer and winter, so by carefully selecting the species, the apiarist may have the flowers at will.

The catalpa, a very rapid-growing tree, throws its large, showy blossoms to the breeze and bees in July. It is rapidly growing in favor as a shade-tree, and is incomparable for posts. It lasts for a great many years when imbedded in the earth. But "the noblest Roman of them all" is the cabbage palmetto—Chamœrops palmetto (Fig. 247). As Mr. Hart, of Florida, says, this is the linden of the South. It yields abundant honey, which, as all who saw and tasted it at the last convention at Cincinnati, can vouch, is unsurpassed in flavor. Mr. Muth well said that he wished no finer. This tree grows to the height of seventy feet. The trunk is leafless nearly to

FIG. 250.

Golden-Rod.—From A. I. Root Co

the top. The small, white blossoms nestle among the long palm leaves in profusion, and are rich in both nectar and pollen, from June 1st till August. The tree is found from the Carolinas to the Gulf. The various palms, as already stated— Chamærops, Phœnix, Cocus, etc.—add not only to the beauty but to the honey-resources of California. The true date-palm (Phœnix dactylifera) bids fair to become an important fruit-tree of Arizona. If it does, it will be very valuable for honey, and add further to the excellent reputation of that section for bees.

At the same time with the above, the white blossom of the black mangrove (Avicennia tomentosa), and its near relative, Avicennia oblongifolia, come forth with their abundant and incomparable nectar, which hangs in drops. The honey from this and the cabbage palmetto is clear, and as fine and beauti-

FIG. 251.

Aster.

ful as that of white clover. This tree is confined to the Peninsula of Florida, where it is regarded as the best honey-plant that grows in that locality.

Here we see the danger of common names. This is not a mangrove at all; though the leaves resemble those of the true mangrove, they are more tomentose or hairy, and, like that tree, it grows down to the very water's edge, so it is not affected by drouth. This is an evergreen, and forms an impenetrable thicket on the muddy shores of the sea. It belongs to the same family as our verbenas—the vervain family.

The true mangrove (Fig. 248) has yellow blossoms, and like the renowned banyan tree, sends numerous stems to the earth, each of which takes root. This tree belongs to the mangrove family, and is Rhizophora mangle.

AUGUST AND SEPTEMBER PLANTS.

The cultivated buckwheat (Fagopyrum esculentum), (Fig. 249), usually blooms in August, as it is sown the first of July— three pecks per acre is the amount to sow—but by sowing the first of June, it may be made to bloom the middle of July, when there is generally, in most localities, an absence of nectar-secreting flowers. Farmers have often grown oats, then raised a crop of buckwheat which matures in two months from sowing, and then have sown to wheat all the same season, and have secured good crops of each, all on the same ground. It often fails to give a crop of honey, though even then it may serve to keep the bees at work and breeding. The bees rarely work on buckwheat after eleven o'clock. Their visits are always a benefit, and never an injury, to the grain. The honey is inferior in color and flavor, though some people prefer this to all other honey. It usually sells for much less than clover or linden honey. The silver-leaf buckwheat blooms longer, has more numerous flowers, and thus yields more grain than the common variety. The Japan buckwheat is much superior to either the common or silver-hull. The grain is larger, and one thousand have been taken from a single stalk. Eighty bushels have been grown on a single acre. Buckwheat is often plowed under to enrich the soil. It is good to loosen the soil and furnish humus, but does not add nitrogen, and so is not equal to clover, peas, lupines, or other legumes. Sown on ground infested with wire-worms, *it* flourishes, and the insects disappear. Heartsease, or western smart-weed (Polygonum persicaria), is a close relative of the buckwheat. It grows very luxuriantly along the Mississippi River. The white or purple flowers hang in great clusters. Mr. T. R. Delong reported at the Lincoln, Nebr., convention that each of two colonies gathered 450 pounds of honey from this plant; and that his entire apiary averaged 250 pounds, all from heartsease. The honey is quite light-colored, and very excellent in quality.

The odd shrub, Hercules' club (Aralia spinosa), is grown as a curiosity North, but is indigenous in Kentucky and Tennessee, and yields abundant nectar. It blooms at Lookout Mountain early in August, just after the sourwood.

Now come the numerous goldenrods. The species of the genus Solidago (Fig. 250), in the Eastern United States, number nearly two score, and occupy all kinds of soils, and are at home on upland, prairie and morass. These abound in all parts of the United States. They yield abundance of rich, golden honey, with flavor that is unsurpassed by any other. Fortunate the apiarist who can boast of a thicket of Solidagos in his locality.

The many plants usually styled sunflowers, because of their resemblance to our cultivated plants of that name, which deck the hillside, meadow and marsh land, now unfurl their showy involucres, and open their modest corollas, to invite the myriad insects to sip the precious nectar which each of the clustered flowers secretes. Our cultivated sunflowers, I think, are indifferent honey-plants, though some think them big with beauty, and their seeds are relished by poultry. But the numerous species of asters (Fig. 251), so wide-spread, the beggar-ticks and Spanish-needles, Bidens, of our marshes, the tick-seed, Coreopsis, also, of the low, marshy places, with hundreds more of the great family Compositæ, are replete with precious nectar, and with favorable seasons make the apiarist who dwells in their midst jubilant, as he watches the bees which fairly flood the hives with the rich and delicious honey. The Hon. J. M. Hambaugh found Spanish-needles—Coreopsis —very abundant in the low flats of Illinois. Almost every year it gave much very thick and excellent honey. It weighed twelve pounds to the gallon. Often the bees took over twelve pounds daily for more than a week at a time. For several years, also, those fifty colonies of bees stored over a ton of this most excellent honey each season. In all of this great family, the flowers are small and inconspicuous, clustered in compact heads, and when the plants are showy with bloom, like the sunflowers, the brilliancy is due to the involucre, or bracts, which serve as a frill to decorate the more modest flowers.

The great willow-herb, or fireweed (Epilobium angusti-folium), (Fig. 252), is often the source of immense honey-har-vests. The downy seeds blow to great distances, and, finding

FIG. 252.

Great Willow-Herb, after Gray.

A Flower with ripe stigma.
St Unripe stamens.
P Petal.
T Pollen-tube.

S Ripe stigma.
R Flower with ripe pollen.
Po Pollen-grain.

a lodgment, their vitality makes them burst forth whenever brush is burned or forest fires rage. Hence the name, fire-weed. This handsome plant often covers acres of burnt lands in northern Michigan with its beautiful pink bloom. Unlike most nectar from late bloom, the honey from this flower is white as clover honey. It often gives a rich harvest to the apiarist of northern Michigan.

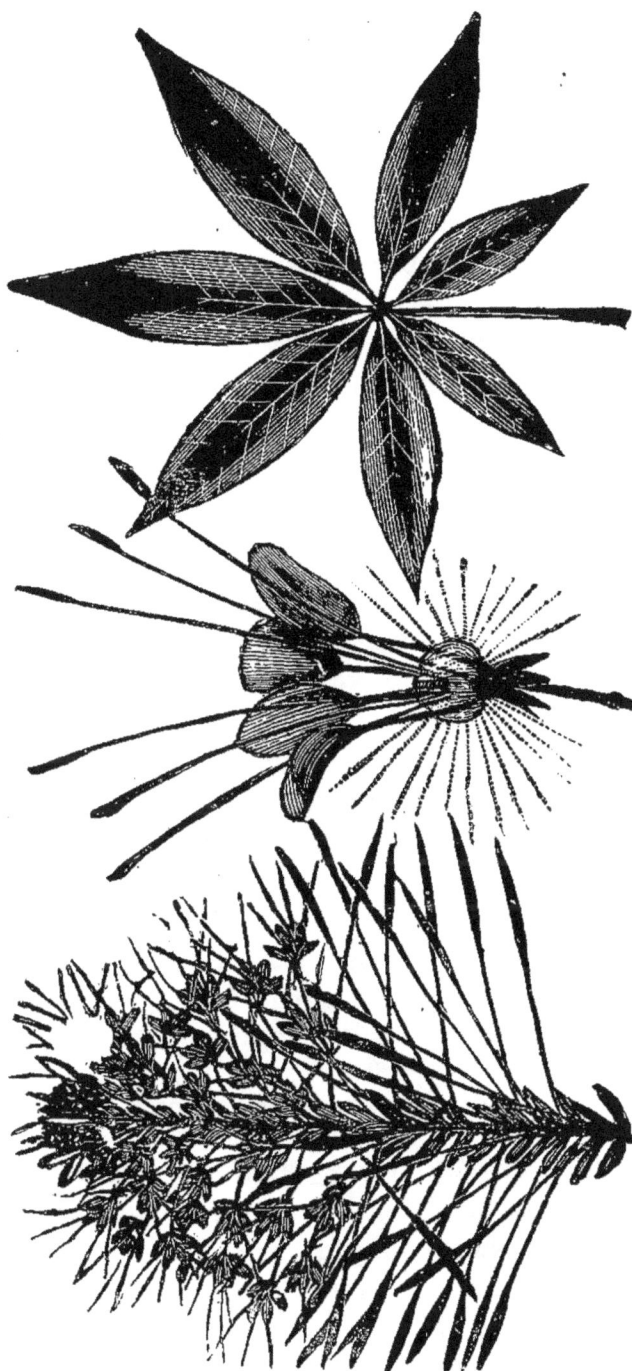

FIG. 253.

Spider-Plant.

Another excellent fall honey-plant of wide range is the coral-berry or Indian currant (Symphoricarpus vulgaris). The honey-product of this plant is worthy its name. The closely related snow-drop (S. racemosus), common in cultivation, is also a honey-plant. I close this account with mention of another, Cleome, the famous spider-plant (Cleome spinosa), (Fig. 253). This plant thrives best in rich, damp, clay soil. It is only open for a little time before nightfall and at early dawn, closing by the middle of the forenoon ; but when open its huge drops of nectar keep the bees wild with excitement, calling them up even before daylight, and enticing them to the field long after dusk. It is a native of the tropics, and is found now from south New Jersey to Florida and Louisiana.

I have thus mentioned the most valuable honey-plants of our country. Of course, there are many omissions. Let all apiarists, by constant observation, help to fill up the list.

BOOKS ON BOTANY.

I am often asked what books are best to make apiarists botanists. I am glad to answer this question, as the study of botany will not only be valuable discipline, but will also furnish abundant pleasure, and give important practical information. Gray's Lessons and Manual of Botany, in one volume, published by Ivison, Phinney, Blakeman & Co., New York, is the most desirable treatise on this subject. A more recent work by Prof. C. E. Bessey, and published by Henry Holt & Co., is also very excellent. Coulter's and Atkinson's Botanies are also most excellent. The first treats of systematic, the second of physiological botany, while the last two are up to date and very fascinating.

PRACTICAL CONCLUSIONS.

It will pay well for the apiarist to decorate his grounds with soft and silver maples, for their beauty and early bloom. If his soil is rich, sugar-maples and lindens may well serve a similar purpose. Indeed, every apiarist should strive to have others plant the linden. No tree is so worthy a place by the roadside. The Judas and tulip trees, both North and South, may well be made to ornament his home. For vines, obtain the wistarias, where they are hardy. In California, encourage long avenues of eucalyptus, the graceful peppers, and the incomparable date-palms.

Sow and encourage the sowing of alsike clover and silver-leaf or Japanese buckwheat in your neighborhood. Be sure that your wife, children, and bees can often repair to a large bed of the new giant or grandiflora mignonettte, and remember that it, with figwort, spider-plant, Rocky Mountain bee-plant, and borage, bloom till frost. Study the bee-plants of your region, and then study the foregoing table, and provide for a succession, remembering that the mustards, rape and buckwheat may be made to bloom almost at pleasure, by sowing at the proper time. Do not forget that borage and the mustards seem comparatively indifferent to wet weather. Be sure that all waste places are stocked with motherwort, catnip, pleurisy-root, figwort, cleome, viper's-bugloss, asters, etc.

The foregoing dates, unless specially mentioned, are only correct for Michigan, northern Ohio, and similar latitudes, and for more Southern latitudes must be varied, which, by comparison of a few, as the fruit-trees, becomes no difficult matter.

CHAPTER XVIII.

WINTERING BEES.

This is a subject, of course, of paramount importance to the apiarists of the Northern States, as this is the rock on which some of even the most successful have split. Yet I come fearlessly to consider this question, as from all the multitude of disasters I see no occasion for discouragement. If the problem of successful wintering has not been solved already, it surely will be, and that speedily. So important an interest was never yet vanquished by misfortune, and there is no reason to think that history is now going to be reversed. Of course this chapter has no practical value to the apiarists of the South and Pacific Coast. There safe wintering is assured, except as the careless bee-keeper permits starvation.

THE CAUSES OF DISASTROUS WINTERING.

I fully believe (and to no branch of this subject have I given more thought, study, and observation) that all the losses may be traced to either unwholesome food, extremes of temperature, or protracted cold. I know from actual and widespread observation, that the severe loss of 1870 and 1871 was attended in Michigan with unsuitable honey in the hive. The previous autumn was unprecedentedly dry. Flowers were rare, and the stores were largely honey-dew, collected from scale insects, and consequently were unwholesome. I tasted of honey from many hives only to find it nauseating. Cider, if collected too freely, will also work ruin in winter. We must remember that bees do not void their intestines for long months, so good food is absolutely imperative.

Extremes of heat and cold are also detrimental to the bees. If the temperature of the hive becomes too high, the bees become restless, eat more than they ought, and if confined to their hives are distended with their feces, become diseased, besmear their comb and hives, and die. If, when they become thus disturbed, they could have a purifying flight, all would be well. Again, if the temperature become extremely low, the bees, to keep up the animal heat, must take more food; they are uneasy, exhale much moisture, which may settle and freeze on the outer combs about the cluster, preventing the bees from getting the needed food, and thus in this case both dysentery and starvation confront the bees. That able and far-seeing apiarist, the lamented M. Quinby, was one of the first to discover this fact; and here, as elsewhere, gave advice that, if heeded, would have saved great loss and sore disappointment. Dr. Miller is doubtless correct in the belief that he has cured and prevented dysentery by use of a coal-stove in the cellar. Of course, Dr. Miller's good judgment and caution were coupled with the artificial heat. I have little doubt, in fact I know from actual investigation, that in the past severe winters, those bees which under confinement have been subject to severe extremes, were the ones that invariably perished. Had the bees been kept in a uniform temperature, ranging from 40 to 45 degrees, F., the record would have been materially changed. Bees do not hibernate in the sense that other insects do, though if the temperature is just right, from 40 to 45 degrees F., they are very quiet and eat but little. Yet that they are even then functionally active is readily shown by the high, independent temperature in the hive and their frequent change of position in the cluster.

Excessive moisture, especially in cases of protracted cold, is always to be avoided. Bees, like all other animals, are constantly giving off moisture, which, of course, will be accelerated if the bees become disturbed and are thus led to consume more food. This moisture not only acts as explained above, but also induces fungous growths. The mouldy comb is not wholesome, though it may never cause death. Hence, another necessity for sufficient warmth to drive this moisture from the hive, and some means to absorb it without opening the hive above and permitting a current which will disturb the bees, and cause the greater consumption of honey. It is probable that, with the proper temperature, moisture will do little harm.

THE REQUISITE TO SAFE WINTERING—GOOD FOOD.

To winter safely, then, demands that the bees have thirty pounds by weight, not guess—I have known many cases where *guessing* meant *starvation*—of good, capped honey (granulated sugar is just as good). With the extractor the temptation is ever with us, to take too much honey from the hive. It is always safest to leave enough, thirty to sixty pounds of the best honey—the best is none too good—for a year's, or in California for two years', stores. It is now proved that it is even safer to feed a syrup made of granulated sugar. We thus are sure that the stores are good and suitable. Often it pays to do this, as we get enough for the extracted honey to pay well for the sugar and our time and trouble. If desired, this may be fed as previously explained, which should be done so early that all will be capped during the warm days of October.

The bees should be able to pass over or through the combs. Hill's device—bent pieces placed above the frames so as to raise the cloth cover—will permit the first, while small holes cut through the combs will enable the bees to pass from one comb to another without having to pass around. In a good cellar it is not necessary to do more at most than so to arrange that the bees can pass over the frames. I used to cut holes, but do so no more. This preparatory work I always do early in October, when I extract all uncapped honey, take out all frames after I have given each colony the thirty pounds, *by weight*, of honey, confine the space with a division-board, cover with the quilt and chaff, and then leave undisturbed till the cold of November calls for further care. We must most carefully exclude honey-dew from scale insects, and must see that cider is not stored for winter food. I prefer that the combs have no pollen in them, and that they be so full of honey that six or eight will be enough. Pollen usually does no harm, though sometimes it is injurious. If the bees can fly often, or if kept in a uniform temperature at from 40 to 45 degrees F., the pollen will do no harm. The combs may well be one-half inch apart. If the bees have been neglected, and mid-winter finds them destitute of stores, then they should not be fed liquid honey, though this has been done with success, but either the Good or Viallon, or some other solid candy, should be placed on the frames just above the cluster. Or we may run the candy into a frame and hang it in the hive. (See Candy, page 318.)

SECURE LATE BREEDING.

Keep the bees breeding till the first of September. Except in years of excessive drouth, this will occur without extra care. Failure may result from the presence of worthless queens. Any queens which seem not to be prolific should be superseded whenever the fact becomes evident. *I regard this as most important.* Few know how much is lost by tolerating feeble, impotent queens in the apiary, whose ability can only keep the colonies alive. Never keep such queens about. Here, then, is another reason for always keeping extra queens on hand. Even with excellent queens, a failure in the honey-yield may occasionally cause breeding to cease. In such cases, we have only to feed as directed under the head of Feeding. It is not true that very large colonies will winter better than smaller ones. Yet it is important that the bees be normal in age and condition.

TO SECURE AND MAINTAIN THE PROPER TEMPERATURE.

We ought also to provide against extremes of temperature. It is desirable to keep the temperature about the hive between 38 and 50 degrees F., through the entire winter, from November to April. If no cellar or house is at hand, this may be partially accomplished as follows : Some pleasant, dry day in late October or early November, raise the stand and place straw beneath ; then surround the hive with a box a foot outside the hive, with movable top, and open on the east ; or else have a long wooden tube, opposite the entrance, to permit flight ; this tube should be six or eight inches square to permit easy examination in winter. The same end may be gained by driving stakes and putting boards around. Then we crowd between the box and the hive either cut straw, chaff or shavings. After placing a good thickness of cut straw above the hive, lay on the cover of the box, or cover with boards. This preserves against changes of temperature during the winter, and also permits the bees to fly, if it becomes necessary from a protracted period of warm weather. I have thus kept all our bees safely during two of the disastrous winters. This plan usually succeeds well, but will fail in a very severe winter like that of 1880–81. As some may wish to try, and possibly to

FIG. 254.

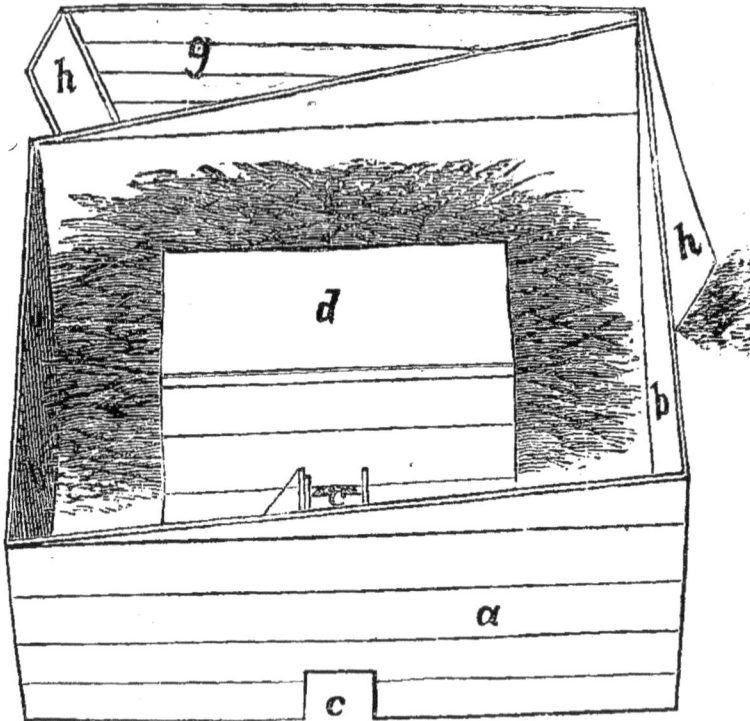

Packing Box.—Original.

adopt it, I will describe the box used at our College, which costs but one dollar, and is convenient to store away in summer.

BOX FOR PACKING.

The sides of this (Fig. 254, *a*, *a*) facing east and west are three and a half feet long, two feet high at the south end, and two and a half feet at the north. They are in one piece, which is secured by nailing the matched boards which form them to cleats, which are one inch from the ends. The north end (Fig. 254, *b*) is three feet by two and a half feet, the south (Fig. 254, *b*) three feet by two feet, and made the same as are the sides. The slanting edges of the side (Fig. 254, *a, a*) are made by using for the upper boards the strips formed by sawing diagonally from corner to corner a board six inches wide and three feet long. The cover (Fig. 254, *g*), which is removed in the figure, is large enough to cover the top and project one inch at both ends. It should be battened, and held in one piece by cleats (Fig. 254, *h*) four inches wide, nailed on to the ends. These will drop over the ends of the box, and thus hold the cover in place, and prevent rain and snow from driving in. When in place this slanting cover permits the rain to run off easily, and will dry quickly after a storm. By a single nail at each corner the four sides may be tacked together about the hive, when it can be packed in with cut straw (Fig. 254) or fine chaff, which should be carefully done, if the day is cold, so as not to disquiet the bees. At the center and bottom of the east side (Fig. 254, *c*) cut out a square eight inches each way, and between this and the hive place a bottomless tube (the top of this tube is represented as removed in figure to show entrance to hive), before putting in the cut straw or chaff and adding the cover. This box should be put in place before the bleak, cold days of November, and retained in position till the stormy winds of April are passed. This permits the bees to fly when very warm weather comes in winter or spring, and requires no attention from the apiarist. By placing two or three hives close together in autumn—*yet never move the colonies more than three or four feet* at any one time. as such removals involve the loss of many bees—one box may be made to cover all, and at less expense. This will also be more trustworthy in very cold winters. Late in spring these boxes may be removed and packed away, and the straw or chaff carried away, or removed a short distance and burned.

CHAFF-HIVES.

Messrs. Townley, Butler, Root, Poppleton, and others, prefer chaff-hives, which are simply double-walled hives, with the four-inch or five-inch chambers filled with chaff. The objections to these I take to be : First, they are not proof against severe and long-continued cold, like the winter of

1880–81: second, such cumbrous hives are inconvenient to handle in summer; and, third, they are expensive. That they would in part supply the place of shade, is, perhaps, in their favor, while Mr. A. I. Root thinks they are not expensive.

Mr. O. O. Poppleton, one of our most intelligent bee-keepers, shows practically that the first objection given above is not valid. So, very likely, the failure in so many apiaries in 1880–81 was rather due to improper use. Mr. Poppleton claims numerous advantages for these hives:

1st. In his hands, success.

2d. They permit early preparation for winter.

3d. They give entire freedom from care of the bees from September till March.

4th. Preparation for winter requires only slight labor.

5th. We can easily get at the bees at any time.

6th. The bees are not excited by a slight rise in temperature, and so are not lost by flying on cold days; do not breed in winter and spring when they need quiet, and do not "dwindle" in spring.

7th. They are valuable aids in building up nuclei and weak colonies at cold periods at any one time of the year.

8th. They are specially desirable to protect the bees in April and May, and prevent "spring dwindling."

RULES FOR THEIR USE.

Mr. Poppleton urges the following important points:

1st. Pack early in autumn before cold weather, and do not remove the packing till the warm weather has come to stay.

2d. Have five or six inches on *all sides* of the bees, of *fine chaff*—timothy is best—entirely freed from straw.

3d. Be sure and have the chaff below the bees, as well as above and on the sides.

4th. Do not put the chaff above the bees on loose, but confine in sacks. This is for convenience and neatness.

5th. Have as much empty space as possible inside the hive and outside the packing; and never let the cover to the hive rest immediately on the packing.

6th. Crowd the bees on to a few frames—never more than eight—and the packing close to the bees.

7th. Winter passages should be made through all the combs.

Mr. Jones prefers that the outer wall of the chaff-hive should be of narrow boards so as to be more impervious to dampness. He also uses fine, dry sawdust instead of chaff. Mr. Root, in his two-story hives (Fig. 255), uses a thicker layer

FIG. 255.

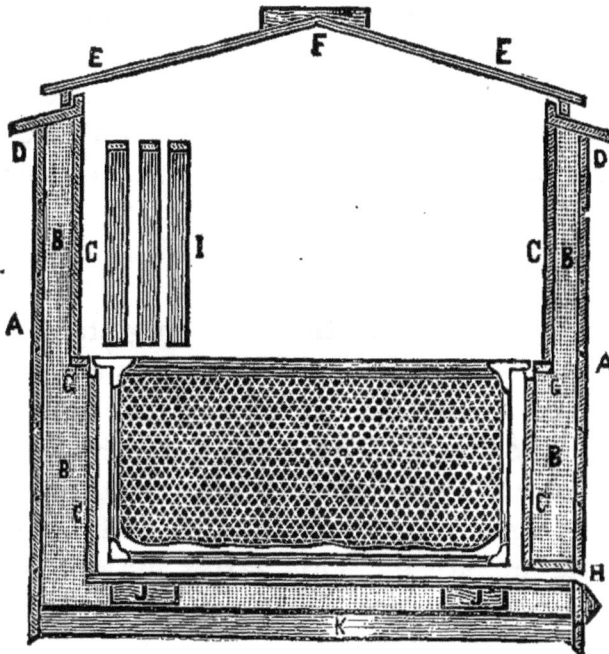

Section of a Chaff-Hive.—From A. I. Root Co.

of chaff below, but carries it to the top. Of course, the double wall need not extend on the sides of the frames. The division-boards on the sides of the frames may make the double wall.

WINTERING IN A BEE-HOUSE.

As Mr. D. A. Jones has tested bee-houses on a very large scale, and met with success, I will quote directly from him :

"The house should be so constructed that the outdoor temperature can not affect that of the bee-house ; and in order to accomplish this its walls should be packed tightly with two feet of dry sawdust or three feet of chaff packing, over-head the same thickness, and the bottom so protected that no frost can penetrate. Next, it should have a ventilating tube at the top, of not less than one square inch to each colony of bees. It should have sub-earth ventilation by means of a tube laid below the depth frost will penetrate, and from one to three hundred feet in length, coming in contact with outside atmosphere at the other end ; as air passes through this tube it is tempered by the distance through the earth, and comes into the house at an even temperature. By means of slides at these ventilators, the temperature can be arranged in the bee-house, which should stand from 43 to 46 degrees, and in no case should it fall lower than 42 degrees. There should be tight-fitting, triple doors, which will make two dead-air spaces.

"When the bee-house is filled, and during warm weather in the spring, the bees should not be set out on the summer stands until the first pollen appears (which is generally from the tag alder or black willow)—it is necessary that the temperature of the room be kept at the wintering standpoint. This may be done by means of an ice-box or refrigerator, filled with ice or snow, and suspended at the top of the room in close proximity to the ceiling. The bottom of the box must be so constructed that while the warm air may be allowed to pass up through the refrigerator, the drippings will not drop to the floor and create moisture. This latter may be prevented by means of a tube running from the box down through the floor."

The rules for removing and storing in the house are the same as those for cellar. From expense and difficulty in maintaining a uniform temperature, I think the house less desirable than the cellar.

WINTERING IN A CELLAR.

North of the latitude of Central (and I think we may say Southern) Ohio, I think a good cellar is not only the safest, but the best place in which to winter bees. I have kept bees for many years in such a cellar with no loss. The great point is to have perfect control of the temperature. This must be kept between 38 degrees F. and 50 degrees F., and should never vary suddenly. It were best if it were always at 45 degrees F. With a cellar all is under ground, and we are thus fortified against the effects of our sudden changes of temperature. The sub-earth ventilator, as described above, though not necessary, as the experience of many has fully proved, is a help. It is still better if the vertical shaft or pipe connect with a stove above which is much used in winter. This creates a draft, and as the air is brought underground through the long sub-earth pipe, the air is warmed. The pipe should connect with the stove-pipe above at quite a height above the stove, or the stove may smoke. I found at the Michigan Agricultural College that we got quite a draft, especially on windy days, even if there was no fire, but the vertical pipe—a common stove-pipe served excellently well—connects simply with a chimney which projects above the house. Such an arrangement not only controls the temperature but ventilates the cellar. A large cistern full of water, or water running through a cellar deep under ground, is a wonderful moderator, and will surely keep the temperature at the proper point. It is imperative that every bee-keeper have a thermometer in his cellar, and by frequent examination KNOW that the temperature is at the proper point. Unless he finds that he can not control the temperature without, he would better not go to the expense of either sub-earth ventilation or a cistern.

Dr. C. C. Miller keeps a small coal-stove burning with an open stove-door in each cellar, and thus keeps the temperature just as he desires. My brother keeps as many bees in his house-cellar with no such pains or labor, and yet is as successful as is Dr. Miller. The thing to remember is, *we must control the temperature.*

I commence preparation for winter as soon as the first frost shows that the harvest is over. I then put five Langstroth or seven Gallup frames at one side or end of the hive, where they are to remain for the winter. If these have not enough food I feed till they have. If other frames have brood I put these close beside, and remove them as soon as the brood has all matured, and close up the other frames by use of a division-board. I now cover all with a cloth and with a super of chaff or dry sawdust. For the past two years I have left all the combs in very strong colonies, and covered simply with a board, and these colonies have done well. In a good cellar bees need no packing about or above the brood-chamber.

Before cold weather—any time from the first to the middle

of November—the bees are carried into the cellar. This would better be done carefully, so as not to disturb the bees. Yet I am not sure that such disturbance is any special injury. To prevent the bees from coming out in case of disturbance, the entrance-blocks must close the entrances. Dr. Miller uses wet cloths to effect this.

In the cellar the hives should rest a foot from the bottom, and may rest on each other, breaking joints, the weakest colonies at the top. When all are in, and quiet, the entrances are opened wide. I would (if it were not for the expense, and I had loose bottom-boards so that I could) place a rim under each hive so as to raise it two or three inches above the bottom-board. Except for the open entrance, I give no special ventilation to each hive. Now we shut our two or three doors, and if our cellar is right we have no more care for the bees till the succeeding April. Should the bees become uneasy and soil their hives about the entrance—they will not if the food is all right and the temperature keeps at the right point (from 38 degrees to 50 degrees F.)—then it may be well to put the bees out for a flight in February or March, in case a warm day affords opportunity. In case there is snow, a little straw may be scattered over it. The day must be quite warm. It is far wiser to have our cellar right so we shall not need to do this.

If the bees get short of stores in winter—this would show great neglect on the part of the bee-keeper—they should be fed "Good candy," cakes of which may be laid on the frames and covered with cloth. Frames of honey or syrup, filled as already described, may be given bees in mid-winter. The idea that bees can not be examined in winter is incorrect. Frames may be taken out or added, though it were doubtless better to leave the bees undisturbed. The cellar should be dark and quiet. If everything is just right, light does no harm; but if it gets pretty cold or too warm then the bees become uneasy and fly out, never to return. Some bees always leave the hive in winter. These are veterans, and are ready to die. Thus, with 100 colonies of bees in a cellar, we need not be anxious even if a good many quarts come out to die.

In spring, when the flowers have started, so that the bees can gather honey and pollen, they may be set out. This better be too late than too early. In Central Michigan, April 15th is usually early enough. I repeat : *Better too late than too early.* The colonies are put each on its own stand, and each hive well cleaned out. Each colony should have plenty of honey. Scant stores in spring always bring loss, if not ruin. We now take away extra frames of comb, giving the bees simply what they will cover, but always a good amount of honey. A frame of pollen taken away the previous autumn may also be added. We close up about the bees with a division-board, and cover warmly above by adding a chaff-filled super.

If we give abundant stores, I am not sure but for strong colonies a full set of frames and board above, which, however, must fit very snugly, is as good as a chaff covering or chaff-hive. For the simple Heddon-Langstroth hive, however, I think a warm cloth under the cover is very desirable. I tried some colonies in this way in two springs, and was pleased with the results. I am not yet sure but it is always better to cover with chaff, sawdust or leaves ; *but we must give plenty of honey,* and perhaps we must cover warmly and snugly, to win the best success. I always thought so in the past, but now I am in doubt. Even if better, it may still prove more profitable to give plenty of honey, and let the hives alone, with a full set of combs in each. This saves much time. Geo. Grimm and my brother have long practiced this and have succeeded.

Perhaps I ought to say that all colonies should be strong in autumn ; but I have said before, never have weak colonies. As before stated, a colony need not be very large to winter well ; but they should be strong, in possession of a good queen, and the proper proportion of young and vigorous bees. Yet for fear some have been negligent, I remark that weak colonies and nuclei should be united in preparing for winter. To do this, approximate the colonies each day, four or five feet, till they are side by side. Now remove the poorest queen, then smoke thoroughly, sprinkle both colonies with sweetened water scented with essence of peppermint, put a sufficient number of the best frames, alternating them as taken from the hives, and put all the bees into one of the hives, and then set this midway between the position of the hives at the commencement of the uniting. Shaking the bees in front of the hive also tends to make the union more complete. The bees will unite peaceably, and make a strong colony. In case of nuclei I usually unite three for winter. Uniting colonies may pay at other seasons.

It may seem rash to some, yet I fully believe that if the above suggestions are carried out in full, I may guarantee successful wintering. But if we do lose our bees, having all our hives, combs and honey, we can buy colonies in the spring with a perfect certainty of making a good percent on our investment. Even with the worst condition of things, we are still ahead, in way of profit, of most other vocations.

BURYING BEES, OR WINTERING IN CLAMPS.

In principle this is the same as cellar-wintering. There are two serious objections to it. First, we do not know that the temperature is just right, and, secondly, if aught goes wrong we know nothing of it—the bees are away out of sight. If this is practiced, the ground should be either sandy or *well drained*. If we can choose a side-hill it should be done. Beneath the hives, and around them, straw should be placed. I should advise leaving the entrance well open, yet secure against mice. *The hives should all be placed beneath the surface level* of the earth, and a mound should be raised above them sufficient to preserve against extreme warmth or cold. A trench about the mound to carry the water off quickly is desirable. In this arrangement the ground acts as a moderator. I would urge the suggestion that no one try this with more than a few colonies, for several years, till repeated successes show that it is reliable in all seasons. I tried burying very successfully for a time, then for two winters lost heavily. These last winters the bees would have wintered well on their summer stands, as the weather was very warm. The bees became too warm, and were worried to death.

SPRING DWINDLING.

In the early years, before the forests were cleared away, the winters were less severe and disastrous, wintering or spring dwindling were seldom experienced. The warmer winters, and possibly better honey in the hive, were the reasons.

As already suggested, spring dwindling is not to be feared

if we keep our bees breeding till autumn, prepare them well and early for winter, and use a good cellar for wintering. It may be further prevented by forbidding late autumn flights, frequent flights in winter, when the weather is warm, and too early flying in spring.

I am aware that this matter of spring dwindling is most stoutly urged as an objection to cellar-wintering, and as an argument in favor of chaff-hives. I have had excellent success in cellar-wintering, and never yet lost a colony by "spring dwindling." Crowd the bees on a few frames when taken from the cellar; give them abundant food; cover warmly above and at the sides of division-boards with generous bags of sawdust, and leave these on the hives if the weather remains cool, until we wish to place the section supers or extracting second story on the hives, and bees from the cellar—a good cellar—will come through the spring in excellent condition. In the winter of 1881–82, I put some chaff-hives into my cellar alongside of my single-walled hives, arranged as just described, and the bees in them did no better in spring after removal from the cellar than in other hives. Be sure in early spring that the bees have no more combs than they can cover, and cover warmly, and spring dwindling will lose its terror. Good wintering, and ample spring stores, are the antidote to spring dwindling. Never set bees permanently on their summer stands from the cellar till the flowers and warmth will enable them to work. Below 60 degrees F. in the shade is too cold for bees to fly. At 70 degrees F. we may safely handle our bees without chilling the brood. When not clustered, bees chill at about 55 degrees.

I have little doubt but that bees will do better if no breeding takes place in winter. Perfect quiet should be our desire. If the bees have no pollen, of course no breeding will take place, and so I advised its removal. It is not for winter use.

CHAPTER XIX.

THE HOUSE-APIARY AND BEE-HOUSE.

The house-apiary (Fig. 256) is a frost-proof house in which the bees are kept the year through. The entrances to the hives are through the sides of the house, and all manipulation of the bees is carried on inside. From what I have said about

FIG. 256.

House-Apiary.—From A. I. Root Co.

wintering, it at once appears that such a house should preserve a uniform temperature. As many such houses were built a few years ago, and are now, with very few exceptions, used for other purposes, I would advise all to study the matter well before building a house-apiary. Where queen-rearing is carried on extensively, or where little room is at command, they may be desirable. Several excellent bee-keepers are now using them with success, and great satisfaction. The old-time objection, of bees collecting in houses while working with them, is now removed, as are the bees by aid of bee-escapes. If the bee-escapes are put on the hives the night before, the extracting or comb honey supers will be practically free of bees in the morning, and all work can be done in the house with very slight annoyance from the presence of the bees. As we all know, cross colonies lose their pugnacity if placed in a house-apiary. They seem cowed by the enclosure. The walls,

FIG. 257.

House-Apiary on Wheels.—From A. I. Root Co.

of course, should be double, and filled in with shavings, and the hives should be the same as are used out-of-doors. A movable house-apiary, on wheels (Fig. 257), has been used, and in some cases may be desirable.

BEE-HOUSES.

As a good and convenient bee-house is very desirable in every apiary of any considerable size, I will give a few hints in reference to its construction.

First, I should have a good cellar under the house, entirely under ground so as certainly to be frost-proof, mouse and rat proof, thoroughly grouted, and ventilated as already described,

I would have three doors to this from the east, the outer one inclined. In our college apiary we had a vestibule to the cel-

FIG. 258.

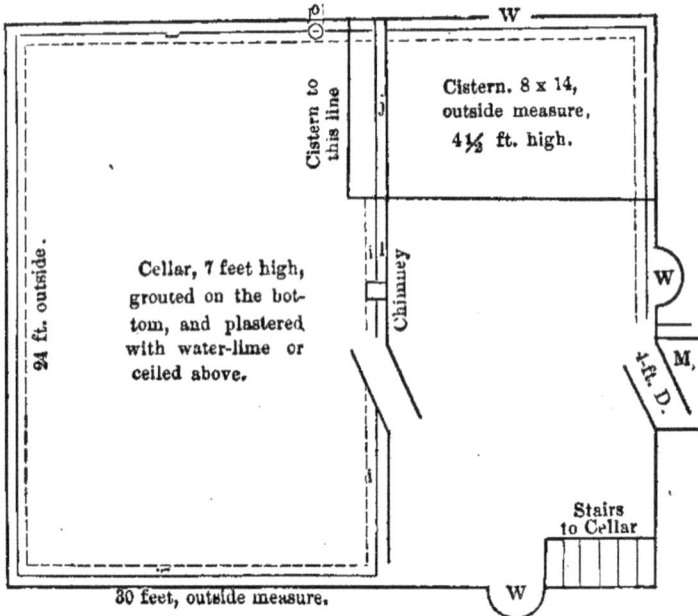

Cistern to this line

Cistern. 8 x 14, outside measure, 4½ ft. high.

24 ft. outside.

Cellar, 7 feet high, grouted on the bottom, and plastered with water-lime or ceiled above.

Chimney

W

4 ft. D.

M,

Stairs to Cellar

30 feet, outside measure.

W

Diagram of Cellar.

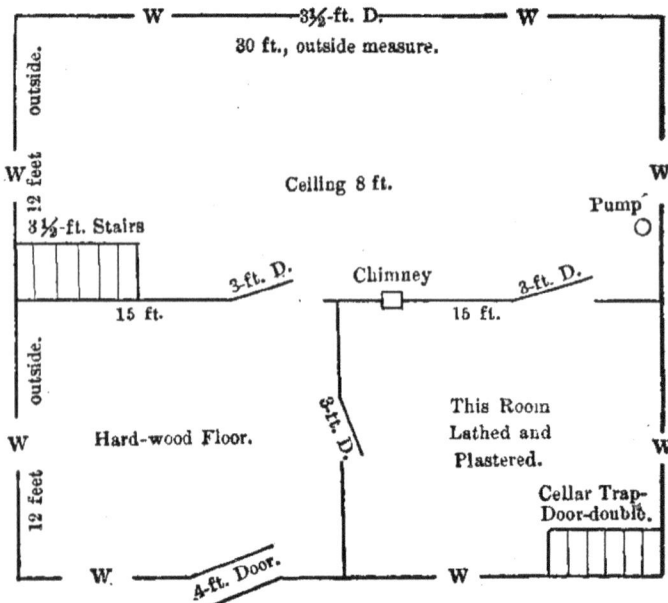

W   3½-ft. D.   W

30 ft., outside measure.

outside.

12 feet

W

3½-ft. Stairs

Ceiling 8 ft.

Pump

W

3-ft. D.

Chimney

3-ft. D.

15 ft.

15 ft.

outside.

3-ft. D.

This Room Lathed and Plastered.

W   Hard-wood Floor.

W

12 feet

Cellar Trap-Door-double.

W.   4-ft. Door.   W

Diagram of First Floor.—Original.

lar, and four doors beside the slanting one, two to the inner one or bee-cellar, and two to the outer or vestibule. I should have the entrance an inclined plane, which, especially if the apiary is large, should be so gradual in its descent that a car could pass down it into the cellar on a temporary track. The cellar should be well drained, or if water be permitted to pass through it, this should be kept in prescribed channels. In our cellar we have a large cistern. This is mostly in the outer cellar, but partly in the inner or bee-cellar. A tight partition separates the two rooms except at bottom of the cistern. In case of large apiaries the track and car make the removal of the bees to and from the cellar an easy matter. The first floor I should have, if my apiary was large, on a level with the ground. This (Fig. 258) should contain three rooms, one on the north for a shop, one on the southeast for comb honey, and one on the southwest for extracting, and storing extracted honey and brood-combs. For 100 colonies of bees, this building need not be more than 20x24 feet. A chimney should pass from the attic at the common angle of these three rooms through the roof. Wide doors on the south, if the apiary is large, should permit the car to enter either of the rooms on an extemporized track, whenever extracting or taking off comb honey is in operation.

The house should be so constructed as to be always free from rats and mice. In summer, wire-gauze doors should be used, also wire-gauze window-screens made to swing out like common window-blinds. Ours are single, not double, light, and so hung that when opened they remain so till shut. At the top the gauze extends outside the upper piece of the frame, and is separated from it by a bee-space width. At the top a few three-eighths inch round holes are made. This permits all bees to leave the house, while the character of the opening precludes outside bees from entering. Inside doors should permit our passing directly from any of these rooms to the others. The position of the chimney makes it easy to have a fire in any of the rooms. This would be desirable in the shop, in winter, when hive-making, etc., is in operation, or when visit-'ng with other bee-keepers is in progress. The ripening of honey or late extracting make it often desirable to have a fire in the extracting-room. If comb honey is kept in the designated room late in the season, it is desirable to warm that room. Of course, a large stove in the shop might be made to heat any or all of the rooms. I would have the comb-honey room very tight, and ventilated by an easily regulated slide into the chimney for the purpose of easy fumigation.

The extractor-room should have close, moth-proof cupboards for receiving brood-combs. Those in our house are high enough for three rows of frames, and wide enough just to receive the top-bar of a frame crosswise. Cleats nailed on to the inside hold the frames, which are turned diagonally a little to pass them to the lower tier. This room ought also to have a table for work, uncapping-box (Fig. 156), and large open tanks, open barrels, or extractor-cans, to hold the honey while it ripens. If the building is painted dark, this room will be warmer in summer. The warmer it becomes the more rapidly the honey thickens.

A chamber above costs but little, and serves admirably as a place for storage. This may be entered by stairs from the shop.

A neat bench and sharp tools, all conveniently placed, make the shop a very desirable fixture to every apiary.

I have spoken of a car and track in large apiaries; such an arrangement, which costs but little, is exceedingly desirable. The tracks run close to the rows of hives, and by means of simple switches, the car can be run anywhere in the apiary.

CHAPTER XX.

EVILS THAT CONFRONT THE APIARIST.

There are various dangers that are likely to vex the apiarist, and even to stand in the way of successful apiculture. Yet, with knowledge, most, if not all, of these evils may be wholly vanquished. Among these are: Robbing among the bees, disease, and depredations from other animals.

ROBBING.

This is a trouble that often very greatly annoys the inexperienced. Whenever bees leave the hives, except at a time of swarming, with the honey-stomach full, we may be sure robbing is in the air. Bees only rob at such times as the general scarcity of nectar forbids honest gains. When the question comes: Famine or theft? like many another, they are not slow to choose the latter. It is often induced by working with the bees at such times, especially if honey is scattered about or left lying around the apiary. It is especially to be feared in spring, when colonies are apt to be weak in both honey and bees, and thus are unable to protect their own meager stores. The remedies for this evil are not far to seek :

First.—Strong colonies are *very rarely* molested, and are almost sure to defend themselves against marauders ; hence, it is only the weaklings of the apiarist's flock that are in danger. Therefore, regard for our motto, "Keep all colonies strong," will secure against harm from this cause.

Second.—Italians—the Cyprians and Syrians are even more spirited in this work of defense than are the Italians—as before stated, are fully able, and quite as ready, to protect their rights against neighboring tramps. Woe be to the thieving bee that dares to violate the sacred rig ts of the home of our beautiful Italians, for such temerity is almost sure to cost the intruder its life.

But weak colonies, like our nuclei, and black bees, are still easily kept from harm. Usually, the closing of the entrance, so that but a single bee can pass through, is all sufficient. Mr. Jones closes the entrance by use of wet grass, straw, or shavings. Mr. Hayhurst places a frame six inches by eighteen inches covered by wire-gauze over the entrance. This keeps the robbers out, and still affords ventilation.

Another way to secure such colonies against robbing is to move them into the cellar for a few days. This is a further advantage, as less food is eaten, and the strength of the individual bees is conserved by the quiet, and as there is no nectar in the fields no loss is suffered. Mr. Root recommends "quiet" robbing at such times to cure robbing. He places hives containing honey near by, with the entrances so contracted that only one bee can enter at a time. The bees seem to prefer this quiet, unresisted robbing, and cease from the other. This, of course, would be expensive in case other apiaries were near by. It is a good way to get partially-filled sections or combs emptied. It works very well in case we give them access to a larger quantity of honey, else robbing may still be kept up.

In all the work of the apiary at times of no honey-gathering, we can not be too careful to keep all honey from the bees unless placed in the hives. The hives, too, should not be kept open long at a time. Neat, quick work should be the watchword. Mr. Root does necessary work at such times by night, using a lantern. I do not like night work; the bees crawl about one's clothes, and often reach quite objectionable places. During times when robbers are essaying to practice their nefarious designs, the bees are likely to be more than usually irritable, and likely to resent intrusion; hence, the importance of more than usual caution, if it is desired to introduce a queen. *Working under the bee-tent* (Figs. 158 and 166) *prevents all danger of inciting the bees to rob.* Dr. Miller inserts a funnel-shaped (Fig. 159) bee-escape in the top of the tent. Such a tent might be placed over the colony being robbed. Mr. Doolittle prizes highly a common sheet in the apiary. In case of robbing he covers the entire hive being robbed with this sheet.

DISEASE.

The common dysentery—indicated by the bees soiling their hives, as they void their fæces within instead of without —which so frequently works havoc in our apiaries, is, without doubt, I think, consequent upon wrong management on the part of the apiarist, poor honey, like cider, rotten apple-juice, rank honey-dew, or burnt sugar, or bad wintering, usually the result of severe weather, as already suggested in Chapter XVIII. As the methods to prevent this have already been sufficiently considered, we pass to the terrible

FOUL BROOD.

This disease, said to have been known to Aristotle— though this is doubtful, as a stench attends common dysentery—though it has occurred in our State as well as in States about us, is not very familiar to me. Of late I receive many samples of this affected brood each season. It is causing sad havoc in many regions of our country. No bee-malady can compare with this in malignancy. By it Dzierzon once lost his whole apiary of 500 colonies. Mr. E. Rood, first President of the Michigan Association, lost all his bees two or three times by this terrible plague.

The symptoms are as follows: Decline in the prosperity of the colony, because of failure to rear brood. The brood seems to putrefy, becomes "brown and salvy," and gives off a stench which is by no means agreeable. With a slight attack, the bad smell is not apparent. In a close box very little of the brood gives the characteristic odor. I often detect it in boxes received by mail before I open them. Later the caps are concave instead of convex, and many will have little holes through them. Holes will often be found in healthy brood-cells. As the cappings were never completed, such holes are smooth at the margins, while those of foul brood are jagged. The most decided symptom is the salvy, elastic mass in the brood-cell. With a pin-head we never draw forth a larva or pupa, but this brown, stringy mass which afterwards dries down in the cell, when it lets go of the pin-head, because of its elasticity, it flies or springs back. This is sometimes less marked.

FIG. 259.

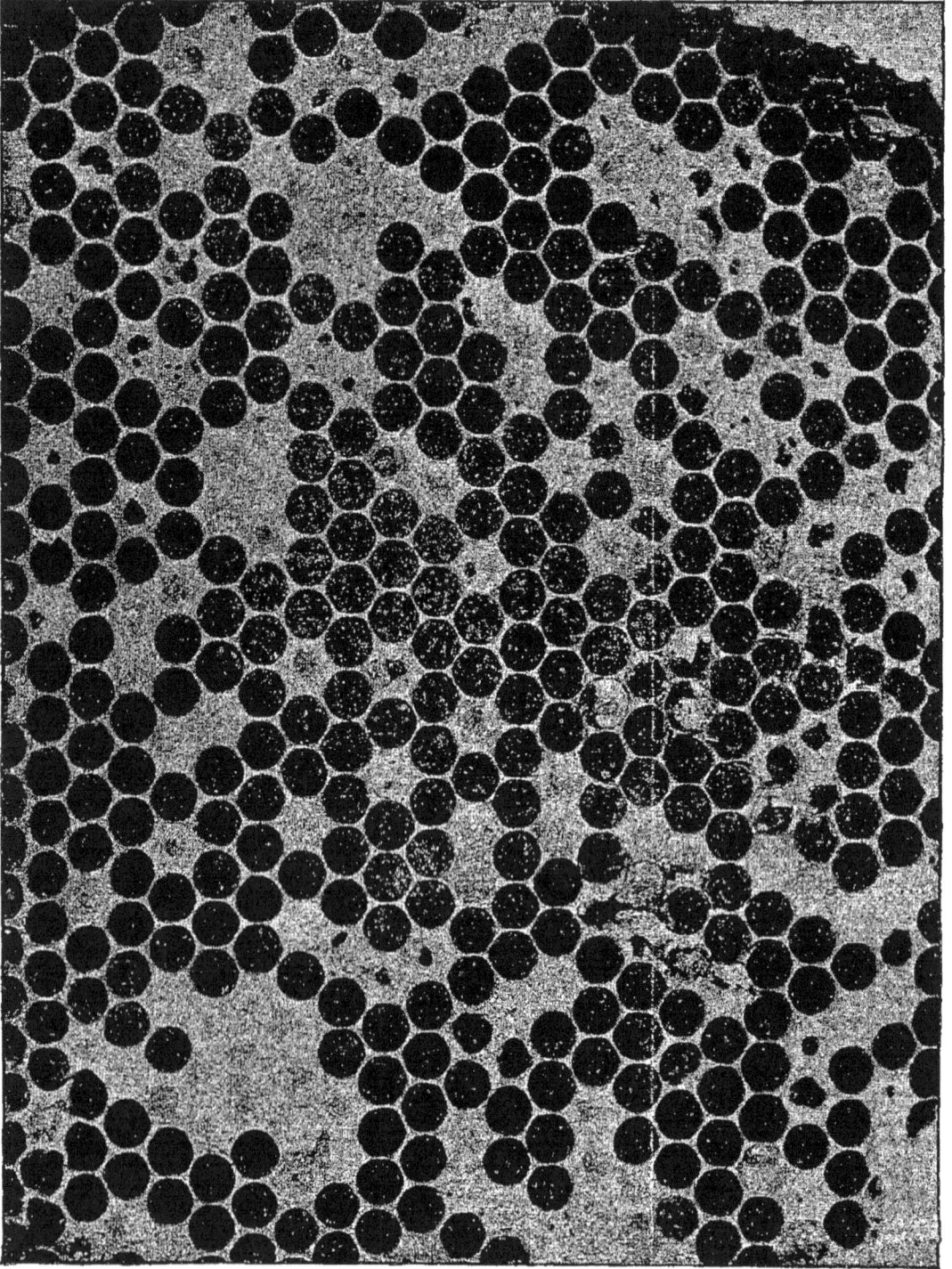

Foul Brood Photographed.—From A. I. Root Co.

There is no longer any doubt as to the cause of this fearful plague. Like the fell "Pebrine," which came so near exterminating the silk-worm, and a most lucrative and extensive industry in Europe, it, as conclusively shown by Drs. Preusz and Schonfeld, of Germany, is the result of minute parasitic organisms. Schonfeld not only infected healthy bee-larvæ, but those of other insects, both by means of the putrescent foul brood and by taking the spores. Professor Cohn discov-

FIG. 260.

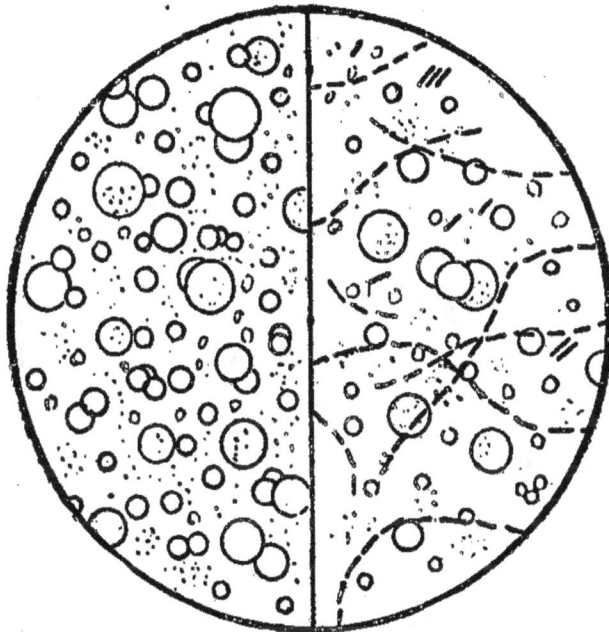

Healthy Stage.   Early Stage.

Foul Brood—From A. I. Root Co.

ered, in 1874, that the cause of foul brood was a microbe, Bacillus alveolaris. Mr. Hilbert, the following year, showed that these micro-organisms existed in the mature bees as well as in the brood. Later Mr. Cheshire gave the microbe the name of Bacillus alvei.

Fungoid growths are very minute, and the spores are so infinitesimally small as often to elude the sharp detection of the expert microscopist. Most of the terrible contagious diseases that human flesh is heir to—like typhus, diphtheria,

cholera, smallpox, etc.—are now known to be due to microscopic germs, and hence to be spread from home to home, and
from hamlet to hamlet, it is only necessary that the germs or
the contained spores, the minute seeds, either by contact or by
some sustaining air current, be brought to new soil of flesh,
blood, or other tissue—their garden-spot—when they at once
spring into growth, and thus lick up the very vitality of their
victims.   The huge mushroom will grow in a night.   So, too,
these other plants—the disease-germs—will develop with mar-

FIG. 261.

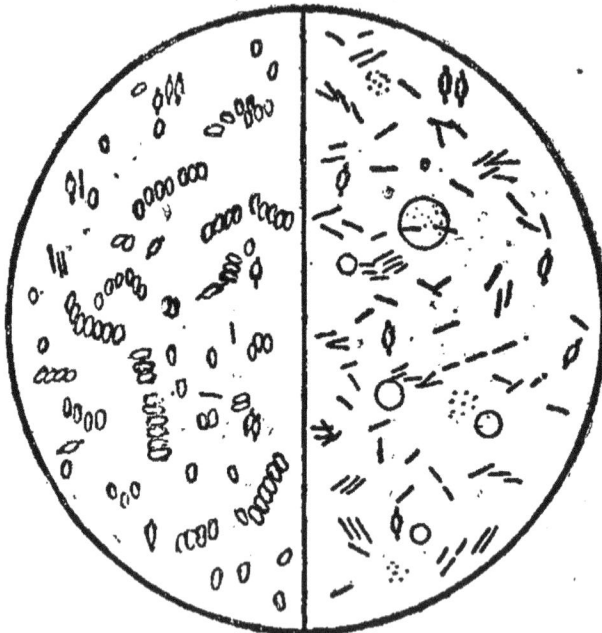

Middle Stage.   Late Stage.

Foul Brood.—From A. I. Root Co.

velous rapidity; and, hence, the horrors of yellow fever, scar-
letina and cholera.   The foul-brood Bacillus, like all bacilli,
is rod-shaped (Fig. 261).   The spore develops in one end, which
becomes slightly enlarged.

To cure such diseases the microbes must be killed.   To
prevent their spread they must be destroyed, or else confined.
But as these are so small, so light, and so invisible—easily
borne and wafted by the slightest zephyr of summer—this is
often a matter of the utmost difficulty.

In "foul brood" these germs feed on the larvæ of the bees, and thus convert life and vigor into death and decay. If we can kill this miniature forest of the hive, and destroy the spores, we shall extirpate the terrible plague. The spores resist heat, are more tenacious of life, and more difficult to kill than are the bacilli themselves.

Some of the facts connected with "foul brood" would lead us to think that the germs or spores of this fungus are only conveyed in the honey. This supposition, alone, enables us to understand one of the remedies which some of our ablest apiarists hold to be entirely sure.

REMEDIES.

"Prevention is better than cure." In case foul brood, black brood, or any suspected germ malady is in the neighborhood or apiary, it will always be wise to feed medicated syrup. Beta napthol is now preferred, as it is non-odorous, and not offensive to the bees. Mr. Thos. W. Cowan uses this successfully as follows : One ounce of the powder is put into a half-pint bottle ; just enough wood alcohol is added to dissolve it fully, when the bottle is filled with water. This will medicate 280 pounds of syrup, made by mixing 140 pounds each of water and granulated sugar. The solution and mixing can best be performed by use of the extractor. Gentle turning soon dissolves the sugar, and thoroughly mixes the beta napthol. Thus we use no heat. (See page 266.)

TO CURE.

No doubt Mr. Hilbert, of Germany, cured foul brood by use of salicylic acid. Mr. Muth did the same, and rendered the solution more easy by adding borax. That this extract of the willow is a powerful germicide is well known. In the cure of foul brood it has so often proved a partial or complete failure, that no one, except for experiment, can afford to use it in this warfare at all.

In 1874, Bontleroff, of Russia, suggested the use of carbolic acid or phenol as a cure of foul brood. Dr. Preusz also thought very highly of it. There is no doubt that this is also a very excellent bacillicide. Only the purest crystals of the acid should be used. To use this to medicate the syrup—one-fortieth of an ounce to a pound of syrup—would be wise as a prevention except that, as stated above, beta napthol is preferable. But, like salicylic acid, these carbolic acid derivatives are too uncertain. So many have failed to cure with these remedies.

So long as we have a safe, sure remedy which works in the hands of all, we can illy afford to risk our success with remedies that so generally fail.

Mr. D. A. Jones, and scores of others, are successful with what is termed the starvation method : The bees are drummed into an empty hive, placed in a cellar, and given no food for three or four days, till they have digested all honey in their stomachs. They are then given foundation and food, and the combs melted for wax, the honey scalded, and the hives scalded thoroughly before being again used. It would seem that the spores are in the honey—we know surely that they are in the chyle, though Schonfeld finds that they are not in the blood of the bee—and by taking that, the contagion is administered to the young bees. The honey may be purified from these noxious germs by subjecting it to the boiling temperature, which is generally, if not always, fatal to the spores of fungoid life. The microbe is killed surely by a temperature even less than the usual boiling, 212 degrees F. The spores, however, are only killed by prolonged boiling. So we better add water to the honey and then boil for an hour to make it safe, after which the honey may be safely fed. Some of wide experience say that it is safe to use the hives, even though they have not been boiled. Mr. McEvoy, of Ontario, after his very extensive experience, urges this. The combs are melted for wax. The disease is probably spread by robber-bees visiting affected hives, and carrying with them in the honey the fatal germs. Mr. Doolittle, after some experience, agrees with the lamented Quinby, that it is not necessary to cause the bees to fast as described by Mr. Jones. They can at once be hived safely on foundation, In this case, all honey is used up before any brood is present to be fed. To secure this, they are after four days changed again on to new foundation. We must in all this be most careful not to scatter honey, or to permit a single robber-bee to get at it. Great care, and the wisest exercise of judgment, is all important. A wee blunder, or little carelessness, may spread the evil rather than effect a cure.

From this remedy it would seem certain that the germs are in the honey.

It should be remembered that it is easy to scatter these fatal germs, and whatever cure is adopted, too great care can not be exercised. Mr. R. L. Taylor tells me that after an experience of two years he does not greatly fear this malady. He finds it easy, by means of the fasting cure, and free use of carbolic acid, to hold it in check or to cure it. Yet he admits that without much care and judgment it might work fearful havoc.

(I have found that a paste made of gum tragacanth and water is very superior, and I much prefer it for either general or special use to gum arabic. Yet it soon sours—which means that it is nourishing these fungoid plants—and thus becomes disagreeable. I have found that a very little salicylic acid will render it sterile, and thus preserve it indefinitely.)

BEE-PARALYSIS.

This is a common malady, more serious, it is claimed, in the warmer parts of the country. The bees become black, show a curious trembling motion, and are often dragged from the hive. Often so many die that the colony is seriously depleted. Change of queen is often a cure. Spraying with salt water has been thought to be of service. I believe this to be a fungoid disease, and, if so, feeding the medicated syrup (page 479) will be a wise practice. I have often seen this trouble in my apiary, but it always disappeared with no serious harm.

NEW BEE-DISEASE.

In California and some other sections, the brood dies without losing its form. We use the pin-head, and we draw forth a larva much discolored, often black, but not at all like the salvy mass that we see in foul brood. This is doubtless a germ disease, which I have greatly mitigated by simply feeding. I believe with this and the similar, if not identical black brood, and all kindred maladies, we should feed freely with the medicated syrup. The removal of old combs and honey, forcing the bees to build new, thus to remove germs would also abet the cure.

Black brood is not ropy like foul brood, and the brood shows affection earlier. It is serious in New York, and is treated precisely as is foul brood. The bees are transferred to other hives on starters of foundation, and this repeated in four days.

ENEMIES OF BEES.

Swift was no mean entomologist, as is shown in the following stanza:

> " The little fleas that do us tease,
> Have lesser fleas to bite them,
> And these again have lesser fleas,
> And so *ad infinitum*."

Bees are no exception to this law, as they have to brave the attacks of reptiles, birds, and other insects. In fact, they are beset with perils at home and perils abroad, perils by night and perils by day.

THE BEE-MOTH—GALLERIA MELLONELLA.

This insect, formerly known as G. cereana, belongs to the family of snout-moths, Pyralidæ. This snout is not the tongue, but the palpi, which fact was not known by Mr. Langstroth, who was usually so accurate, as he essayed to correct Dr. Harris, who stated correctly that the tongue was "very short and hardly visible." This family includes the destructive hop-moth, and the noxious meal and clover moths, and its members are very readily recognized by their usually long palpi, the so-called snouts. The family is now more restricted, and named Galleriidæ.

The eggs of the bee-moth are white, globular, and very small. These are usually pushed into crevices by the female moth as she extrudes them, which she can easily do by aid of her spy-glass-like ovipositor. They may be laid in the hive, in the crevice underneath it, or about the entrance. Soon these eggs hatch, when the gray, dirty-looking caterpillars, with brown heads, seek the comb on which they feed. To protect themselves better from the bees, they wrap themselves in a silken tube (Fig. 262), which they have power to spin. They remain in this tunnel of silk during all their growth, enlarging it as they eat. The noise, as they eat, can be heard plainly by holding the comb to the ear. As they tunnel

FIG. 262.

FIG. 263.

Tunnel of Bee-Moth Larva.—Original.

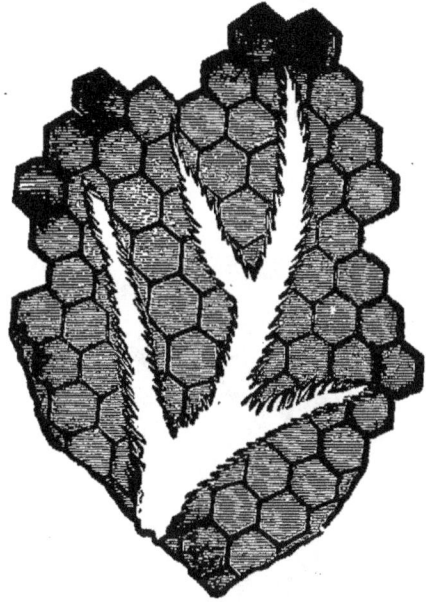

Tunnel in Comb.—Original.

among the larvæ in brood-combs, the larvæ are destroyed, and will be removed from the hives. Thus, the presence of dead larvæ in front of the hive is often a sign of the presence of insects in the hive. By looking closely, the presence of these

FIG. 264.

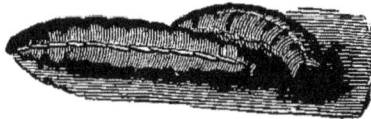

Larvæ of Bee-Moth.—Original.

larvæ may be known by this robe of glistening silk, as it extends in branching outlines (Fig. 263) along the surface of the comb. A more speedy detection, even, than the defaced comb, comes from the particles of comb, intermingled with

the powder-like droppings of the caterpillars, which will
always be seen on the bottom-board in case the moth-larvæ
are at work. Soon, in three or four weeks, the larvæ are full
grown (Fig. 264). Now the six-jointed and the ten prop-legs—
making sixteen in all, the usual number possessed by caterpillars—are plainly visible. These larvæ are about an inch
long, and show by their plump appearance that *they at least*
can digest comb. However, though these are styled wax-
moths they must have either pollen or dead bees to mingle
with their wax. While it is true that there is a little nitrogenous material in wax, there is not enough so that even the wax-

FIG. 265.

FIG. 266.

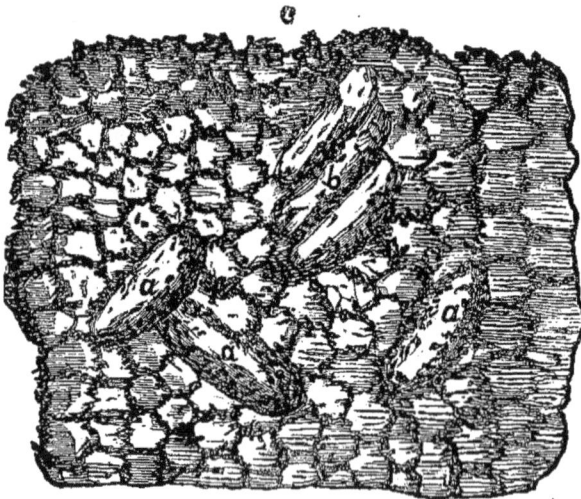

Cocoons.—*Original.*

Bee-Moth.—*Original.*

moth larva could thrive on it alone. They now spin their
cocoons, either in some crevice about the hive, or, if very
numerous, singly (Fig. 265, *a*) or in clusters (Fig. 265, *b*) on the
comb, or even in the drone-cells (Fig. 265, *c*), in which they
become pupæ, and in two weeks, even less sometimes, during
the extreme heat of summer, the moths again appear. In
winter they may remain as pupæ for months. The moths or
millers—sometimes incorrectly called moth-millers—are of an
obscure gray color, and thus so mimic old boards that they are
very readily passed unobserved by the apiarist. They are
about three-fourths of an inch long, and expand (Fig. 266)

nearly one and one-fourth inches. The females are darker than the males, possess a longer snout, and are usually a little larger. The wings, when the moths are quiet, are flat on the back for a narrow space, then slope very abruptly. They rest by day, yet, when disturbed, will dart forth with great swiftness, so Reaumur styled them "nimble-footed." They are active by night, when they essay to enter the hive and deposit their one or two hundred eggs. If the females are held in the hand they will often extrude their eggs; in fact, they have been known to do this even after the head and thorax were severed from the abdomen, and, still more strange, while the latter was being dissected.

It is generally stated that these are two-brooded, the first moths occurring in May, the second in August. Yet, as I have seen these moths in every month from May to September, and as I have proved by actual observation that they may pass from egg to moth in less than six weeks, I think under favorable conditions there may be even three broods a year. It is true that the varied conditions of temperature—as the moth-larvæ may grow in a deserted hive, in one with few bees, or one crowded with bee-life—will have much to do with the rapidity of development. Circumstances may so retard growth and development that there may be, not more than two, and possibly, in extreme cases, not more than one brood in a season.

It is stated by Mr. Quinby that a freezing temperature will kill these insects in all stages, while Mr. Betsinger thinks that a deserted hive is safe; neither of which assertions is entirely correct. Still, I believe exposure of combs to cold the winter through would kill most, if not all, of the bee-moth larvæ. I believe, in very mild winters, the moth and the chrysalids might be so protected as to escape unharmed, even outside the hive. It is probable, too, that the insects may pass the winter in any one of the various stages, though they generally exist as pupæ during the cold season.

HISTORY.

These moths were known to writers of antiquity, as even Aristotle tells of their injuries. They are wholly of Oriental origin, and are often referred to by European writers as a terrible pest. The late Dr. Kirtland, the able scientist, and first president of our American bee-association, once said in a letter to Mr. Langstroth, that the moth was first introduced into America in 1805, though bees had been introduced long before. They first seemed to be very destructive. It is quite probable, as has been suggested, that the bees had to learn to fear and repel them; for, unquestionably, the bees do grow in wisdom. In fact, may not the whole of instinct be inherited knowledge, which once had to be acquired by the animal? Surely bees and other animals learn to battle new enemies, and vary their habits with changed conditions, and they also transmit this knowledge and their acquired habits to their offspring, as illustrated by setter and pointer dogs. In time, may not this account for all those varied actions, usually ascribed to instinct? At least I believe the bee to be a creature of no small intelligence.

REMEDIES.

In Europe, late writers give very little space to this moth. Once a serious pest, it has now ceased to alarm, or even to disquiet the intelligent apiarist. In fact, we may almost call it a blessed evil, as it will destroy the bees of the heedless, and thus prevent injury to the markets by their unsalable honey, while to the attentive bee-keeper it will work no injury at all. Neglect and ignorance are the moth-breeders.

As already stated, Italian bees are rarely injured by moths, and strong colonies never. As the enterprising apiarist will possess only these, it is clear that he is free from danger. The intelligent apiarist will also provide not only against weak but queenless colonies as well, which, from their abject discouragement, are the surest victims to moth invasion. Knowing that destruction is sure, they seem, if not to court death, to make no effort to delay it.

As my friend, Judge W. H. Andrews, asserts, no bees, black or Italian, will be troubled with these insects so long as all the combs are covered with bees.

In working with bees an occasional web will be seen glistening in the comb, which should be picked out with a knife till the manufacturer—the ruthless larva—is found, when it should be crushed. Any larva seen about the bottom-board, seeking place to spin its cocoon, or any pupæ, either on comb or in crack, should also be killed. If, through carelessness, a colony has become thoroughly victimized by these filthy wax-devourers, then the bees and any combs not attacked should be transferred to another hive, after which the old hive should be sulphured by use of the smoker, as before described; then by giving one or two each of the remaining combs to strong colonies, after killing any pupæ that may be on them, they will be cleaned and used, while by giving the enfeebled colony brood, and if necessary a good queen, if it has any vigor remaining it will soon be rejoicing in strength and prosperity.

We have already spoken of caution as to comb honey and frames of comb (page 380), and so need not speak further of them.

THE WEE BEE-MOTH.

In 1887 another smaller moth attacked comb in New York and Michigan. Mr. W. J. Ellison, of South Carolina, wrote me that this insect does much harm in his State. It is Ephestia

FIG. 267.

Wee Bee-Moth.—Original.   Wing.—Original.

interpunctella, Hub., and belongs to the same family of moths, Pyralidæ, or snout-moths, that contains the old bee-moth. I shall call this (Fig. 267) the Wee bee-moth. The moths lay eggs in July and August, upon the comb. The larvæ feed in August, September and October upon the pollen, and do mischief by spreading a thin layer of silk over the combs. Mr. Ellison says the web on the comb honey is no small injury. Very likely there is an early summer brood.

REMEDIES.

The only suggestion I can offer at present is that the combs shall not be exposed. Fumigation, of course, either with the bisulphide of carbon or sulphur fumes, will destroy these also, and might be desirable in case comb honey is injured.

TWO DESTRUCTIVE BEETLES.

There are two destructive beetles that often work on the comb, more, however, for the pollen and dead bees than for the wax. One of these, Tenebrio molitor, Linn., is the common flour or meal beetle. It is dark brown in color, and five-eighths of an inch (16 mm.) long. The larva or grub is of a lighter color, and when fully developed is one inch (25 mm.) long. It resembles very closely the larva of our Elater beetles —the wire-worms. The other is the bacon beetle, Dermestes lardarius, Linn. (Fig. 268), which is a sore pest in museums,

FIG. 268.

D. Lardarius.—Original.

as it feeds on all kinds of dried animal tissues. The beetle is black, while nearly one-half of the wing-covers, next to the thorax, are yellowish-gray, lined in the middle with black. The beetle is three-eighths of an inch (10 mm.) long. The larva is some longer, very hairy, and ringed with brown and black bands. These beetles are not very troublesome in the apiary, and can be readily destroyed by use of bisulphide of carbon. Care is necessary, however, in the use of this very inflammable and explosive liquid. It is no more to be feared than would be gasoline. We have only to keep the match or lighted cigar away. There are other beetles and moths of similar habits, which are likely at any time to invade the apiary.

ROBBER-FLIES.

There are several of these flies that prey upon bees. The most common is Asilus missouriensis, Riley. This is a two-winged fly, of the predacious family Asilidæ, which attacks and takes captive the bee and then feeds upon its fluids. It is more common in the southern part of our country. The fly (Fig. 269) has a long, pointed abdomen, strong wings, and is very powerful. I have seen an allied species attack and overcome the powerful tiger-beetle, whereupon I took them both with my net, and now they are pinned, as they were captured, in the college cabinet. These flies delight in the warm sunshine, are very quick on the wing, and so are not easily captured. It is to be hoped that they will not become very numer-

FIG. 269.

Robber-Fly.—Original.

ous. If they should, I hardly know how they could be kept from their evil work. Frightening them or catching with a net might be tried, yet these methods would irritate the bees, and need to be tried before they are recommended. I have received specimens of this fly from nearly every Southern State. During the summer-time these flies are usually well employed in Michigan. They have been observed to kill the cabbage butterfly by scores. The Asilids are very common in California, yet I am persuaded that they do far more good than harm.

I have also a fly of the same family, with the same bee-destroying habits, a species of Erax (Fig. 270). In form it resembles the one referred to above. The wing (Fig. 269), as will be seen, is quite different in its venation. I received this

FIG. 270.   FIG. 271.

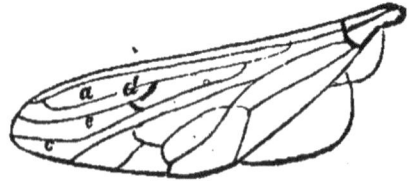

Robber-Fly and Wing.—Original.

species from Louisiana. Fig. 272 shows the antennæ magnified. The Nebraska bee-killer, Promachus bastardi, is the

FIG. 272.   FIG. 273.

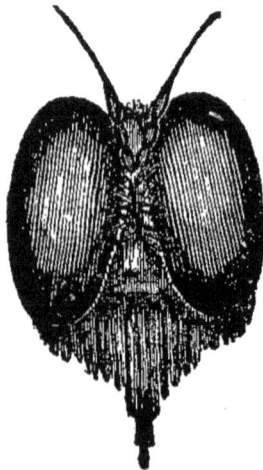

Head and Tarsus of Robber-Fly.—Original.

same in general appearance as the above. The second vein of the primary wing, not the third, as in the case of Asilus, *forks*. In Erax, as seen in Fig. 271, this branch is disconnected.

There are two other insects of this family, Mallophora orcina and Mallophora bomboides, which differ greatly in form from those mentioned above; they look more like bumble-bees, for which they have been mistaken.

I have received these insects from several of our enterprising bee-keepers of the South—Tennessee, Georgia, and Florida—with the information that they dart forth from some convenient perch, and with swift and sure aim grasp a bee, and bear it to some bush, when they leisurely suck out all but the mere crust, and cast away the remains.

The insects in question, which in form, size, and color much resemble bumble-bees, belong to Loew's third group,

FIG. 274.

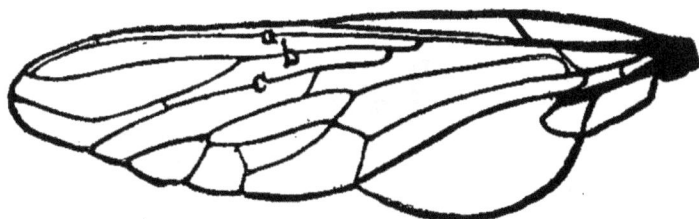

Wing of Mallophora.—Original.

Asilina, as the antennæ end in a bristle (Fig. 272), while the second longitudinal vein of the wing (Fig. 274, b) runs into the first (Fig. 274, a).

The genus is Mallophora. The venation of the wings much resembles that of the genus Promachus, though the form of these insects is very different.

In Mallophora and Promachus the venation is as represented in Fig. 274, where, as will be seen, the second vein (Fig. 274, b) forks, while in the genus Asilus (Fig. 269) the third vein is forked, though in all three genera the third joint of the antennæ (Fig. 272) ends in a prolonged bristle.

One of the most common of these pests, which I am informed by Dr. Hagen, is Mallaphora orcina, Weid., is one inch long, and expands one and three-fourths inches (Fig. 275). The head (Fig. 272) is broad, the eyes black and prominent, the antennæ three-jointed, the last joint terminating in a bristle, while the beak is very large, strong, and, like the eyes and antennæ, coal-black. This is mostly concealed by the light yellow hairs, which are crowded thick about the mouth and between the eyes.

The thorax is prominent and thickly set with light yellow hairs. The abdomen is narrow, tapering, and covered with yellow hairs, except the tip, which is black, though there are scattering hairs of a grayish yellow color on the black legs. The pulvilli, or feet-pads (Fig. 273, *b*), are two in number, bright yellow in color, surmounted by strong, black claws (Fig. 273, *a*), while below and between is the sharp spine (Fig. 273, *c*), technically known as the empodium.

The habits of the flies are interesting, if not to our liking. Their flight is like the wind, and, perched near the hive, they

FIG. 275.

M. oricina.—Original.

rush upon the unwary bee returning to the hive with its full load of nectar, and grasping it with their hard, strong legs, they bear it to some perch near by, when they pierce the crust, suck out the blood, and drop the carcass, and are then ready to repeat the operation. A hole in the bee shows the cause of its sudden taking off. The eviscerated bee is not always killed at once by this rude onslaught, but often can crawl some distance away from where it falls, before it expires.

Another insect nearly as common is Mallophora bomboides, Weid. This fly might be called a larger edition of the one just described, as in form, habits, and appearance it closely resembles the other. It belongs to the same genus, possessing all the generic characters already pointed out. It is very difficult to capture this one, as it is so quick and active.

This fly is one and five-sixteenths inches long, and expands two and a half inches. The head and thorax are much as in the other species. The wings are very long and strong, and, as in other species, are of a smoky brown color. The abdomen is short, pointed, concave from side to side on the under surface, while the grayish yellow hairs are abundant on the legs and whole under portion of the body. The color is a lighter yellow than in the other species. These insects are powerfully built, and if they become numerous must prove a formidable enemy to the bees. I believe all of the robber-flies are our friends. They destroy few bees, comparatively, and hosts of our insect enemies.

Another insect very common and destructive in Georgia, though it closely resembles the two just described, is of a different genus. It is the Laphria thoracica, of Fabricius. In this genus the third vein is forked, and the third joint of the antenna is without the bristle, though it is elongated and tapering. The insect is black, with yellow hair covering the upper surface of the thorax. The abdomen is wholly black, both above and below, though the legs have yellow hairs on the femurs and tibiæ. This insect belongs to the same family as the others, and has the same habits. It is found North as well as South.

THE STINGING BUG—PHYMATA EROSA, FABR.

This insect is very widely distributed throughout the United States. I have received it from Maryland to Missouri on the South, and from Michigan to Minnesota on the North. The insect will lie concealed among the flowers, and upon occasion will grasp a bee, hold it off at arm's length, and suck out its blood and life.

This is a Hemipteron, or true bug, and belongs to the family Phymatidæ, Uhler. It is the Phymata erosa, Fabr., the specific name erosa referring to its jagged appearance. It is also called the " stinging bug," in reference to its habit of repelling intrusion by a painful thrust with its sharp, strong beak.

The " stinging bug " is somewhat jagged in appearance, about three-eighths of an inch long, and generally of a yellow color, though this latter seems quite variable. Frequently there is a distinct greenish hue. Beneath the abdomen, and on the back of the head, thorax, and abdomen, it is more or less specked with brown ; while across the dorsal aspect of the broadened abdomen is a marked stripe of brown (Fig. 277, *d, d*). Sometimes this stripe is almost wanting, sometimes a mere patch, while rarely the whole abdomen is very slightly marked, and as often we find it almost wholly brown above and below. The legs (Fig. 277, *b*), beak and antennæ (Fig. 278, *a*) are greenish yellow. The beak has three joints (Fig. 278, *a, b, c*), and a sharp point (Fig. 278, *d*). This beak is not only the great weapon of offense, but also the organ through which the

FIG. 276.   FIG. 277.   FIG. 278.

Side view, natural size.   *Magnified twice.*   *Beak much magnified.*
—Original.   *—Original.*   *—Original.*

food is sucked. By the use of this, the insect has gained the sobriquet of "stinging bug." This compact, jointed beak is peculiar to all true bugs, and by observing it alone we are able to distinguish all the very varied forms of this group. The antenna is four-jointed. The first joint (Fig. 279, *a*) is short, the second and third (Fig, 279, *b* and *c*) are long and slim, while the terminal one (Fig. 279, *d*) is much enlarged. This enlarged joint is one of the characteristics of the genus Phymata, as described by Latreille. But the most curious structural peculiarity of this insect, and the chief character of the genus Phymata, are the enlarged anterior legs (Figs. 280 and 281). These, were they only to aid in locomotion, would seem like awkward, clumsy organs, but when we learn that they are used to grasp and hold their prey, then we can but appreciate and admire their modified form. The femur (Fig. 281, *b*) and the tarsus (Fig. 281, *a*) are toothed, while the latter is greatly

FIG. 279.

Antenna, much magnified.

FIG. 280. - FIG. 281.

Interior view.   *Exterior view.*

Anterior Leg, magnified.—Original.

enlarged. From the interior lower aspect of the femur (Fig. 282) is the small tibia, while on the lower end of the tarsus (Fig. 281, *d*) is a cavity in which rests the single claw. The other four legs (Fig. 283) are much as usual.

FIG. 282.   FIG. 283.

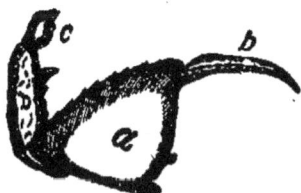

Claw, enlarged.—Original.   *Middle Leg, much magnified.—Original.*

This insect, as already intimated, is very predaceous, lying in wait, often almost concealed, among flowers, ready to capture and destroy unwary plant-lice, caterpillars, beetles, butterflies, moths, and even bees and wasps. We have already noticed how well prepared it is for this work by its jaw-like anterior legs, and its sharp, strong, sword-like beak.

It is often caught on the golden-rod. This plant, from its color, tends to conceal the bug, and from the character of the plant—being attractive as a honey-plant to bees—the slow bug is enabled to catch the spry and active honey-bee.

As Prof. Uhler well says of the "stinging bug": "It is very useful in destroying caterpillars and other vegetable-feeding insects, but is not very discriminating in its tastes, and would as soon seize the useful honey-bee as the pernicious saw-fly." And he might have added that it is equally indifferent

FIG. 284.

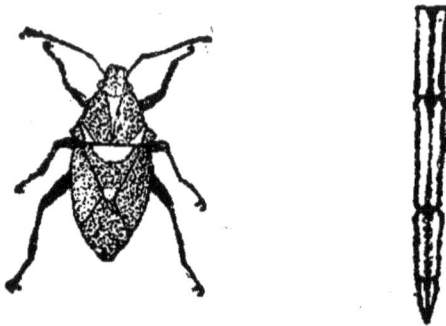

Bee-Stabber, and Beak.—Original.

to the virtues of our friendly insects, like the parasitic and predaceous species.

We note, then, that this bug is not wholly evil, and as its destruction would be well-nigh impossible, for it is as widely scattered as are the flowers in which it lurks, we may well rest its case, at least until its destructiveness becomes more serious than at present.

THE BEE-STABBER.

In the Southern States there is another bug, Euthyrhyn-chus floridanus, Linn. (Fig. 284), which I have named the bee-stabber. This bug places itself at the entrance of the hive and stabs and sucks the bees till they are bloodless. As will be seen, its powerful four-jointed beak fits it well for this purpose. This bug is purplish or greenish blue, with dull, yellowish markings, as seen in the figure. It is also yellowish beneath. It is one-half of an inch long. Other similar bugs may also learn that bees with their ample honey-sac full of nectar are most toothsome.

BEE-HAWK—LIBELLULA.

These large, fine, lace-wings (Fig. 285) are Neuropterous insects. They work harm to the bees mostly in the Southern States, and are called mosquito-hawks. Insects of this genus are called dragon-flies, devil's darning-needles, etc. They are

FIG. 285.

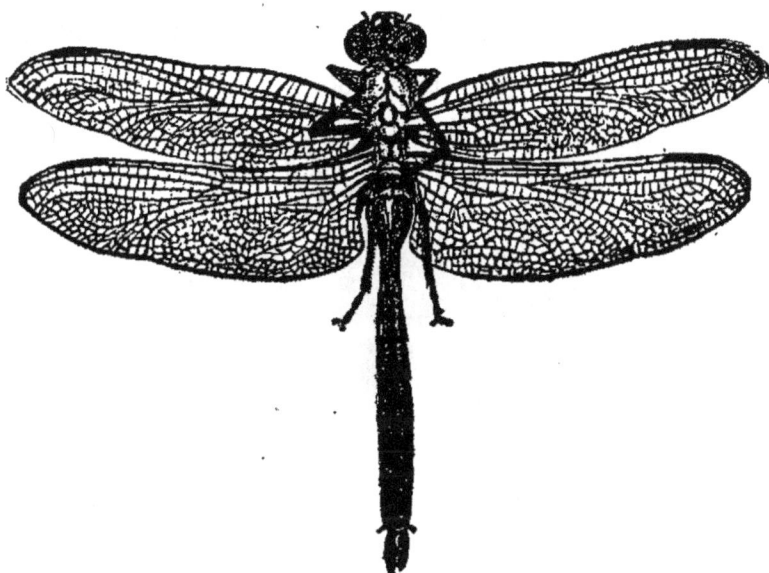

Bee-Hawk.—Original.

exceedingly predaceous. In fact, the whole order is insectivorous. From its four netted veined wings, we can tell it at once from the Asilids, before mentioned, which have but two wings. The bee or mosquito hawks are resplendent with metallic hues, while the bee-killers are of sober gray. The mosquito-hawks are not inaptly named, as they not only prey upon other insects, swooping down upon them with the dexterity of a hawk, but their graceful gyrations, as they sport in the warm sunshine at noonday, are not unlike those of our graceful hawks and falcons. These insects are found most abundant near water, as they lay their eggs in water, where the larvæ live and feed upon other animals. The larvæ are peculiar in breathing by gills in the rectum. The same water that bathes these organs and furnishes oxygen, is sent out in a jet, and thus sends the insect darting along. The larvæ also possess enormous jaws, which formidable weapons are masked till it is desired to use them, when the dipper-shaped mask is dropped or unhinged, and the terrible jaws open and close upon the unsuspecting victim, which has but a brief time to bewail its temerity.

A writer from Georgia, in Gleanings in Bee-Culture, Vol. IV, page 35, states that these destroyers are easily scared away, or brought down by boys with whips, who soon become as expert in capturing the insects as are the latter in seizing

FIG. 286.

Tachina-Fly.—Original.

the bees. One of the largest and most beautiful of these (Fig. 285) is Anax junius. It has a wide range in the United States (North and South), and everywhere preys upon the honey-bee.

TACHINA-FLY.

From descriptions which I have received, I feel certain that there is a two-winged fly, probably of the genus Tachina (Fig. 286), that works on bees. I have never seen these, though I have repeatedly requested those who have to send them to me. My friend, J. L. Davis, put some sick-looking bees into a cage and hatched the flies, which, he told me, looked not unlike a small house-fly. It is the habit of these flies, which are closely related to our house-flies, which they much resemble, to lay their eggs on other insects. Their young, upon hatching, burrow into the insect that is being victimized, and grow by eating it. It would be difficult to cope with this evil, should it become of great magnitude. We may well hope that this habit of eating bees is an exceptional one with it. The affected bees will be found dead at early dawn in front of the hives.

BEE-LOUSE—BRAULA CŒCA, NITSCH.

This louse (Fig. 287) is a wingless Dipteron, and one of the uniques among insects. It is a blind, spider-like parasite, and serves as a very good connecting link between insects and spiders, or, still better, between the Diptera, where it belongs, and the Hemiptera, which contains the bugs and most of the lice. It assumes the semi-pupa state almost as soon as

FIG. 287.

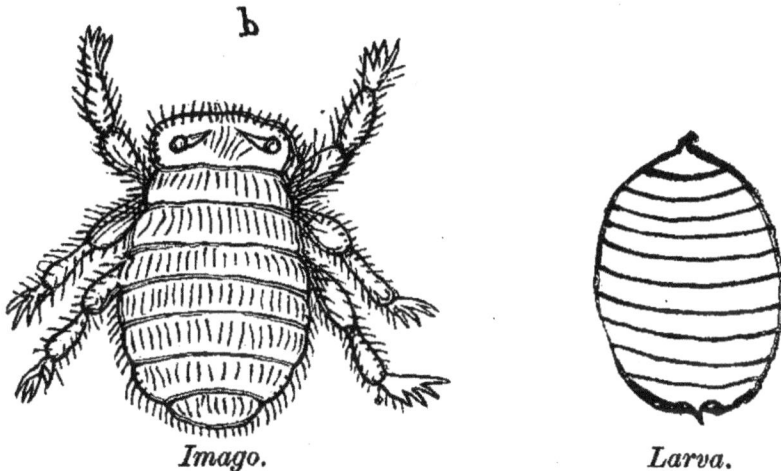

Imago. *Larva.*

B. cœca.—Original.

hatched, and, strangest of all, is, considering the size of the bee on which it lives, and from which it sucks its nourishment, enormously large. Two or three, and sometimes as many as ten, are found on a single bee. When we consider their great size, we cannot wonder that they soon devitalize the bees.

These have done little damage, except in the south of Continental Europe, Cyprus, and other parts of the Orient. The reason that they have not been naturalized in other parts of Europe and America may be owing to climate, though I think more likely it is due to improved apiculture. Mr. Frank Benton, who has had much experience with these bee-lice in Cyprus, writes me that the Braula is no serious pest if the bees are properly cared for. "In fact, if hives are kept clean inside, and colonies supplied with young queens and kept strong, the damage done by the Braula is very slight, if anything. In old, immovable-comb hives, where the combs are black and thickened, and in case the queens are old; or where through some extraneous cause the colonies have become weak, these lice are numerous on queens and workers. I have not noticed them on the drones. Since they are found on workers as well as the queen, their removal from the latter will bring temporary relief. About ten is the greatest number that I have seen on one queen. I have only thought it necessary to remove them in case there were three or more on a queen. The only way to remove them is to pick them off with a knife, scissors, forceps or similar instrument. They are quick-footed, and glide from one place to another like the wax-moth. I hold the queen between the thumb and first finger of the left hand, and with pocket-knife or clipping scissors shave off the parasite. It is no easy matter to get them the first time, as when you attempt their removal they glide around to the other side of the queen so adroitly that you have to turn the queen over to try again." Mr. Benton says that it is not practicable to remove these lice by lessening the size of the entrance to the hive. He thinks that, with the attention given to bees in America, the Braula will never become a serious pest, if introduced here. While these lice have been imported to America several times, they seem to disappear almost at once, which verifies Mr. Benton's prophecy.

ANTS.

These cluster about the hives in spring for warmth, and seldom, if ever, I think, do any harm in our cold climates, though in California and the South they do much harm. Should the apiarist feel nervous, he can very readily brush them away, or destroy them by use of any of the fly-poisons which are kept in the markets. As these poisons are made attractive by adding sweets, we must be careful to preclude the bees from gaining access to them. As we should use them in spring, and as we then need to keep the quilt or honey-board close above the bees, and as the ants cluster above the brood-chamber, it is not difficult to practice poisoning. One year I tried Paris-green with success. There are several reports of ants entering the hives and killing the bees; even the queen is said to have been thus destroyed.

I learn from Mr. H. E. Hill, of Florida, of a large, red ant peculiar to that section (Fig. 288), which is a terror to bees. It has destroyed nineteen nuclei in one week, and hundreds of dollars worth of bees, for Mr. Hill. It hides and burrows in rotten wood, above and below ground, in hive-covers, in parts of hives separated by the division-boards—anywhere where

FIG. 288.

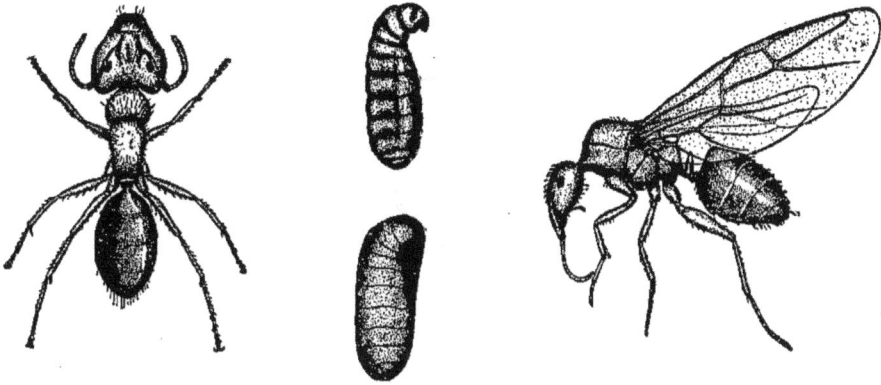

Florida Ant, in all stages.—Original.

concealment is possible. So numerous are they that Mr. Hill thinks there may be thousands in a colony, and he has destroyed hundreds of colonies within the past two years. Weak and queenless colonies suffer most, but none are exempt. Scouts are sent out to locate the prey in the early twilight. Later the chosen victims are stormed by the ant army and routed, though many ants die in the conflict. This ant (Fig. 288) is known as the bull-dog ant in Florida. It is known to science as Camponotus esuriens. (See American Bee Journal, Vol XLI, page 72.) Mr. Hill finds only one way—burning—to destroy them, and only one to keep them at bay. The legs of the hive stands are cut with a basin (Fig. 289), which is waxed and kept filled with carbolic acid. This is not satisfactory, as it evaporates quickly. I would suggest mixing kerosene and lard, both of which are very obnoxious to insects, and fill the basins with this.   Ants in California are killed by saturating the runs with gasoline, and then burning all.   It is a quick remedy, but hard for the ants.

This ant is red except the eyes and abdomen, which may be nearly all black, large soldiers, or tipped with black—common workers.   There are many hairs on the abdomen.

In such cases, if they occur, it is best to put a sweet poisonous mixture in a box and permit the ants to enter through

FIG. 289.

Leg of Hive-Stand.—Original.

an opening too small to admit bees, and thus poison the ants. Or we may find the ant's nest, and, with a crowbar, make a hole in it, turn in this an ounce of bisulphide of carbon, and quickly plug it up by packing clay in the hole and on the nest. The liquid will kill the ants.   This better be done when the ants are mostly in their nest.

THE COW-KILLER.

This ant-like insect, Mutilla coccinea (Fig. 290), has been sent me from Illinois and the South as far as Texas.   It is a formidable enemy of the bees.   The male has wings and no sting.   The female has no wings, but is possessed of a powerful sting.   She is an inch (25 mm.) long, very hairy, and black, except the top of her head and thorax, and a broad basal band and the tip of the upper part of her abdomen, which are bright red. A central band of black divides the red spaces of the abdomen. The entire under part of the body and all the members are black. There are several species of varying size and color in California. Grayish white species are nearly as common as the red and black ones. Some are as large as a worker-bee.

So hard and dense is the chitinous crust of these insects that they enter the hives fearlessly, and, unmindful of stings,

FIG. 290.

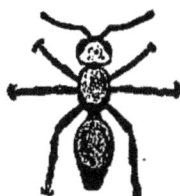

Cow-Killer.—Original.

deliberately kill the bees and feed on the young. The males are said to sting. This is certainly a mistake. The sting is a modified ovipositor—an organ not possessed by males. These insects belong to the family Mutillidæ, so called because the females are wingless. They are closely allied in structure to the ants, which they much resemble.

THE PRAYING MANTIS.

This strange insect I have received from Indiana and other Southern and Western States. Its scientific name is Mantis carolina, Linn. A similar species I often take in Los Angeles County, Calif. It is very predaceous, and the female has been known to eat up her mate immediately after the sexual act. No wonder that they make our friends of the hive contribute to their support. This insect (Fig. 291) is a sort of nondescript. In the South it is known as devil's race-horse. It is a corpulent "walking-stick" with wings. In fact, is closely related to the "walking-sticks" of the North. Its anterior legs are very curious. As it rests upon them, it appears as if in the attitude of devotion, hence the name, praying mantis. It also raises these anterior legs in a supplicating attitude, which would also suggest the name. It might well be preying mantis. These peculiar anterior legs, like the same in Phymata erosa, are used to grasp its victims. It is reported to move with surprising rapidity, as it grasps its prey.

FIG. 291.

Mantis.—Original.

Its eggs (Fig. 292) are glued to some twig, in a scale-like mass, and covered with a sort of varnish. Some of these hatched out in one of my boxes, and the depravity of these insects was manifest in the fact that those first hatched fell to

FIG. 292.

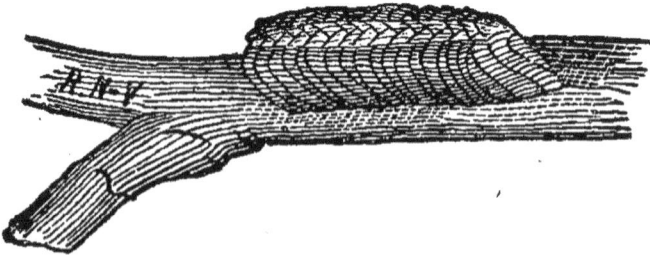

Eggs of Mantis.— Original.

and ate the others. They do much good in destroying our insect enemies.

BLISTER-BEETLES.

I have received from Mr. Rainbow, of San Diego Co., Calif., the larvæ (Fig. 293, *a*) of some blister-beetles, probably Meloe barbarus, Lec., as that is a common species in California. Mr. Rainbow took as many as seven from one worker-bee. Fig. 293, *B*, represents the female of Meloe angusticollis, a common species in Michigan and the East. I have also received larvæ from Mr. Hammond, of New York, who took them from his bees. He says they make the bees uncomfortable. These are likely M. angusticollis. As will be seen, the wing-covers are short, and the beetle's abdomen fairly drags with its weight of eggs. The eggs are laid in the earth. The larvæ, when first hatched, crawl upon some flower, and, as occasion permits, crawl upon a bee and thus are borne to the hive, where they feast on eggs, honey and pollen. These

FIG. 293.

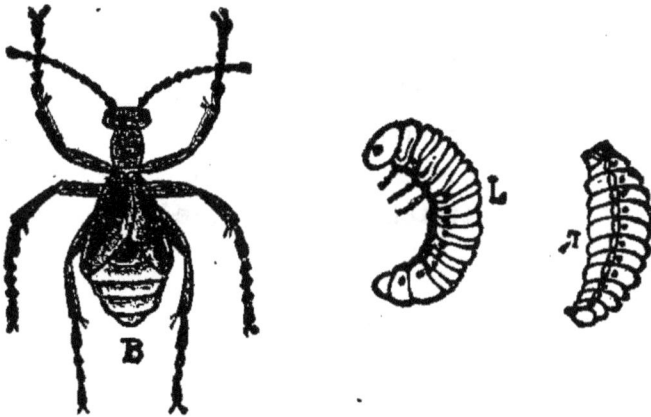

Blister-Beetle and Larvæ.—Original.

insects undergo what M. Faher styles hyper-metamorphosis, as the larva appears in four different forms instead of one. Two of these forms show in the figure. The Spanish-fly—Cantharides of the shops—is an allied insect. Some of our common blister-beetles are very destructive to plants. Girard, in his excellent work on bees, gives illustrations of all the forms of this insect.

WASPS.

I have never seen bees injured by wasps. In the South, as in Europe, we hear of such depredations. I have received wasps, sent by our Southern brothers, which were caught destroying bees. The wasp sent me is the large, handsome Stizus speciosus, Drury. It is black, with its abdomen imperfectly ringed with yellow. The wasps are very predaceous, and do immense benefit by capturing and eating our insect pests. I have seen wasps carry off "currant worms" with a celerity that was most refreshing.

As the solitary wasps are too few in numbers to do much damage—even if they ever do any—any great damage which may occur would doubtless come from the social paper-makers. In this case, we have only to find the nests and apply the torch, or hold the muzzle of a shot-gun to the nest and shoot. This should be done at nightfall, when the wasps have all gathered home. Let us not forget that the wasps do much good, and so not practice wholesale slaughter unless we have strong evidence against them.

A BEE-MITE.

It has long been known to chicken fanciers that our poultry often suffer serious injury from a small mite. These little arachnids often enter houses in countless thousands, much to the annoyance of the owners. Kerosene may be used to repel them. Other mites attack the cow, the horse, the sheep, etc.

FIG. 294.

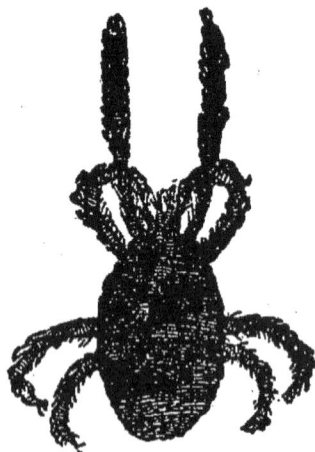

Mite.—Original.

The Texas cattle-tick—Boophalus bovis—which so often worries horses and cattle, and which carries the minute protozoan (Pyrosoma bigeminum) that causes the terrible Texas fever, is a colossal mite.

One spring a lady bee-keeper of Connecticut discovered these mites in her hives while investigating to learn the cause of their rapid depletion. She had noticed that the colonies were greatly reduced in number of bees, and upon close observation she found that the diseased or failing colonies were covered with these mites. A celebrated queen-breeder of New York State sent me these same mites in 1887, with the report that they killed his queens while yet in the cell. I found great numbers in a cell sent by this gentleman. The strong and prosperous colonies were exempt from the annoyance. So small.are these little pests that a score could take possession of a single bee, and not be near neighbors, either. The lady states that the bees roll and scratch in their vain attempts to rid themselves of these annoying stick-tights, and, finally, worried out, either fall to the bottom of the hive or go forth to die outside.

The bee-mite (Fig. 294) is very small, hardly more than five mm. (1–50 of an inch) long. The female is slightly larger than the male, and somewhat transparent. The color is black, though the legs and more transparent areas of the females appear yellowish.

REMEDIES.

The fact that what would be poison to the mite would probably be death to the bees, makes this question of remedy quite a difficult one. I can only suggest what Mrs. Squire has tried—frequent changing of the bees from one hive to another, after which the hive can be freed from the mites by scalding. Of course, the more frequent the transfer the more thorough the remedy.

I would suggest placing pieces of fresh meat, greased or sugared paper, etc., in the hives, in hopes to attract the pests, which, when massed on these decoys, could easily be killed.

CALIFORNIA BEE-KILLER.

Mr. J. D. Enas, of Napa Co., Calif., sent me specimens of a curious bee-enemy (Fig. 295), which he finds quite a serious enemy of bees. I have taken many of these here at Claremont, but have not known of their disturbing bees.

This is a Datames, possibly D. Californicus, Simon, though it does not quite agree with the description of that species. It, like the mites just described, belongs to the sub-class Arachnida or spiders, and is related to the scorpions. The group of animals is known as the family Solpulgidæ. As will be seen, the head and thorax are not separate, as they are in true insects.

The abdomen is long and segmented, a shield-like plate covers the head, and the eyes are far forward, small and globular. The most peculiar organs are the jaws or falces, which are immense, and armed with formidable teeth, spines, hairs, etc. The family is small, little known, and, except in one case,

FIG. 295.

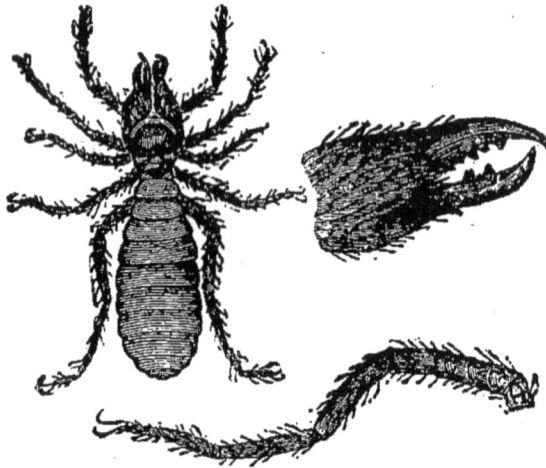

California Bee-Killer.—Original.
(Jaws and falces, and posterior leg.)

Datames pallipes, Say., which is said to live in houses in Colorado, and to feed on bed-bugs; the habits have not been described.

Mr. Enas finds this species in his hives, killing and eating the bees. The remedy must be hand-picking, which will not be very difficult.

SPIDERS.

These sometimes spread their nets so as to capture bees. If porticos—which are, I think, worse than a useless expense—are omitted, there will very seldom be any cause for complaint against the spiders, which, on the whole, are friends. As the bee-keeper who would permit spiders to worry his bees would not read books, I will discuss this subject no further.

THE KING-BIRD—TYRANNUS CAROLINENSIS.

This bird, often called the bee-martin, is one of the fly-catchers, a very valuable family of birds, as they are wholly insectivorous, and do immense good by destroying our insect pests. The king-birds are the only ones in the United States that deserve censure. The species in California is Tyrannus verticalis, or Western king-bird; that of the East, Tyrannus tyrannus. Another, the chimney swallow of Europe, has the same evil habit. Our chimney swallow has no evil ways. I am sure, from personal observation, that these birds capture and eat the workers, as well as drones, as I have taken worker-bees from their stomachs; and, I dare say, they would pay no more respect to the finest Italian queen. They perch on a tree or post and dart with the speed of an arrow as their poor victim comes heavily laden towards the hives. How is it that the bird is not stung? Some say that they pull the bees apart and simply eat the honey-stomach. Do they handle the bee so as to avoid the stings? Who will determine this point? King-birds killed close by an apiary here at Claremont had only rob-ber-flies in their stomachs; thus it was befriending the bees. Yet, in view of the good that these birds do, unless they are far more numerous and troublesome than I have ever observed them to be, I should certainly be slow to recommend the death warrant.

TOADS.

The same may be said of toads, which may often be seen sitting demurely at the entrance of the hives, and lapping up the full-laden bees with the lightning-like movement of their tongues, in a manner which can but be regarded with interest, even by him who suffers loss. Mr. Moon, the well-known api-arist, made this an objection to low hives; yet, the advantage of such hives far more than compensates, and with a bottom-board, such as described in the chapter on hives, we shall find that the toads do very little damage. In case of toads, the bees sting their throats, as I have taken, on several occasions, the stings from the throats of the toads, after seeing the latter capture the bees. As the toads make no fuss, it seems prob-able that their throats are callous against the stings.

MICE.

These little pests are a consummate nuisance about the apiary. They enter the hives in winter, mutilate the combs, especially those with pollen or old combs that have been long used for breeding, irritate, perhaps destroy, the bees, and create a very offensive stench. They often greatly injure comb which is outside the hive, destroy smokers, by eating the leather off the bellows, and, if they get at the seeds of honey-plants, they never retreat till they make complete the work of destruction.

In the house and cellar, unless they are made as they should always be—mouse-proof—these plagues should be, by use of cat or trap, completely exterminated. If we winter bees on the summer stands, the hive-entrance should be so con-tracted that mice can not enter the hive. In case of packing, as I have recommended, I should prefer a more ample opening, which may be safely secured by taking a piece of wire-cloth or perforated tin or zinc, and, tacking it over the entrance, letting it come within one-fourth of an inch of the bottom-board. This will give more air, and still preclude the entrance of these miserable vermin.

SHREWS.

These are mole-like animals (Fig. 295), and look not unlike a mouse. They have a long, pointed nose like the moles, to which they are closely related. They are insectivorous, and

FIG. 295.

Shrew.—Original.

have needle-shaped teeth, quite unlike those of the Rodentia, which includes the true mice. I have received from Illinois and Missouri species of the short-tailed shrews—Blarina—which enter the hives in winter and eat the bees, only refusing the head and wings. They injure the combs but little. As they will pass through a space three-eighths of an inch wide, it is not easy to keep them out of hives where the bees are wintering on their summer stands. I have received a short-tailed shrew—Blarina brevicauda, Gray—which was taken in the hives by Mr. Little, of Illinois.

SKUNKS.

Skunks sometimes annoy bee-keepers. They disturb the bees at nightfall, and as the bees come out of the hive they gulp them down. Of course, they can be poisoned or trapped. But as insect-destroyers they do great good, and I doubt if we can ever afford to kill the skunks. The small, striped skunk in Southern California depredates on our poultry. Yet I would use wire-netting and keep them out of the poultry-house, and not kill them.

CHAPTER XXI.

CALENDAR AND AXIOMS.

WORK FOR DIFFERENT MONTHS.

Though every apiarist will take one, at least, of the several excellent journals relating to this art, printed in our country, in which the necessary work of each month will be detailed, yet it may be well to give some brief hints in this place.

These dates are arranged for the Northern States, where the fruit-trees blossom about the middle of May. By noting these flowers, the dates can be easily changed to suit any locality.

JANUARY.

During this month the bees will need little attention. Should the bees in the cellar or depository become uneasy, which will not happen if the requisite precautions are taken, and there comes a warm day, it were well to set them on their summer stands, that they may enjoy a purifying flight. At night, when all are again quiet, return them to the cellar. While out I would clean the bottom-boards, especially if there are many dead bees. This is the time to read, visit, study, and plan for the ensuing season's work.

FEBRUARY.

No advice is necessary further than that given for January, though if the bees have a good flight in January, they will scarcely need attention in this month. The presence of snow on the ground need not deter the apiarist from giving his bees a flight, providing the day is warm and still. It is better to let them alone if they are quiet, as they should and will be if all is right. In California we must be sure the stores are sufficient.

MARCH.

Bees should be kept housed, and those outside still retain about them the packing of straw, shavings, etc. Frequent flights do no good, and wear out the bees. Colonies that are uneasy and besmear their hives are not wintering well, and may be set out and allowed a good flight and then returned. In California we do the April work of the East.

APRIL.

Early in this month the bees may all be put out. It will be best to feed all, and give all access to flour, when they will work at it, though usually they can get pollen as soon as they can fly out to advantage. Keep the brood-chamber contracted so that the frames will all be covered, and cover well above the bees to economize heat.

The colony or colonies from which we desire to rear queens and drones should now be fed to stimulate breeding. By careful pruning, too, we may and should prevent the rearing of drones in any but the best colonies. If from lack of care the previous autumn any of our colonies are short of stores, now is when it will be felt. In such cases feed either honey, sugar, or syrup, or place candy on top of the frames beneath the oil-cloth cover. Remember that plenty of stores insures rapid breeding. In California we will do the May work of the East in April.

MAY.

Prepare nuclei to start extra queens. Feed sparingly till bloom appears. Give room for storing. Extract if necessary, and keep close watch for swarms. Now, too, is the best time to transfer.

JUNE.

Keep all colonies supplied with vigorous, prolific queens. Divide the colonies or allow swarming as may be desired. Extract if necessary, or best, adjust frames or sections, if comb honey is desired, and be sure to keep all the white clover honey, in whatever form taken, separate from all other. Now is the best time to Italianize.

JULY.

The work this month is about the same as that of June. Keep the basswood honey by itself, and tier up sections as soon as the bees are well at work in them. Be sure that queens and workers have plenty of room to do their best, and do not suffer the hot sun to strike the hives.

AUGUST.

Do not fail to supersede impotent queens. Between basswood and fall bloom it may pay to feed sparingly. Give plenty of room for queen and workers, as fall storing commences.

SEPTEMBER.

Remove all surplus boxes and frames as soon as storing ceases, which usually occurs about the middle of this month. See that all colonies have enough stores for winter. If necessary to feed honey or sugar for winter, it should be done at this time.

OCTOBER.

If not already done, prepare colonies for winter. See that all have at least 30 pounds, by weight, of good, capped stores, and that all are strong in bees. If the bees are to be packed, it should be done early in October.

NOVEMBER.

Before the cold days come, remove the bees to the cellar or depository.

DECEMBER.

Now is the time to make hives, honey-boxes, etc., for the coming year. Also labels for hives. These may contain just the name of the colony, in which case the full record will be kept in a book; or the label may be made to contain a full register as to time of formation, age of queen, etc. Slates are also used for the same purpose.

I know from experience that any who heed all of the above may succeed in bee-keeping—may win a double success—receive pleasure and make money. I feel sure that many experienced apiarists will find advice that it may pay to follow. It is probable that errors abound, and certain that much remains unsaid, for of all apiarists it is true that what they *do* not know is greatly in excess of what they do know.

AXIOMS.

The following axioms, given by Mr. Langstroth, are just as true to-day as they were when written by that noted author:

There are a few *first principles* in bee-keeping which ought to be as familiar to the apiarist as the letters of the alphabet.

First.—Bees gorged with honey never volunteer an attack.

Second.—Bees may always be made peaceable by inducing them to accept of liquid sweets.

Third.—Bees, when frightened by smoke or by drumming on their hives, fill themselves with honey and lose all disposition to sting, unless they are hurt.

Fourth.—Bees dislike any *quick* movements about their hives, especially any motion which *jars* their combs.

Fifth.—In districts where forage is abundant only for a short period, the largest yield of honey will be secured by a *very* moderate increase of colonies.

Sixth.—A moderate increase of colonies in any one season will, in the long run, prove to be the easist, safest, and cheapest mode of managing bees.

Seventh.—A queenless colony, unless supplied with a queen, will inevitably dwindle away, or be destroyed by the bee-moth or by robber-bees.

Eighth.—The formation of new colonies should ordinarily be confined to the season when bees are *accumulating* honey; and if this or any other operation must be performed when forage is scarce, the greatest precaution should be used to prevent robbing.

The essence of all profitable bee-keeping is contained in Oettl's Golden Rule: KEEP YOUR COLONIES STRONG. If you can not succeed in doing this, the more money you invest in bees the heavier will be your losses; while, if your colonies are strong, you will show that you are a *bee-master* as well as a bee-keeper, and may safely calculate on generous returns from your industrious subjects.

" Keep all colonies strong."

GLOSSARY.

Abdomen—The third or last part of bee's body, p. 54, 65.

Absconding Swarm—Swarm that has separated from cluster and is going to its new home, p. 305.

Adulteration—Making impure, as mixing glucose with honey, p. 175.

After-Swarms—Swarms that issue within a few days after the first swarms, p. 168.

Air-Tubes—Tracheæ; Lungs of insects, p. 86.

Albino—Usually applied to animals with no pigment in skin, hair, etc. In bee-culture it refers to a variety of Italians with white rings, p. 55.

Alighting-Board—Board in front of entrance, on which bees alight as they return to their hives, p. 214.

American Hive—Langstroth hive with frames one foot square.

Antennæ—Horn-like organs of insects, p. 70.

Antenna Cleaner—Organ on anterior leg of bees, wasps, etc., to dust antennæ, p. 148.

Apiarian—Adjective, as apiarian implements; incorrectly used as a noun for apiarist.

Apiarist—One who keeps bees.

Apiary—Place where bees are kept, including bees and all.

Apiculture—Art of bee-keeping.

Apidæ—Family of bees, p. 38.

Aphis—Plant-lice, p. 390.

Apis—Genus of the honey-bee, p. 44.

Arthropada—Branch or phylum of insects, p. 31.

Articulata—Old name for branch containing insects, p. 31.

Artificial Fecundation or Impregnation—Fecundation in confinement (?).

Artificial Heat, Swarms, Pasturage, etc.—Furnished by man; not natural.

Atavism—Inheriting from a remote ancestor.

Balling of Queen—Bees gathering snugly about the queen in form of a sphere, p. 312.

Bar-Hives—Hives with bars across the top to which the combs are attached, p. 210.

Barren—Sterile; not able to produce eggs or young, p. 118.

Bees—Insects of the Family Apidæ, p. 38.

Bee-Bird or Bee-Martin—A fly-catcher that captures bees, p. 508.

Bee-Bread—The albuminous food of bees, usually pollen, p. 186.

Bee-Culture—Keeping bees.

Bee-Dress—Special suit worn by apiarist while working with bees, p. 345.

Bee-Escape—Device for clearing upper story of hive or section-case of bees, pp. 330, 341, 469.

Bee-Glue—Propolis, p. 190.

Bee-Gum—Section of hollow tree used as a bee-hive.
Bee-Hat—Hat so arranged as to prevent bees from stinging the face, p. 344.
Bee-Hawk—Dragon fly, p. 497.
Bee-Hive—Box for bees. See bee-gum and skep, p. 207.
Bee-House—House where bees are kept, where bee-work is done, or bees
 wintered, p. 468.
Bee-Keeper—One who keeps bees; apiarist.
Bee-Line—Straight line, like the route of bee from field to hive, p. 262.
Bee-Louse—Braula Cœca, p. 499.
Bee-Martin—King or bee bird, p. 508.
Bee-Master—English, bee-keeper.
Bee-Moth—Galleria mellonella, formerly G. cereana, moth that feeds on
 wax, etc., p. 482.
Bee-Pasturage—Honey-plants, p. 389.
Bee-Plants—Plants which secrete nectar, and so are visited by bees, p. 389.
Bee-Space—Space that will just allow a bee to pass: it is three-sixteenths
 of an inch. A double bee-space, three-eighths of an inch minus, is
 the space that bees do not fill with brace-combs or glue.
Beeswax—Secretion of the bee from which comb is fashioned, p. 176.
Bee-Tent—Tent covering hive and bee-keeper, pp. 332, 351. In England,
 tent for lectures on bees.
Bee-tree—A hollow tree in which bees breed and store, p. 262.
Bee-Veil—Veil for protecting face while working with bees, p. 344.
Bell-Glass—Glass vessel used for surplus comb-honey storing.
Bingham-Knife—Uncapping knife with beveled edge, p. 325.
Bingham-Smoker—Bee-smoker with open draft, p. 348.
Bisulphide of Carbon—Valuable insecticide, pp. 380, 487.
Black Bee—Common or German race of bees, p. 52.
Black Brood—Diseased brood, but not foul brood, p. 482.
Bottom-Board—Floor of hive, pp. 215, 217, 226.
Box-Hive—Plain box in which bees are kept, p. 207.
Box-Honey—Comb honey stored in boxes.
Brace-Combs—Incorrectly called "burr-combs." Small columns of wax
 connecting brood-combs, p. 219.
Brain—Nerve mass in head of insects, p. 82.
Breed—Race; Italian breed, p. 53.
Breeding-In—Close breeding, as when a queen is fecundated by one of
 her own drones.
Bridal Trip—Flight of queen to meet drone, p. 112.
Brimstoning—Killing bees with sulphur. Now happily obsolete, pp. 380,
 487.
Brimstone—The same as sulphur,   pp. 380, 487.
Broad-Frame—Wide frame for holding sections, p. 244.
Brood—Immature bees, or bees yet in the cell, p. 98.
Brood-Comb—Comb used for breeding, p. 179.
Brood-Nest—Space in hive used for breeding.
Brood-Rearing—Rearing of brood.
Brown Bee—A supposed variety of the common black bee, p. 52.
Bumble-Bee—Our large wild bee or humble-bee, p. 40.
Burr-Combs—Small pieces of wax built above the top-bars of the frames,
 p. 219.

Candied Honey—Honey crystallized or granulated, p. 175.
Cane Sugar—Common sugar, or the sugar of nectar, p. 17·
Cap—Box to shut over top of a hive, p. 220.

Cap—To seal or close a cell.

Capped Brood—Brood sealed.

Capped Honey—Honey sealed.

Cappings or Caps—Thin wax sheets cut off in extracting.

Card—Frame of comb. Rare.

Cardo—Part of maxilla, p. 66.

Carniolans—Same as Krainer. Race of black bees from Krain, Austria, pp. 57, 310, 346.

Carton—Paper box to hold comb honey, p. 382.

Casts—After-swarms. Rare.

Caterpillar—Larva of butterfly or moth.

Caucasian Bee—Variety of black bee, from Caucasian Mountains, pp. 48, 52.

Cell—Opening in comb for brood, honey or bee-bread, p. 179.

Chaff-Hive—A double-walled hive with space filled with chaff, pp. 215, 459.

Chitine—Substance which makes crust of insects hard, p. 32.

Chyle—Digested food; probable food of larva; p. 141.

Chyme—Partially digested food; word of doubtful use, p. 141.

Chrysalid or Chrysalis—Pupa of butterflies. Sometimes applied to other papæ.

Clamp—Hives placed close together and covered, p. 466.

Cleansing Flight—Removing bees from cellar that they may fly, p. 464.

Closed End or Top Frames—Where end-bars of frames and ends of top-bars are close fitting, p. 233.

Cluster—Bees in compact mass, pp. 166, 167.

Clustering—Many bees hanging together, pp. 166, 167.

Clypeus—Portion of head of insects below the eyes, p. 66.

Cocoon—Case, often containing silk fibers, which surrounds a pupa; cup lining cells of comb, pp. 90, 101, 162, 184.

Collateral System—Side-storing. English.

Colon—Part of intestine, rectum, pp. 89, 145.

Colony—The bees of one hive.

Comb—The fabric which holds the brood and honey, p. 179.

Comb-Basket—The frame of an extractor which holds the comb, p. 323.

Comb-Carrier—Box for carrying combs; most used in extracting, p. 329.

Comb foundation—Thin sheets of impressed wax, like the foundation of real comb, p. 353.

Comb Foundation Machine—Machine for making comb foundation, p. 354.

Comb-Guide—Strip of wood, comb or foundation on the bottom of top-bar of frame, to induce bees to build comb in proper place, p. 361.

Comb-Holder—Device for holding combs, 324.

Comb Honey—Honey in comb, p. 335.

Compound Eyes—Large eyes of insects, so called as they consist of many simple eyes, p. 73.

Corbicula—Pollen-basket on hind leg of worker-bee, p. 154.

Cover—Lid of hive, or cover of brood-frames, pp. 220, 223, 233.

Coxa—First part or joint of the insect's leg, p. 79.

Crate—Box for sections on the hive, or for shipping comb honey, pp. 247, 381.

Cushion—Quilt or bag for covering bees, p. 223.

Cyprian Bees—A yellow race from the Isle of Cyprus, p. 55.

Dalmation Bees—A variety of black bees from Dalmatia, the Southwestern Province of Austria, p. 58.

Darts—Lancets of sting, p. 157.

Decoy Hive—Hives set to catch absconding swarms.

Diarrhea—Dysentery, p. 475.

Dipping-Board—Board for securing thin wax sheets in making foundation, p. 358.

Dividing—Forming colonies artificially, p. 316.

Division-Board—Board for reducing the size of the brood-chamber, p. 222.

Dollar Queens—Queen sold for one dollar, p. 361.

Driving Bees—Causing the bees to pass out of a hive into a box placed above by rapping on the hive, 258.

Drone—Male bee, p. 121.

Drone-Brood—Brood which produces drone-bees, p. 126.

Drone-Comb—Comb with large cells, in which drones may be reared, p.183.

Drone-Eggs—Eggs that produce drones, p. 126.

Drone-Trap—Trap for catching drones, p. 285.

Drumming Bees—Forcing bees from one hive to another hive or box by rapping on the first with a stick or hammer, p. 258.

Dry Feces—Supposed dry excreta of bees.

Ductus Ejaculatorus—Part of male apparatus, p. 92.

Dummies—Division-boards, p. 222.

Dysentery—Winter disease of bees, p. 475.

Dzierzon Theory—Parthenogenesis; agamic reproduction; theory that unfecundated eggs will develop, and in bees such eggs always produce drones, p. 126.

Egg—The initial or first stage of all the higher animals, pp. 95, 101.

Egyptian Bee—Yellow bee from Egypt, p. 57.

Eke—Rim to raise and enlarge the hive; often a half hive.

Embryo—The young animal while yet in the egg or before birth.

Entrance—Opening of the hive where the bees enter, p. 217.

Entrance-Blocks—Pieces of wood, usually triangular, for contracting or closing the entrance of hive, p. 217.

Entrance-Guard—Perforated zinc to prevent drones or queen from leaving the hive, p. 285.

Epicranium—Part of head between and above the eyes, p. 66.

Epipharynx—Part of mouth.

Extracted Honey—Honey thrown from comb by use of extractor, pp.281, [333, 376.

Extractor—Machine for extracting, p. 321.

Exuvium—Cast-skin of larva. Substance left in cell when bee emerges, pp. 89, 98, 184.

Eyes—Organs of sight in insects; there are usually two large compound and three small simple or Ocelli, p. 73.

Feces—Intestinal excreta of animals.

Farina—Flour; incorrectly used for pollen.

Fecundate—Union of sperm and germ cells; to impregnate, p. 103.

Feeder—Device for feeding bees, p. 266.

Femur—Third and largest joint of an insect's legs, p. 79.

Fence—Separator to be used with plain sections, p. 242.

Fertile—Productive; often used for impregnated or fecundated. A queen that can lay eggs is fertile; after mating she is fecund.

Flagellum—Outer part of antenna, p. 69.

Foul Brood—Malignant disease of a fungoid character which attacks bees, p. 475.

Foundation, Fdn.—Stamped wax sheets, p. 353.

Frame—Device for holding comb in the hive, p. 227.

Fumigate—To surround with fumes. We fumigate the bees with smoke and the combs with sulphur fumes, pp. 380, 487.

Gallup Frame—Frame 11¼ inches square, p. 229.
Ganglia—Knots of nerve matter like the brain, p. 81.
Gastric Juice—Digestive ferment secreted by stomach.
Gena—Cheek of insects.
German Bee—Common black bee, p. 52.
Glands—Tubular or sack-like organs which form from elements taken from the blood a liquid called a secretion. Bees have several pairs of glands, p. 134.
Glassing—Covering or protecting sections of comb honey with glass.
Glucose—Reducing sugar, p. 172.
Good Candy—Candy made by mixing sugar and honey, p. 318.
Grafted Cells—Queen-Cells with the larva replaced by another, p. 278.
Grafting Cells—Taking small larvæ from cells and placing them in queen-cells, p. 278.
Granulated Honey—Honey that has crystallized or candied, p. 175.
Green Honey—Unripe honey, p. 327.
Grub—Larva of beetle, p. 98.
Guide Comb—Narrow piece of comb or starters fastened to top-bar of frame or section, p. 295.
Gullet—Œsophagus, pp. 89, 142.

Hatch—To issue from egg; egg hatches, the brood develops and emerges from cell.
Hatching Brood—Incorrectly used to refer to bees coming from cells.
Heart—Circulating organ; in insects a tube along the back, p. 84.
Heath Bees—Variety of German bees from Luneberg Heath, Europe, p. 57.
Heddon Hive—Hive with divided brood-chamber, the division being horizontal, p. 223.
Heddon-Langstroth Hive—Langstroth hive as used by Heddon, p. 215.
Hexapoda—Class insects, p. 32.
Hill's Device—Curved sticks used to raise cloth a little from the frames in winter. p. 456.
Hive—Box or receptacle for bees, p. 207.
Hiving—Removing a swarm of bees from cluster to hive, p. 297.
Hiving Basket or Box—Basket or box used in hiving swarms, p. 297.
Holy-Land Bees—Yellow bees from Southern Palestine, p. 48.
Honey—Nectar digested by the bees, p. 171.
Honey-Bee—Apis Mellifera, the domestic bee, p. 52.
Honey-Bag—Honey stomach, pp. 89, 143.
Honey-Board—Board between brood-chamber and section-case, p. 219.
Honey-Box—Box for surplus comb honey.
Honey-Comb—Fabric that holds the honey and brood, p. 179.
Honey-Dew—Nectar from insects like Aphides and bark-lice, or from extra floral glands, pp. 392, 393.
Honey-Extractor—Machine for extracting honey, p. 321.
Honey-Gate—Faucet to draw extracted honey from an extractor or barrel. It is closed instantly by a slide or gate.
Honey-Knife—A knife for uncapping honey, p. 325.
Honey-Sac—Honey stomach, pp. 89, 143.
Honey-Slinger—Honey extractor, p. 321.
Honey-Stomach—Honey-sac where bee carries honey, pp. 89, 143.
House-Apiary—Building frost-proof where bees are kept continually, p. 468.
Hungarian Bee—Variety of the black bee from Hungary, p. 58.

Hybrid—Properly an animal which is a cross between two different species. A hybrid bee is a cross between two different races; all the
bees except the drones from an Italian queen mated to a black
drone will be hybrids; the drones will be pure if the queen is (see
Dzierzon theory).

Hymenoptera—Order of insects which includes bees, ants and wasps, p. 35.

Hymettus—A mountain of Greece famed for its delicious honey.

Hypopharnyx—Membrane or curtain connecting the base of the mouth
organs.

Ileum—Small intestine, pp. 89, 145.

Imago—The mature insect; the last or winged stage of an insect, p. 101.

Insects—Hexapoda—Class of bee, p. 32.

Intestine—Digestive tube beyond the stomach, p. 145.

Introducing—Method of making bees accept a strange queen. p. 311.

Introducing-Cage—Cage for introducing a queen, p. 312.

Inverting—Turning a hive, section, case or frame bottom up. Reversing
is also used, p. 230.

Italian Bee—A yellow race from Italy. Every worker-bee has three well
marked yellow bands, pp. 53, 307.

Italianizing—Changing bees from some other race to Italians, p. 306.

Jaws—Same as mandibles, p. 146.

Krainer Bees—Bees from Krain, Austria; same as Carniolans, pp. 57,
310, 346.

Labium—Under lip of an insect, pp. 66, 131.

Labrum—Upper lip of an insect, p. 66.

Lamp-Nursery—Tin double-walled box used for rearing queens. p. 286.

Langstroth Frame—Adopted by Mr. Langstroth for his hive; size 17⅜ by
9⅛, pp. 215, 227.

Langstroth Hive—L. Hive; hive with frame suspended in a case or box;
invented by Rev. L. L. Langstroth, p. 210.

Larva—plu. Larvæ—Immature bees, p. 98.

Laying Worker—Worker-bee that lays eggs, p. 130.

Ligula—End of labium; the tongue in bees, pp. 66, 131.

Ligurian Bee—Same as Italian; named from Liguria, a province in Italy,
pp. 53, 307.

Lining Bees—Noting direction of flight to find bee-tree, etc., p. 262.

Loose Frames—Frames not fixed, p. 233.

Lora—Part of labium, p. 132.

Maggot—Footless larva of two-winged flies; often applied to any footless
larvæ.

Maiden Swarm—First swarm.

Malpighian Tubules—Renal tubules attached to the stomach, p. 90.

Mandibles—Main jaws of insects, p. 146.

Manipulation—Handling.

Marriage Flight—Mating of queen, p. 112.

Mat—Flexible cover to place over brood-frames, made of slats, straw, etc.

Maturing Brood—Where the bees are just emerging from the cells.

Maxillæ—The second or under jaws of insects, pp. 66, 131.

Mel Extractor—Honey extractor, p, 321.

Meliput—Honey extractor, p, 321.

Mentum—Second joint of labium or under lip, p. 131.

Meso-Thorax—Second joint of thorax, p. 78.

Meta-Thorax—Third joint of thorax, p, 78.

Metal Corners—Tins to fasten and unite corners of frames.

Micropyle—Openings in eggs where sperm-cells enter, p. 101.

Midrib of Comb—Center partition of comb, p. 182.

Miller—Moth, which is the more proper word, p. 482.

Mismated—Not purely mated.

Moth—All scale-winged insects except butterflies.

Moth-Larva—Immature moth, p. 483.

Moth-Miller—Incorrect term often used for moth, p. 484.

Moth-Trap—Trap for catching moths.

Movable-Frame Hive—Langstroth hive, p. 210.

Muscles—Organs that produce motion, p. 80.

Nadir—The under story of a two-story hive; a wide eke, p. 213,

Nectar—Sweet substance, as the liquid in nectaries of flowers, p. 171.

Nectaries—Nectar-glands of flowers.

Nerves—White threads which connect organs to convey impressions or impulses, p. 81.

Nervures of Wings—Same as veins, p. 45.

Neuter—Incorrect name for worker-bees; they are not neuters, but undeveloped females, p. 129.

New Idea Hive—Long one-story hive with many frames.

Non-Swarming Hive—A purely ideal hive, supposed to prevent swarming.

Normal—Usual; regular.

Nucleus—plural, nuclei; miniature colony of bees for queen-rearing, p.281.

Nurse-Bees—Young bees or ones that feed the brood, p. 164.

Nursery—Device for rearing queens. See lamp-nursery, p. 286.

Nymph—An insect in the pupa state; the immature bee in cell that is the form of adult bee is a nymph, p. 99.

Observatory Hive—Hive with glass sides, so that bees can be seen without disturbing them, p. 238.

Ocelli—Simple eyes on epicranium, usually three, p. 73.

Œsophagus—Tube leading from pharynx to honey-stomach, pp. 89, 142.

Open Sections—Sections that do not touch on sides, p. 240.

Ovary—Essential organs of the female, where the eggs grow, p. 94.

Over-stocking—Where more bees are kept than a locality can supply with a full harvest of nectar.

Oviduct—Tube for passage of egg from ovary, p. 94.

Ovipositor—Same as oviduct, p. 94.

Ovum—Egg, pp. 95, 101.

Palestine Bees—Race of yellow bees found in Southern Syria; the so-called Holy-Land bees, p. 48.

Paraffine—Wax-like crystalline substance used to coat barrels and prevent leakage; one of the products of crude petroleum.

Parasite—An organism that feeds upon another, p. 37.

Parent Colony—The colony from which a swarm has issued.

Paraglossæ—Short appendages at base of tongue, pp, 67, 132.

Parthenogenesis—Reproduction without males, p. 126.

Pasturage—Plants from which food is secured, p. 389.

Pecten of Legs—Fringe or comb of hairs.

Perforated Zinc—Zinc with holes cut so worker-bees can pass, but drones and queens can not, p. 219.

Pharynx—Throat or back of the mouth, p. 89.

Phenol—Pure carbolic acid, p. 479.

Pincers—Wax jaws of hind legs, p. 153.

Piping of Queens—Noise made by young queens when one has emerged from cell and others have not, p. 168.

Plain Sections—Sections with no inset or bee-way; the edges are straight, p. 241.

Planta—Soles or bottom of feet, p. 150.

Poison-Sac—Sac at base of sting to hold the poison, p. 157.

Pollen—Male cell or element of flowers; bee-bread.

Pollen-Basket—Corbicula; cavity on posterior leg for carrying pollen, pp. 152, 186.

Pollen-Combs—Rows of hairs on first tarsus of second and third pairs of legs of worker, on the inside, also pecten, p. 153.

Pollen-Hairs—Compound or webbed hairs of bees, used for collecting pollen, p. 79.

Portico—Porch to hive, p. 210.

Pound Section—Section 4¼ inches square, p. 242.

Prime Swarm—First swarm.

Prize Section—Section 6¼ by 5¼ inches, p. 242.

Propolis—Bee-glue.

Propolize—To cover with propolis, p. 190.

Prothorax—First joint of thorax, p. 78.

Prune—To cut out undesirable comb, as drone or old.

Puff-Ball—A large fungus, which, when pressed, sends out myriads of spores; it is sometimes used to subdue bees.

Pulvilli—Adhesive disks on the last joint of an insect's leg, p. 150.

Pupa—Third stage of insects, that between larva and imago; also called nymph, p. 99.

Pygidium—Last joint of abdomen.

Queen—Mother-bee, p. 102.

Queen-Cage—Cage for introducing queen, p. 312.

Queen-Cell—Cell in which queen is reared, pp. 100, 111.

Queenless—Having no queen.

Queen-Rearing—Rearing of queens, p. 273.

Queen Register—Card to show state of hive as to queen, p. 291.

Queen's Voice—Noise made by queen like piping; true voice, p. 168.

Queen-Yard—Box with perforated zinc, to keep a clipped queen from being lost when she comes out with a swarm; also called queen-trap.

Quilt—Cover for brood-frames, consisting of two cloths containing wool or cotton sewed together, p. 223.

Quinby Hive—Large Huber style of hive, p. 235.

Quinby Frame—Large frame 18½ by 11¼ inches, p. 227.

Quincunx—Where things in rows alternate, thus, . · . · .

Rabbet—Where one side of the edge of a board is planed down for a short distance, p. 216.

Race—Breed. Where a variety has been closely bred so long as to transmit its peculiarities to its offspring. Race is a natural breed, p. 52.

Rack—Crate or case; section-rack.

Rectal Glands—Glands in the rectum, p. 146.

Rectum—Large intestine, p. 146.

Rendering Wax—Melting and cleaning wax, 367.

Reversing—Inverting; turning bottom up, pp. 229, 339.

Rhomb—Four equal sided figure, two of whose opposite angles are equal and acute, the others equal and obtuse.

Ripe Honey—Honey that has cured or evaporated, so it is thick, p. 327.

Robbing—When bees steal honey from another colony, p. 473.

Royal Jelly—Food fed to queen-larvæ, p. 108.

Saliva—Secretion of the mouth, p. 91.

Scape—Base of antenna, p. 69.

Scouts—Bees that go forth just before swarming to find and prepare the new home, p. 166.

Seal—To close.

Sealed Brood—Brood in cells that the bees have capped, p. 162.

Sealed Honey—Honey in cells that are capped, p. 183.

Section—Small frame for comb honey, 239.

Seminal Vesicle—Sac to hold sperm-cells or semen, p. 93.

Separator—Wood or tin strip, very thin, for separating sections, so that bees will build straight and true combs, p. 250.

Septum—Base between cells of comb; incorrectly called midrib, p. 182.

Sholtz Candy—Good candy; sugar and honey mixed; invented years ago by Sholtz, a German, p. 318.

Skep—Straw hive, such as were used in olden times.

Smell—Sense located in antennæ of insects, p. 70.

Smoker—Instrument used to smoke or quiet bees, p. 348.

Smyrnian Bees—A variety or race of bees from a province—Smyrna—in Asiatic Turkey, p. 58.

Species—Animals so long bred as to have distinctive characteristics more fixed, p. 52.

Spent Queen—One sterile with age, p. 118.

Spermatheca—The sac off oviduct of queen that holds the sperm, p. 104.

Spermatozoæ—Sperm-cells; the male element or fecundating principle, p. 124.

Spring Dwindling—Rapid dying of bees in the spring, p. 466.

Stand—Support of hive. Incorrectly used for colony.

Starter—A small piece of comb or foundation fastened to the top-bar of a hive, 295.

Sterile Queen—One that does not lay, or whose eggs do not hatch, p. 118.

Sting—The organ of defense of bees, wasps, etc., p. 156.

Stock—Wrongly used for colony; if used at all it should refer to bees, hive and all.

Stomach—Where the food is mainly digested, pp. 90, 143.

Stomach-Mouth—Organ at base of honey-stomach, p. 142.

Storify—Used in England for adding upper stories to hives.

Storifying—English, tiering up.

Strain—A variety, as a strain of bees, developed by the bee-keeper.

Strained Honey—Honey strained through a cloth, not extracted honey.

Sulphur—A yellow mineral used to fumigate honey.

Super—Upper story, either for extracted honey or honey in sections, p. 214.

Supersede—To replace with another.

Swarm—Bees that leave hive in natural division, p. 166.

Swarming-Basket—Basket to convey swarm from place of clustering to hive, 297.

Swarming Impulse or Fever—Desire of the bees to swarm.

Swarming Season—Season of year when bees are likely to swarm.

Syrian Bee—Race of yellow bees from Northern Palestine, p. 55.

Taking up Bees—Destroying bees to get the honey. Rare now.

Tarsus—Last one to five joints of insect leg; foot, p. 79.

Tested Queen—One proved pure by examination of her offspring.

Thorax—Second part of insect's body, p. 64.
Tibia—Fourth joint of an insect's leg, from the body, p. 79.
Tibial Spur—Spur at end of tibia, p. 79.
Tier Up—Setting additional stories or supers of sections on a hive.
Tongue—Sucking tube of bee, p, 132.
Tracheæ—Air-tubes or turbular lungs of insects, p. 81.
Transferring—Removing colony of bees from one hive to another, p. 258.
Transformations—Changes from larva to pupa to imago, p. 96.
Travel-Stain—Soil of comb when left long in hive.
Trochanter—Second joint of insect's leg, 79.

Uncapping—Cutting caps from comb-cells, p. 325.
Unfertile—Queen or eggs that can not produce young.
Uunicomb Hive—Hive with one comb and glass sides; observatory hive, p. 238.
Uniting—To put two or more colonies into one, p. 465.
Unqueening—Removing queen from colony.
Unripe—Thin honey; honey not cured or evaporated, p. 327.
Unsealed—Applied to honey and brood when not capped.
Untested Queen—One whose purity has not been demonstrated.
Urinary Tubules—Tubes attached to the stomach of a bee, p. 90.

Variety—Division of a race; a strain, p. 52.
Veil—Protection for face, p. 344.
Velum—Part of antenna cleaner, p. 148.
Ventilation—Changing the air so it shall be constantly pure.
Virgin—Unmated queen.

Wax—Secretion formed between the abdominal segments of worker-bees, p. 155.
Wax-press—Press for expressing wax, p. 371.
Wax-extractor—Device for separating the wax from comb, p. 367.
Wax Plates or Pockets—Place where the wax-scales form on the underside of a worker-bee, p. 155.
Wedding Flight—Flight of queen to mate with the drone, p. 112.
Wild Bees—Bees in the forest, etc., with no owner.
Wind-Break—High fence or evergreen hedge to protect from wind, p. 253.
Winter-Passages—Holes through the center of combs so bees can pass through, p. 456.
Wired-Frames—Frames with opposite sides connected with fine wire, pp. 230, 364.
Worker-Bees—The undeveloped females; the bees that do the work except that of egg-laying.
Worker-Eggs—Eggs that develop into workers, p. 129.
Worm—Term usually applied to a larva; really a footless cylindrical animal like an angle-worm, p. 31.

INDEX.

www.ingramcontent.com/pod-product-compliance
Lightning Source LLC
Chambersburg PA
CBHW061738210326
41599CB00034B/6718